# Approximation Theory

ACADEMIC PRESS RAPID MANUSCRIPT REPRODUCTION

*Proceedings of an International Symposium*
*Conducted by the University of Texas and*
*the National Science Foundation*
*at Austin, Texas, January 22-24, 1973*

# Approximation Theory

Edited by

G. G. Lorentz

*Department of Mathematics*
*University of Texas*
*Austin, Texas*

*in cooperation with H. Berens,*
*E. W. Cheney, L. L. Schumaker*

ACADEMIC PRESS, INC.
NEW YORK AND LONDON 1973
A Subsidiary of Harcourt Brace Jovanovich, Publishers

ACADEMIC PRESS, INC.
111 Fifth Avenue, New York, New York 10003

*United Kingdom Edition published by*
ACADEMIC PRESS, INC. (LONDON) LTD.
24/28 Oval Road, London NW1

**Library of Congress Cataloging in Publication Data**
Main entry under title:

Approximation theory.

"Proceedings of an international symposium conducted by the University of Texas and the National Science Foundation, at Austin, Texas, January 22-24, 1973."

1. Approximation theory–Congresses. 2. Spline theory–Congresses. 3. Functional analysis–Congresses. I. Lorentz, G. G., ed. II. United States. National Science Foundation. III. Texas. University at Austin.
QA221.A65 515'.7 73-2083
ISBN 0-12-455750-3

PRINTED IN THE UNITED STATES OF AMERICA

# CONTENTS

Asterisk denotes the author who presented the paper at the symposium.

CONTENTS

CONTENTS

# CONTRIBUTORS

**Dan Amir**, Department of Mathematical Sciences, Tel Aviv University, Tel Aviv, Israel

**John Warren Baker**, Department of Mathematics, Florida State University, Tallahassee, Florida 32306

**M. W. Bartelt**, Department of Mathematics, Rensselaer Polytechnic Institute, Troy, New York 12181

**H. Berens**, Department of Mathematics, University of Texas, Austin, Texas 78712

**J. Blatter**, Institut für Angewandte Mathematik, Universität Bonn, 5300 Bonn, West Germany

**R. Bojanic**, Department of Mathematics, The Ohio State University, Columbus, Ohio 43210

**Carl de Boor**, Mathematics Research Center, University of Wisconsin, Madison, Wisconsin 53706

**Dietrich Braess**, Institut für Numerische and Instrumentelle Mathematik, Universität Munster, 4400 Munster, Roxelerstrasse 64, West Germany

**B. Brosowski**, Lehrstühle für Numerische und Angewandte Mathematik, Universität Göttingen, 3400 Göttingen, West Germany

**Herman Burchard**, Department of Mathematics and Statistics, Oklahoma State University, Stillwater, Oklahoma 74074

**P. L. Butzer**, Lehrstuhl A für Mathematik, Technological University of Aachen, 5100 Aachen, West Germany

**M. P. Carroll**, Department of Mathematics, Virginia Polytechnic Institute and State University, Blacksburg, Virginia 24061

**A. S. Cavaretta, Jr.**, Department of Mathematics, Kent State University, Kent, Ohio 44240

**E. W. Cheney**, Department of Mathematics, University of Texas, Austin, Texas 78712

**Chin-Hung Ching**, Department of Mathematics, University of Melbourne, Melbourne, Victoria, Australia

**Charles K. Chui**, Department of Mathematics, Texas A & M University, College Station, Texas 77843

**Stephen Demko**, Department of Mathematics, Kent State University, Kent, Ohio 44240

**Frank Deutsch**, Department of Mathematics, Pennsylvania State University, University Park, Pennsylvania 16802

**Ronald DeVore**, Department of Mathematics, Oakland University, Rochester, Michigan 48063

**Charles B. Dunham**, Department of Computer Science, University of Western Ontario, London, Canada

**Rui J. P. de Figueiredo**, Mathematics Research Center, University of Wisconsin, Madison, Wisconsin 53706

**L. Flatto**, Department of Mathematics, Belfer Graduate School of Science, Yeshiva University, New York, New York 10033

**T. W. Gamelin**, Department of Mathematics, University of California, Los Angeles, California 90024

**Tord H. Ganelius**, Department of Mathematics, University of California at San Diego, La Jolla, California 92037

**M. v. Golitschek**, Institut für Angewnadte Mathematik, Universität Würzburg, 8700 Würzburg, West Germany

**E. Görlich**, Lehrstuhl A für Mathematik, RWTH Aachen, 5100 Aachen, Templergraben 55, West Germany

**T. N. E. Greville**, National Center for Health Statistics, 5600 Fishers Lane, Rockville, Maryland 20852

**Alfred P. Hallstrom**, Department of Mathematics, University of Washington, Seattle, Washington 98195

**Werner Haussmann**, Institut für Mathematik, Ruhr-Universität,463 Bochum, West Germany

**William Hintzman**, Department of Mathematics, California State University at San Diego, San Diego, California 92115

**K.-H. Hoffman**, Mathematisches Institut, Universität München, 8000 München 2, West Germany

**Joseph W. Jerome**, Department of Mathematics, Northwestern University, Evanston, Illinois 60201

**David W. Kammler**, Department of Mathematics, Southern Illinois University, Carbondale, Illinois 62901

**J. Korevaar**, Department of Mathematics, University of California at San Diego, La Jolla, California 92037

**Werner Krabs**, Fachbereich Mathematik, Technische Hochschule, Darmstadt, Hochschulstr. 1, 6100 Darmstadt, West Germany

**P. J. Laurent**, Mathématiques appliquées, Université Scientifique et Médicale de Grenoble, Cédex 53, 38041 - Grenoble, France

**Y. H. Lee**, Department of Mathematics, The Ohio State University, Columbus, Ohio 43210

**William H. Ling**, Department of Mathematics, Union College, Schenectady, New York 12308

**G. G. Lorentz**, Department of Mathematics, University of Texas, Austin, Texas 78712

**C. A. Micchelli**, Mathematical Sciences Department, IBM Thomas J. Watson Research Center, Yorktown Heights, New York 10598

**R. H. Moore**, Department of Mathematics, University of Wisconsin, Milwaukee, Wisconsin 53201

**P. D. Morris**, Department of Mathematics, Pennsylvania State University, University Park, Pennsylvania 16802

**G. Mühlbach**, Institut für Mathematik, Technische Universität Hannover, 3000 Hannover, West Germany

**Heinz Josef Münch**, Rechenzentrum Ruhr-Universitat, 463 Bochum, West Germany

**M. Z. Nashed**, School of Mathematics, Georgia Institute of Technology, Atlanta, Georgia 30332

**D. J. Newman**, Department of Mathematics, Yeshiva University, New York, New York 10033

**K. H. Price**, Department of Mathematics, Stephen F. Austin College, Nacogdoches, Texas 75961

**David L. Ragozin**, Department of Mathematics, University of Washington, Seattle, Washington 98195

**John R. Rice**, Department of Mathematics, Purdue University, Lafayette, Indiana 46907

**L. A. Rubel**, Department of Mathematics, University of Illinois, Urbana, Illinois 61801

**I. J. Schoenberg**, Mathematics Research Center, University of Wisconsin, Madison, Wisconsin 53706

**Larry L. Schumaker**, Department of Mathematics, University of Texas, Austin, Texas 78712

**Paul Schwartz**, Department of Mathematics, University of Texas, Austin, Texas 78712

**G. L. Seever**, Department of Mathematics, University of Texas, Austin, Texas 78712

**A. Sharma**, Department of Mathematics, University of Alberta, Edmonton, Alberta, Canada

**A. L. Shields**, Department of Mathematics, University of Michigan, Ann Arbor, Michigan 48104

**O. Shisha**, Mathematics Research Center, Naval Research Laboratory, Washington, D. C. 20390

**S. D. Silliman**, Department of Mathematics, Cleveland State University, Cleveland, Ohio 44115

**Wilhelm Sippel**, Institut für Angewandte Mathematik I, Universität Erlangen-Nürnberg, 8520 Erlangen, West Germany

**P. W. Smith**, Department of Mathematics, Texas A & M University, College Station, Texas 77843

**J. J. Swetits**, Department of Mathematics, Old Dominion University, Norfolk, Virginia 23508

**B. A. Taylor**, Department of Mathematics, University of Michigan, Ann Arbor, Michigan 48104

**Gerald D. Taylor**, Computer Science Department, Stanford University, Stanford, California 94305

**J. Edward Tornga**, Department of Mathematics, Union College, Schenectady, New York 12308

**Walter Trebels**, Lehrstuhl A für Mathematik, RWTH Aachen, D-51-Aachen, Templegraben 55, West Germany

**Richard S. Varga**, Department of Mathematics, Kent State University, Kent, Ohio 44240

**A. K. Varma**, Department of Mathematics, University of Florida, Gainesville, Florida 32601

**Manfred Wolff**, Fachbereich Mathematik, Universität Dortmund, 4600 Dortmund-Barop, West Germany

**B. Wood**, Department of Mathematics, University of Arizona, Tucson, Arizona 85721

**Daniel Wulbert**, Department of Mathematics, University of California at San Diego, La Jolla, California 92037

**K. L. Zeller**, Mathematisches Institut, Universität Tubingen, 74 Tübingen, West Germany

**Zvi Ziegler**, Faculty of Mathematics, Israel Institute of Technology, Haifa, Israel

# PREFACE

An International Symposium on Approximation Theory was held at the University of Texas in Austin, January 22-24, 1973. It received support from the National Science Foundation and the Graduate School of the University of Texas. Mathematicians of several countries were able to participate, among them many Germans and Canadians.

Two types of papers appear in the Proceedings: long survey papers, based on one-hour invited lectures, and about fifty shorter research papers, presented at the thirty- and fifteen-minute talks. The purpose of the Organizing Committee was to have a wide coverage of the field, interpreting Approximation Theory in a broad sense, and to let younger mathematicians come forward; the older ones seemed not to object.

The opening of the Symposium coincided with the dedication of the new Physics, Mathematics and Astronomy building on our campus. Funds for the building have been provided by the State of Texas and the National Science Foundation.

The Organizing Committee consisted of H. Berens, E. W. Cheney, Y. Ikebe, G. G. Lorentz, and L. L. Schumaker; Mrs. F. Griffin was our helpful secretary. The typing of the manuscripts was in the hands of Mrs. M. A. Zivley and Becky Tatman. We extend our warmest thanks to these dedicated persons and organizations!

G. G. Lorentz

# Approximation Theory

# THEOREMS OF KOROVKIN TYPE FOR POSITIVE LINEAR OPERATORS ON BANACH LATTICES

H. Berens and G. G. Lorentz[1]

## 1. Introduction

Let X be a compact metric space, and let C(X) be the space of all real-valued, continuous functions f on X under the supremum norm

$$\|f\|_\infty = \sup\{|f(x)|: x \in X\}.$$

By $C^+(X)$ we denote the cone of positive functions in C(X).

Let E be a real Banach lattice with positive cone $E^+$ and norm $\|\cdot\|$. A linear transformation T on C(X) into E is said to be <u>positive</u> if $TC^+ \subset E^+$. Any positive linear transformation T on C(X) into E is continuous with norm

$$\|T\| = \sup\{\|Tf\|: \|f\|_\infty \le 1\} = \|T1\|.$$

We denote by $L^+(C,E)$ the cone of all positive linear transformations in $L(C,E)$.

In the following a transformation in $L(C,E)$, denoted by the capital letter P, will always be a lattice homomorphism on C(X) into E, i.e., P is a linear transformation on C(X) into E such that

$$P(f \vee_C g) = Pf \vee_E Pg.$$

Let E be a Banach lattice and P a lattice homomorphism on C(X) into E. For a non-empty set S in C(X) we define its <u>shadow with respect to</u> $L^+(C,E)$ <u>and</u> P, in notation $S(S,E,P)$, to be the set of all functions f such that

(1) given any sequence $T_n$, n=1,2,..., in $L^+(C,E)$, the relation

$$\lim_n T_n g = Pg \quad \text{in} \quad E \qquad (g \in S)$$

implies

$$\lim_n T_n f = Pf \quad \text{in} \quad E.$$

A simple argument shows that for a given set S in C(X) its shadow $\mathcal{S}(S,E,P)$ is a well-defined linear subspace in C(X). If $\mathcal{S}(S,E,P)$ equals C(X), then S is called a Korovkin set in C(X) with respect to $\mathcal{L}^+(C,E)$ and P.

We define the shadow of S in C(X) with respect to $\mathcal{L}^+(C,E)$ to be the set

$$\mathcal{S}(S,E) = \bigcap_P \mathcal{S}(S,E,P),$$

where the intersection is taken over all lattice homomorphisms P on C(X) into E. Moreover, the total shadow of S is defined by

$$\mathcal{S}(S) = \bigcap_E \mathcal{S}(S,E).$$

Here the intersection is taken over all Banach lattices E. Also, if $\mathcal{S}(S,E)$, resp. $\mathcal{S}(S)$, equals C(X), then we say S is a Korovkin set in C(X) with respect to $\mathcal{L}^+(C,E)$, resp. a total Korovkin set in C(X).

It is the aim of the paper to characterize the various shadows of a given set S in C(X) under the additional assumptions that (i) $1 \in S$, and (ii) S separates points on X. Such a set in C(X) is usually called an admissible set. Under assumption (i), for any Banach lattice E and any P: $C(X) \to E$ the associated shadows of a given set S are closed linear subspaces of C(X).

One can define shadows and Korovkin sets in a more general set-up. For instance, we replace C(X) by an arbitrary, but fixed Banach lattice F and define, in the same way as before, the shadow of a set S in F with respect to $\mathcal{L}^+(F,E)$ and a given lattice homomorphism P in $\mathcal{L}^+(F,E)$. To mention a second example, let F be equal to E; for a set S in E we define its shadow with respect to contractions on E and the identity I.

The present stand of the theory may be briefly described as follows. For F and E equal to C(X) and P equal to I, a very satisfactory theory for Korovkin sets exists, due to Yu. A. Šaškin for positive operators given in [20], and for contractions due to Šaškin in [22] and D. E. Wulbert in [27]. Šaškin mainly treats the case of finite Korovkin sets. An exposition of this theory, also valid for infinite Korovkin sets, has been given in G. G. Lorentz [15].

For positive operators on C(X) into a Banach lattice, more generally, into a locally convex topological vector lattice E, first important results are due to M. A. Krasnosel'skiĭ and E. A. Lifšič [11]. The purpose of the present paper is to outline a parallel theory. Krasnosel'skiĭ and Lifšič give sufficient conditions for a set S in C(X) to be a Korovkin set. The present authors are often able to describe shadows, even if they are not equal to C(X), exactly or by inclusion; this does not seem to be possible with the old technique. Accordingly, these methods of proof are more general. In preliminary form their results appear in G. G. Lorentz [15, Chapter 3]. Thus, the present paper is not expository.

The paper of Krasnosel'skiĭ and Lifšič remained little known, perhaps because of intricacy of the proof of their main theorem. We take the liberty to give a simplified version of their proof in Section 5.

For positive operators on a Banach lattice F into a Banach lattice E, M. Wolff [25]--as we learned during the preparation of the manuscript--gives a satisfactory extension of Šaškin's theorems, see his contribution [26] in these proceedings. Contractions and positive contractions on a Banach lattice E have been studied by D. E. Wulbert [27], for $L^1(0,1)$, and by the authors in [5].

In Section 2 we introduce further definitions and notations

and prove auxiliary results. The main theorems of the paper are in Sections 3 and 4 Theorem 3, for example, characterizes the total shadow of an admissible set S in C(X). In Section 4, we discuss Banach function spaces, rather than Banach lattices; thus the set-up here is the same as for the theorems of Krasnosel'skiĭ and Lifšič, discussed in Section 5. An Appendix discusses convergence sets for positive contractions in Banach lattices.

Historical and critical remarks are concentrated at the end of each section.

## 2. Auxiliary Results

Let E be a Banach lattice and P a lattice homomorphism on C(X) into E. Here are some simple facts about shadows.

Proposition 1. Let $\Phi$ be a lattice isomorphism on E onto itself. Then

$$\mathcal{S}(S,E,\Phi P) = \mathcal{S}(S,E,P).$$

In particular, for $E = C(X)$ and $P = I$

$$\mathcal{S}(S,C,\Phi) = \mathcal{S}(S,E,I).$$

Proof: Obviously, $\Phi P$ is a lattice homomorphism on C(X) into E. By definition, a function f belongs to $\mathcal{S}(S,E,P)$ if, given any sequence $T_n$, $n=1,2,\ldots$, in $\mathcal{L}^+(C,E)$, $T_n g \to Pg (g \in S)$ implies $T_n f \to Pf$. If $T_n'$, $n=1,2,\ldots$, is any sequence in $\mathcal{L}^+(C,E)$ satisfying $T_n' g \to \Phi Pg (g \in S)$, then $T_n' f \to \Phi Pf$. Indeed, setting $T_n = \Phi^{-1} T_n'$, then $T_n g \to Pg (g \in S)$, hence $T_n f \to Pf$ which in turn implies $T_n' f \to \Phi Pf$. The rest of the proof is clear.

Proposition 2. Let $\Psi$ be a lattice isomorphism of C(X) onto itself. Then

$$\mathcal{S}(S,E,P) = \Psi^{-1}(\Psi S,E, P\Psi^{-1}).$$

In particular,

$$S(S,C,I) = \Psi^{-1}S(\Psi S,C,\Psi^{-1}).$$

We omit the easy proof.

Let P be a lattice homomorphism on C(X) into a Banach lattice E. The kernel of P, $N(P) = \{f \in C(X): Pf = 0\}$, is a solid closed sublattice in C(X). Indeed, N(P) is a closed sublattice. Also, if $f \in N(P)$ and if $g \in C(X)$ is such that $|g| \leq |f|$, then the estimate $0 \leq |Pg| = P|g| \leq P|f| = |Pf| = 0$ shows that N(P) is solid.

To each $f \in C(X)$ let us associate the point set $N_f = \{x \in X: f(x) = 0\}$. The <u>support of</u> P is defined to be the intersection of all sets $N_f$, where $f \in N(P)$, in notation

$$\text{supp } P = \bigcap_{N(P)} N_f.$$

By definition, the support of P is a closed subset of X. By a compactness argument one verifies easily that supp P = $\emptyset$ exactly when N(P) equals C(X) or, equivalently, P is the zero transformation. Also, supp P = X holds exactly when N(P) = {0}. More generally, we obtain from Kakutani's characterization of solid closed sublattices in C(X) (cf. M. H. Stone [23, Theorem 7]).

<u>Lemma 1.</u> A function $f \in C(X)$ belongs to N(P) exactly when f(x) vanishes on supp P.

For the reader's convenience let us give the proof.

<u>Proof:</u> We first show that there is a function $f_0 \in N(P)$ with $N_{f_0}$ = supp P. Since X is metrizable, hence separable, the intersection defining supp P reduces to a countable intersection, say, $\bigcap_{k=1}^{\infty} N_{f_k}$. We can assume that $f_k \geq 0$, $\|f_k\| \leq 1$. Then $f_0 = \Sigma_{k=1}^{\infty} 2^{-k} f_k$ is in N(P) and $N_{f_0} = \bigcap N_{f_k}$.

Now we show that $f \in N(P)$ if f vanishes on supp P. Let $\varepsilon > 0$ be arbitrary, then there exists a neighborhood U of this set such that $|f(x)| < \varepsilon$ on U. On $X \setminus U$, $f_0(x) \geq c$ for some

$c > 0$. Hence

$$|f(x)| < \varepsilon + c^{-1}\|f\| f_0(x), \quad x \in X,$$

and

$$|Pf| \leq \varepsilon P1 + c^{-1}\|f\| Pf_0 = \varepsilon P1.$$

Thus $Pf = 0$, proving the lemma.

Lemma 1 implies that the quotient $C(X) / N(P)$ is isometrically isomorphic to $C(\operatorname{supp} P)$. Another consequence, needed below, is given in

Lemma 2. Let $f \in C(X)$ be positive on supp P in X, then Pf is contained in $E^+$.

We omit the proof.

By Riesz representation theorem, the dual of $C(X)$ is isometrically isomorphic to $M(X)$, the space of regular Borel measures on X. Let us set $M^+(x)$ for the cone of positive measures and $P(X)$ for the probability measures on X, $P(X) = \{\mu \in M: \mu(1) = 1 = \|\mu\|\}$.

Let S be an admissible set in $C(X)$ and set $G = \operatorname{lin} S$. For any $x \in X$, $P_x(S)$ denotes the set of all $\mu \in P$ such that

$$\mu(g) = \int g d\mu = g(x) \qquad (g \in S).$$

The set $P_x(S)$ is not empty. In particular, the evaluation functional $\varepsilon_x$: $\varepsilon_x(f) = f(x)$ $(f \in C(X))$, belongs to $P_x(S)$.

Let S be an admissible set in $C(X)$ and let A be a subset of X. We denote by $\hat{G}_A$ the set of all $f \in C(X)$ such that

(2) for each $x \in A$ and each $\mu \in P_x(S)$

$$\mu(f) = f(x).$$

Clearly $\hat{G}_A$ is a closed linear subspace in $C(X)$ containing G. It may be characterized as follows: For an $x \in X$, let $\ell_x$ denote the restriction of $\varepsilon_x$ on G. The set $\hat{G}_A$ is the largest closed linear subspace in $C(X)$ on which $\ell_x$ has a unique extension for all $x \in A$.

The following lemma gives two equivalent characterizations of $\hat{G}_A$. To be able to formulate the lemma we have to introduce one more notion.

As above, let S be admissible and G = lin S. For an $f \in C(X)$, we denote by

$$\overline{f}(x) = \inf \{g(x): f \leq g,\ g \in G\}$$

and

$$\underline{f}(x) = \sup \{g(x): g \leq f,\ g \in G\}$$

the <u>upper</u> and <u>lower</u> G-<u>envelope</u> of f, respectively. Obviously, $\underline{f} = -\overline{(-f)}$, and $\overline{f}$ is an upper semi-continuous function on X. Also, for $f_1, f_2 \in C(X)$, $f_1 \leq f_2$ implies $\overline{f}_1 \leq \overline{f}_2$, and $\overline{f_1 + f_2} \leq \overline{f}_1 + \overline{f}_2$, $\overline{\lambda f} = \lambda \overline{f}$ in case $\lambda \geq 0$.

The G-envelopes of a function f were introduced by H. Bauer [3, Sätze 3 and 4] who also formulated and proved the following

<u>Lemma 3</u>. For an $f \in C(X)$ and a compact subset K of X the following statements are equivalent:

(i) $f \in \hat{G}_K$;

(ii) $\underline{f}(x) = f(x) = \overline{f}(x)$ for all $x \in K$;

(iii) given any $\delta > 0$, there exists finitely many $g_1', g_2', \ldots g_m', g_1'', \ldots, g_m''$ in G such that for the elements

$$\underline{g} = \bigwedge_{i=1}^{m} g_i', \quad \overline{g} = \bigvee_{i=1}^{m} g_i''$$

one has

$$\underline{g} \leq f \leq \overline{g} \quad \text{and} \quad (\overline{g} - \underline{g})|_K < \delta.$$

This lemma is crucial for the proofs of the theorems given in the following two sections. Because of this we shall indicate its proof. An essential part of it rests upon the following two statements which we give without proof.

(a) For each $x \in X$, $\mu \in P_x(S)$ and $f \in C(X)$
$$\underline{f}(x) \leq \mu(f) \leq \overline{f}(x).$$

(b) If $x \in X$, $f_0 \in C(X)$ and $\gamma \in R$ are given such that

$$\underline{f}_0(x) \leq \gamma \leq \overline{f}_0(x),$$

then there exists a $\mu \in P_x(S)$ such that $\gamma = \mu(f_0)$.

<u>Proof:</u> The equivalence (i) $\Longleftrightarrow$ (ii) follows quickly from (a) and (b). To prove the implication (ii) $\Rightarrow$ (iii), let $\delta > 0$ be given. For each $x \in K$, by (ii) there exist $g'_x$ and $g''_x$ in G such that

$$g'_x \leq f \leq g''_x \quad \text{and} \quad g''_x(x) - g'_x(x) < \delta.$$

By continuity, $g''_x(y) - g'_x(y) < \delta$ for some open neighborhood $U_x$ of x. The family $\{U_x: x \in K\}$ forms an open cover of K, and since K is compact, there is a finite subcover, say $\{U_{x_i} : i=1,2, \ldots,m\}$. The associated functions $g'_{x_i}$ and $g''_{x_i}$ $(i=1,2,\ldots,m)$--in simplified notation $g'_i$ and $g''_i$--have the desired properties. Finally, let us show that (iii) $\Rightarrow$ (i). Let $f \in C(X)$ be such that (iii) holds. For an $x \in K$ and $\mu \in P_x(S)$

$$g'_i(x) \leq \int f d\mu \leq g''_i(x), \qquad i=1,2,\ldots,m.$$

Hence $\underline{g}(x) \leq \mu(f) \leq \overline{g}(x)$. Since $\underline{g}(x) \leq f(x) \leq \overline{g}(x)$ for all $x \in X$, $|f(x) - \mu(f)| \leq \overline{g}(x) - \underline{g}(x) < \delta$ for each $x \in K$, proving the implication.

Let S be an admissible set, and $G = \text{lin } S$. If $x \in X$ is such that the evaluation functional $\ell_x$ on G has a unique extension onto $C(X)$, or, equivalently,

$$P_x(S) = \{\varepsilon_x\},$$

then x is said to be an <u>extremal point</u> for G. The set of extremal points is known as the <u>Choquet boundary for</u> G, briefly: $\partial_{Ch}G$. H. Bauer [3] proved that admissible sets have extremal points, or equivalently, that the Choquet boundary is a non-empty subset of X.

Let us mention one further result which is of interest to

us in Section 4. Throughout the paper the compact space X is assumed to be metrizable. In this case, for any admissible subspace G, $\partial_{Ch} G$ is a $G_\delta$-set in X, hence Borel measurable. For these and further results on Choquet theory we refer to H. Bauer [3] and R. R. Phelps [19].

Bauer's Lemma 3 has as a predecessor a theorem of G. G. Lorentz [13], characterizing almost convergent sequences. Analogues of (i) and (ii) appear there, while (iii) has a somewhat different form.

## 3. Theorems for Banach Lattices

We are now in a position to formulate and prove the first theorems.

Theorem 1. Let E be a Banach lattice and let P be a lattice homomorphism on C(X) into E. If S is an admissible set in C(X), then the subspace $\hat{G}_{\text{supp } P}$ is contained in the shadow of S with respect to $L^+(C,E)$ and P, briefly:

(3) $\hat{G}_{\text{supp } P} \subset \mathcal{S}(S,E,P)$, where $G = \text{lin } S$.

Moreover,

(4) $\hat{G}_X \subset \mathcal{S}(S,E)$.

Proof: Let f be a function in $\hat{G}_{\text{supp } P}$. Since supp P is compact, by Lemma 3 (iii) for any given $\delta > 0$ there exist

$$\overline{g} = \bigwedge_{i=1}^{m} g_i'' \quad \text{and} \quad \underline{g} = \bigvee_{i=1}^{m} g_i'$$

such that

$$\underline{g} \leq f \leq \overline{g} \quad \text{and} \quad \overline{g} - \underline{g}|_{\text{supp } P} < \delta.$$

Hence

$$T_n\underline{g} - P\overline{g} \leq T_n f - Pf \leq T_n\overline{g} - P\underline{g}$$

and by Lemma 2

$$T_n\underline{g} - P\underline{g} - \delta P1 \leq T_n f - Pf \leq T_n\overline{g} - P\overline{g} + \delta P1.$$

Furthermore,

$$T_n\overline{g} \leq \bigwedge_{i=1}^{n} T_n g_i'' \quad \text{and} \quad T_n\underline{g} \geq \bigvee_{i=1}^{n} T_n g_i',$$

and

$$T_n\overline{g} - P\overline{g} \leq \sum_{i=1}^{m} |T_n g_i'' - Pg_i''| \quad \text{and} \quad T_n\underline{g} - P\underline{g} \geq -\sum_{i=1}^{m} |T_n g_i' - Pg_i'|,$$

giving

$$|T_n f - Pf| \leq \sum_{i=1}^{m} |T_n g_i'' - Pg_i''| + \sum_{i=1}^{m} |T_n g_i' - Pg_i'| + \delta P1.$$

Consequently,

$$\lim_{n\to\infty} \|T_n f - Pf\| \leq \delta\|P1\|$$

for any $\delta > 0$, proving relation (3). Relation (4) follows immediately from (3).

In general the inclusions (3) and (4) are proper as we shall see in Section 4. However, the following converse is true.

<u>Theorem 2</u>. If S is an admissible set in $C(X)$, then $\hat{G}_X$ contains the shadow of S with respect to $L^+(C)$ and the identity I, briefly:

$$\mathcal{S}(S,C,I) = \hat{G}_X.$$

<u>Proof:</u> We only have to prove $\mathcal{S}(S,C,I) \subset \hat{G}_X$. Let $f_0$ be a function in $C(X)$ which does not belong to $\hat{G}_X$. We shall construct a sequence $T_n$, $n=1,2,\ldots$, in $L^+(C)$ such that $T_n g \to g$ in C for all $g \in S$, while $T_n f_0 \not\to f_0$ in C. Since $f_0 \notin \hat{G}_X$, by Lemma 3(ii) there exists a point $x_0$ in X such that either $\underline{f}_0(x_0) < f_0(x_0)$ or $f_0(x_0) < \overline{f}_0(x_0)$. Let us assume the latter inequality is true. If $d(\cdot,\cdot)$ is the distance function on X, we define functions $L_n \in C(X)$, $n=1,2,\ldots$, as follows: $\varphi_n(x) = 1$ when $d(x,x_0) \leq 1/2n$, $= 0$ when $d(x,x_0) \geq 1/n$, and $0 \leq \varphi_n(x) \leq 1$ elsewhere. The sequence of operators in $L^+(C)$

$$T_n f(x) = \varphi_n(x)\mu_0(f) + \{1 - \varphi_n(x)\}f(x), \qquad n=1,2,\ldots$$

where $\mu_0 \in P_{x_0}(S)$ is such that $\mu_0(f_0) = \overline{f}_0(x_0)$, has the desired properties. The existence of such a $\mu_0$ is guaranteed by property (b), stated below Lemma 3. Indeed, for any $g \in S$

$$T_n g(x) - g(x) = \varphi_n(x)\{g(x_0) - g(x)\},$$

which converges uniformly to zero as $n \to \infty$. On the other hand, for the function $f_0$

$$\lim_{n\to\infty} T_n f_0(x_0) = \overline{f}_0(x_0) \neq f_0(x_0),$$

i.e., $T_n f_0(x)$ does not even converge pointwise to $f_0(x)$.

An immediate consequence of Theorem 1 and 2 is

<u>Theorem 3.</u> Let S be an admissible set in C(X). The total shadow of S equals $\hat{G}_X$, where G = lin S. Moreover, S is a total Korovkin set in C(X) exactly when the Choquet boundary for G equals X.

<u>Remark 1.</u> Theorem 1 remains true for locally convex topological vector lattices (instead of Banach lattices). The proof remains the same, with the norm $\|\cdot\|$ now being replaced by a family of semi-norms, which define the topology.

<u>Remark 2.</u> Bauer's Lemma 3 holds for C(X), where X is a compact Hausdorff space. Consequently, Theorem 1 remains true for these spaces X. Metrizability is needed for Theorem 2, but even here we can omit it if we assume that, for each $x \in X$, $\{x\}$ is a $G_\delta$-set.

<u>Remark 3.</u> One easily shows that a total Korovkin set S in C(X), i.e., a Korovkin set with respect to $L^+(C)$ and I, necessarily has to separate points and to contain a strictly positive function, say $g_0$. The mapping $f \to f/g_0$ defines a lattice isomorphism $\Psi$ on C(X) onto itself, mapping $g_0$ into 1. It follows from Proposition 2 that $S(S,C,I) = \Psi^{-1}S(\Psi S, C, \Psi^{-1})$. Hence

it is no restriction to assume that S itself contains the function 1.

On the other hand, let S be a set in C(X) containing 1. For the development of the theory of shadows for S we need not to assume that S separates points. In this case we have to replace definition (2) by

(2') For an $x \in X$, let $\hat{G}_x$ denote the closed linear subspace of C(X) onto which the evaluation functional $\ell_x$ on $G = \text{lin } S$, i.e., $\ell_x(g) = g(x)$, has a unique linear extension. For a set $A \subset X$ we define

$$\hat{G}_A = \bigcap_{x \in A} \hat{G}_x .$$

If S separates points, then (2') is equivalent to (2). With $\hat{G}_A$ defined as in (2') Lemmas 3 and 4 (in Section 4), as well as their proofs remain valid. This discussion shows that the theorems of this and the next section remain valid if one only assumes that the set S contains a strictly positive function $g_0$.

Remark 4. As an application, we have, for S satisfying the above condition, that the shadow of S lies between the closed linear hull of S, and the closed vector lattice generated by S.

Theorem 3 extends Šaškin's theorem on Korovkin sets in C(X) with respect to $L^+(C)$ and I in essentially two directions. First, we are able to characterize the shadow $S(S,C,I)$ for any admissible set S in C(X), namely,

$$S(S,C,I) = \hat{G}_X = \{f \in C(X): \ \underline{f} = f = \overline{f}\}.$$

Secondly, we prove that for any Banach lattice E and any lattice homomorphism P: $C(X) \to E$

$$S(S,C,I) \subset S(S,E,P).$$

This inclusion contains a great many of the Korovkin type theorems known in the literature, see, for example, Theorem 4 below.

The method of proof of Theorem 1 is also new (although less simple than the one of Šaškin). The idea to use Lemma 3 for the proof of Theorem 1 we owe to an unpublished manuscript of W. B. Arveson [2]. In this paper, Lemma 3 (iii) was used to obtain Korovkin type theorems for function algebras. For Theorem 2 see Yu. A. Šaškin [22].

The concept of a shadow of a set in C(X) is implicitly contained in Yu. A. Šaškin [21], explicitly it is given in M. A. Krasnosel'skiĭ and E. A. Lifšič [11] and also C. Franchetti [8].

Theorems for contractions on C(X), corresponding to Šaškin's results for positive operators, were given independently by D. E. Wulbert [27] and Yu. A. Šaškin [22] (see also the remark about this possibility in [11, p. 467]) and later by C. Franchetti [8] who in addition characterized the shadow of admissible sets in C(X) with respect to contractions and the identity I. For a detailed discussion of these results we refer to the lecture notes [15] of Lorentz.

During the preparation of the paper we learned about further results by H. Bauer [4] and M. Wolff [25], [26]. H. Bauer considers the space C(X), where X is a locally compact Hausdorff space. By use of an extension of Lemma 3 he characterizes the shadow of so-called adapted subspaces of C(X) with respect to $L(C)$ and the identity I. In addition he remarks that the operators under consideration need not be linear but merely <u>monotone</u>, i.e., $Tf \leq Tg$ whenever $f \leq g$. M. Wolff proves Korovkin type theorems for vector lattices. He replaces the space of continuous functions C(X) by a locally convex sequentially complete vector lattice F. Among others Wolff proves that F has a <u>finite</u> total Korovkin system if and only if F is finitely generated. If F is a separable Banach lattice with order continuous norm then F has a Korovkin set of three elements.

As Korovkin's celebrated results show the (total) shadow of an admissible set S in C(X) is usually much larger than its closed linear hull $\overline{G}$. It has been observed by H. Bauer [3] that if $G = \text{lin } S$ is a lattice under the canonical ordering, then $S(S,C,I) = \overline{G}$. As an example the above named author gives the set $S = \{1,x\}$ in $C[0,1]$. The linear functions form a lattice under the canonical order, but not a sublattice of $C(X)$, see also C. Franchetti [8].

As a first application let $E = R$. A continuous, linear functional, not identically zero, is a lattice homomorphism P on $C(X)$ into R exactly when it is a positive multiple of an evaluation functional, i.e.,

$$Pf = \lambda f(x)$$

for some $x \in X$ and some $\lambda > 0$. (An easy consequence of Lemma 1.) In this case the support of P equals $\{x\}$. We have

<u>Proposition 3</u>. Let S be an admissible set in $C(X)$ and $G = \text{lin } S$. If $\mu_n, n=1,2,\ldots$ is any sequence in $M^+(X)$ such that, for some $x \in X$, $\lim_{n\to\infty} \mu_n(g) = g(x)$, $g \in S$, then

$$(5) \quad \lim_{n\to\infty} \mu_n(f) = f(x)$$

for those $f \in C(X)$ for which $\underline{f}(x) = f(x) = \overline{f}(x)$. Also, relation (5) holds for all $f \in C(X)$ if, and only if,

$$P_x(S) = \{\varepsilon_x\}.$$

The second part of Proposition 3 is again essentially Šaškin's [20], while the case $C[0,1]$ goes back to Korovkin. See also Wulbert [27].

As a second example let $E = C(Y)$, where Y is a compact Hausdorff space. If $\Phi$ is any continuous mapping of Y into X and $\lambda(y)$ any function in $C^+(Y)$, then $Pf(y) = \lambda(y)\, f\circ\Phi(y)$, $f \in C(X)$, forms a lattice homomorphism on $C(X)$ into $C(Y)$. More generally, W. A. J. Luxemburg (oral communication) proved:

Let P be a lattice homomorphism of $C(X)$ into $C(Y)$, X and Y compact Hausdorff spaces, there exists a continuous mapping

$\Phi$ of the cozero set $\{y \in Y: [P1](y) > 0\}$ into X such that for all $f \in C(X)$

$$[Pf](y) = [P1](y) \cdot f \circ \Phi(y).$$

For these P, one obtains many meaningful special versions of Theorem 1. We leave their formulation to the reader.

We will conclude with the following observation. Let $X_0$ be a compact subset of X, let $S_0 = S|_{X_0}$ be the set of restrictions to $X_0$ of the function $g \in S$.

Proposition 4.

$$\mathcal{S}(S_0, C(X_0), I) \supset \mathcal{S}(S, C(X), I)|_{X_0}.$$

In particular, if S is a Korovkin set in C(X), so is $S_0$ in $C(X_0)$.

Proof: The inclusion follows from relation (5) of Theorem 1, because the envelopes for f and $f_0 = f|_{X_0}$ satisfy

$$\underline{f} \leq \underline{f}_0 \leq f_0 \leq \overline{f}_0 \leq \overline{f} \quad \text{on } X_0.$$

## 4. Theorems for Banach Function Spaces

A typical application of Theorem 1 is outlined in the following paragraphs.

Let X be a compact metric space, and let $\nu_0$ be, henceforth, a fixed probability measure on the measure space $(X, \mathcal{B})$, where $\mathcal{B}$ is the Borel field. We denote by $M = M(X, \mathcal{B}, \nu_0)$ the space of the (equivalence classes of the) real-valued Borel measurable functions on X which are finite ($\nu_0$-) a.e. Endowed with the topology of convergence in measure, M is a Frechet lattice with positive cone $M^+ = \{f \in M: f(x) \geq 0 \text{ a.e.}\}$. The identification of an element $f \in C(X)$ with its equivalence class in M forms a lattice homomorphism I of C(X) into M. We simply write f again instead of If.

Let $L^\rho = L^\rho(X, \mathcal{B}, \nu_0)$ be a Banach function space in the sense of Luxemburg with norm $\|\cdot\|_\rho = \rho(|\cdot|)$, where $\rho$ is a function norm on $M^+$ having the Fatou property,

(i) $0 \leq \rho(f) \leq \infty$ for all $f \in M^+$; and $\rho(f) = 0$ if and only if $f = 0$.

(ii) $\rho(f + g) \leq \rho(f) + \rho(g)$ for all $f, g \in M^+$; $\rho(\lambda f) = \lambda\rho(f)$ for all $f \in M^+$ and all $\lambda \geq 0$; and $0 \leq g \leq f$ implies $\rho(g) \leq \rho(f)$.

(iii) If $0 \leq f_n \uparrow f$ pointwise, then $\rho(f_n) \uparrow \rho(f)$.

Under these assumptions on $\rho$, $L^\rho$ is a Banach space which is continuously embedded into M, that is:

(6) If $f_n$, $n=1,2,\ldots$ and $f_0$ are functions in $L^\rho$ and if $f_n \to f_0$ in $L^\rho$ as $n \to \infty$ then $f_n \to f_0$ in measure.

(see W. A. J. Luxemburg [17]). In addition, let us assume that $\rho(1)$ is finite or, equivalently,

(7) $L^\infty$ is continuously embedded in $L^\rho$.

The following result is an immediate consequence of Theorem 1.

Theorem 4. Let S be a total Korovkin set in C(X), and let $T_n$ $n=1,2,\ldots$ be a sequence of operators in $L^+(L^\rho)$ such that $\lim_n T_n g = g$ in $L^\rho$ $(g \in S)$. Then, for all $f \in C(X)$

(8) $\lim_n T_n f = f$ in $L^\rho$.

Moreover, if the norms of the transformations $T_n$ are uniformly bounded in n, then (8) holds for the closure of C(X) in $L^\rho$, expecially for all $f \in L^\rho$, if C(X) is dense in $L^\rho$.

Proof: For each $n=1,2,\ldots$, we define on C(X) the positive linear transformation $T'_n = T_n I$ into $L^\rho$, where I is the natural embedding of C(X) into $L^\rho$, and apply Theorem 1.

It is the aim of the following considerations to characterize the shadow of an admissible set S with respect to

$L^+(C,L^0)$ and $I$, the natural embedding of $C(X)$ into $L^0$. To do this we need

Lemma 4. Let $S$ be an admissible set in $C(X)$. For an $f \in C(X)$ the following statements are equivalent.

(i) $f \in \tilde{G}_X$, i.e., for almost all $x \in X$
$$f(x) = \int f d\mu, \qquad \mu \in P_x(G),$$

(ii) $\underline{f}(x) = f(x) = \overline{f}(x)$ a.e.

(iii) given any $\varepsilon$, $0 < \varepsilon < 1$, and $\delta > 0$, there is a compact set $K \subset X$ such that $\nu_0(X \setminus K) < \varepsilon$ and there are functions $g_1', \ldots, g_m', g_1'', \ldots, g_m''$ in $G$ such that
$$\bigvee_{i=1}^m g_i' = \underline{g} \le f \le \overline{g} = \bigwedge_{i=1}^m g_i'' \quad \text{and} \quad \overline{g} - \underline{g}\big|_K < \delta.$$

By taking into account that the measure $\nu_0$ is regular, the proof follows readily from Lemma 3.

Clearly, $\tilde{G}_X$ is a closed linear subspace of $C(X)$. We also have
$$G_{\text{supp } \nu_0} \subset \tilde{G}_X.$$

Moreover,

Proposition 5. Let $\partial_{Ch} G$ be the Choquet boundary for $G$, $G = \text{lin } S$. Then $\nu_0(\partial_{Ch} G) = 1$ holds exactly when $\tilde{G}_X = C(X)$.

Proof: Since $C(X)$ is separable, $\partial_{Ch} G$ is a $G_\delta$-set in $X$ and hence Borel measurable. If $\nu_0(\partial_{Ch} G) = 1$ then trivially $\tilde{G}_X$ equals $C(X)$. Conversely, let $\tilde{G}_X = C(X)$. We take a dense sequence $f_m$, $m=1,2,\ldots$, in $C(X)$. For each $m$, let $N_m$ be the null set of points $x \in X$ for which there exists a $\mu \in P_x(G)$ such that $f_m(x) \ne \mu(f_m)$, and let $N = \bigcup_{m=1}^\infty N_m$. The set $N$ is also a null set. Let $f$ be an arbitrary element in $C(X)$. There is a subsequence $f_{m_k}$, $k=1,2,\ldots$, for which $f_{m_k} \to f$ in $C$. If $x$ is any point in $X \setminus N$ and $\mu$ any measure in $P_x(G)$, then

$$\mu(f) = \lim_k \mu(f_{m_k}) = \lim_k f_{m_k}(x) = f(x).$$

Hence, for all $x \in X \setminus N$ and all $f \in C(X)$ we have $f(x) = \mu(f)$, $\mu \in P_x(G)$. Thus $X \setminus N \subset \partial_{Ch} G$. Since $\nu_0(X \setminus N) = 1$ it follows that $\nu_0(\partial_{Ch} G) = 1$.

Theorem 5. Let S be an admissible set in $C(X)$, and let I be the natural embedding of $C(X)$ into $L^\rho$. Then

(9) $\hat{G}_{\text{supp } \nu_0} \subset S(S, L^\rho, I) \subset \tilde{G}_X$.

Proof: Since the support of I equals supp $\nu_0$, the first inclusion follows directly from Theorem 1. The second inclusion is a corollary of the following lemma.

Lemma 5. Let S be an admissible set. If a function $f_0 \in C(X)$ does not belong to $\tilde{G}_X$, then there exists a positive linear transformation T on $C(X)$ into $L^\infty$ for which

(i) $Tg(x) = g(x)$ on X $(g \in S)$

(ii) $Tf_0(x) \neq f_0(x)$ on a set of positive measure.

Proof: Since $f_0 \notin \tilde{G}_X$, there exists a set of positive measure such that $f_0(x) < \overline{f}_0(x)$ (if not, we replace $f_0$ by $-f_0$).

Let $G_0 = G + Rf_0$, and define the operator $T_0$: $G_0 \to L^\infty$ by

$$T_0 g_0(x) = g(x) + \lambda \overline{f}_0(x)$$

for all $g_0 = g + \lambda f_0$ in $G_0$. Since $\overline{f}_0$ is upper semi-continuous and less or equal to $\|f_0\|$, $T_0$ obviously defines a linear transformation on $G_0$ into $L^\infty$. Moreover, $T_0$ is positive, since $g + \lambda f_0 \geq 0$ implies $g + \lambda \overline{f}_0 \geq 0$, and restricted to G, $T_0$ is the identity, $T_0 g = g$ $(g \in G)$.

The rest of the proof follows from an extension theorem of Kantorovič (see B. Z. Vulikh [24, Theorem X,3.1]): Let F be a vector lattice and $F_0$ a subspace of F which majorizes F (that is, for each $f \in F$ there is an $f_0 \in F_0$ such that $|f| \leq f_0$). Then each positive linear transformation $T_0$ defined on $F_0$ and with values in an order complete vector lattice E, has a positive linear extension on F. Since $1 \in S$, the subspace $G_0$ majorizes $C(X)$, and $L^\infty$ is order complete. By the theorem of Kantorovič $T_0$: $G_0 \to L^\infty$ has a positive linear extension T on $C(X)$, satisfying (i) and (ii). This completes the proof.

Let us conclude with the remark that the operator $T: C(X) \to L^\infty$ forms a measurable selection from the set-valued function $x \to P_x(S)$ on X into $P$.

We can now complete the proof of Theorem 5. From Lemma 5 it follows that if $f \notin \tilde{G}_X$ then there is a $T \in L^+(C,L^\rho)$ such that $Tg = g$ for all $g \in S$ and $Tf \neq f$. Therefore, $f \notin S(S,L^\rho,I)$, proving the second inclusion (9).

While the first inclusion (9) is an immediate consequence of Theorem 1, the second inclusion is a statement parallel to Theorem 2. In the following theorem we give a sufficient condition for the function space $L^\rho$ to insure that the shadow $S(S,L^\rho,I)$ actually equals $\tilde{G}_X$.

The space $L^\rho$ has the dominated convergence property if the following is true:

(10) For a sequence of functions $f_m$, $m=1,2,\ldots$ in $L^\rho$, conditions $|f_m| \leq F_0$ for some $F_0 \in L^\rho$, and $f_m \to f_0$ a e. imply that $f_m \to f_0$ in $L^\rho$.

Theorem 6. Let S be an admissible set in $C(X)$, and let $L^\rho$ have the dominated convergence property. Then

$$S(S,L^\rho,I) = \tilde{G}_X.$$

Proof: We only have to prove the inclusion $\tilde{G}_X \subset S(S,L^\rho,I)$.

First let us recall that the dominated convergence property of $L^\rho$ is equivalent to the absolute continuity of the norm, that is,

(11) For any sequence of Borel sets $E_n$ which decreases to zero and any $f \in L^\rho$, $\|f\chi_{B_n}\|$ decreases to zero.

Here, $\chi_B$ is the characteristic function of B

Let $f \in \tilde{G}_X$. We may assume that $|f| \leq 1$. Given $0 < \varepsilon < 1$ and $\delta > 0$, by Lemma 4 (iii) there exist a compact subset $K \subset X$ with $\nu_0(X\backslash K) < \varepsilon$ and functions $g_1', g_2', \ldots, g_m', g_1'', \ldots, g_m''$ in G such that for $\underline{g} = \bigwedge_{i=1}^m g_i'$ and $\overline{g} = \bigvee_{i=1}^m g_i''$ one has

$$\underline{g} \leq f \leq \overline{g} \quad \text{and} \quad (\overline{g} - \underline{g})|_K < \delta.$$

We may assume that $|\underline{g}|, |\overline{g}| \leq 1$. It follows that

$$\overline{g} \leq \underline{g} + \delta + 2(\overline{g} - \underline{g})\chi_{X\backslash K} \leq \underline{g} + \delta + 2\chi_{X\backslash K}\,.$$

As in the proof of Theorem 1 we obtain

$$T_n\underline{g} - \underline{g} - 2\chi_{X\backslash K} \leq T_n f - f \leq T_n\overline{g} - \overline{g} + \delta + 2\chi_{X\backslash K}$$

and furthermore

$$|T_n f - f| \leq \sum_{i=1}^m \{|T_n g_i'' - g_i| + |T_n g_i' - g_i'|\} + \delta + 2\chi_{X\backslash K}\,.$$

Consequently,

$$\lim_n \|T_n f - f\|_\rho \leq \delta\|1\|_\rho + 2\|\chi_{X\backslash K}\|_\rho.$$

Since the norm $\|\cdot\|_\rho$ is absolutely continuous, we have $\|\chi_{X\backslash K}\|_\rho \leq \delta\|1\|_\rho$, if $\varepsilon > 0$ is sufficiently small. Hence $\overline{\lim}_n \|T_n f - f\|_\rho \leq 3\delta\|1\|_\rho$ for any $\delta > 0$, proving the theorem.

The Lebesque spaces $L^\rho$, $1 \leq \rho < \infty$, have the dominated convergence property, while $L^\infty$ does not. For this space we have

Theorem 7. Let S be an admissible set in $C(X)$. Then

$$\mathcal{S}(S, L^\infty, I) = G_{\operatorname{supp} \nu_0}.$$

Proof: We only have to show that $f_0 \notin G_{\operatorname{supp} \nu_0}$ implies that $f_0$ does not belong to the shadow. We can assume that there is a point $x_0 \in \operatorname{supp} \nu_0$ with the property $f(x_0) < \overline{f}_0(x_0)$. We take $T_n$, $n=1,2,\ldots$ in $L^+(C, L^\infty)$ as in the proof of Theorem 2, then

$$T_n f_0(x) - f_0(x) = \varphi_n(x)\{\mu_0(f_0) - f_0(x)\},$$

where $\mu_0(f_0) - f_0(x_0) = \gamma > 0$ For each n=1,2,... there is an open neighborhood $V_n$ of $x_0$ such that $T_n f_0(x) - f_0(x) > \frac{1}{2}\gamma$, $x \in V_n$. Since $\nu_0(V_n)$ is strictly positive, $\|T_n f_0 - f_0\|_\infty > \frac{1}{2}\gamma$ for each n, proving the theorem.

The validity of Korovkin theorems is not restricted to normed function spaces on $(X,\mathcal{B},\nu_0)$. As an example let us state

Theorem 8. For an admissible set S in C(X)

$$\mathcal{S}(S,M,I) = \tilde{G}_X.$$

We omit the proof and turn the reader's attention to the following Korovkin type theorem for convergence a.e.

Theorem 9. Let S be an admissible set in C(X), and let $T_n$, n=1,2,... be any sequence of positive linear transformations on C(X) into M. Necessary and sufficient that

(12) $\lim_n T_n g(x) = g(x)$ a.e. $(g \in S)$

implies for an $f \in C(X)$

(13) $\lim_n T_n f(x) = f(x)$ a.e.

is that $f \in \tilde{G}_X$.

Proof: Necessity follows directly from Lemma 5. To prove sufficiency, let $f \in \tilde{G}_X$. By Lemma 4 (iii) for any $\varepsilon$, $0 < \varepsilon < 1$, and any $\delta > 0$ there is a compact set $K \subset X$ with $\nu_0(X\setminus K) < \varepsilon/2$ and there are functions $g_1', \ldots, g_m'$, $g_1'', \ldots, g_m''$ in G such that, setting $\underline{g} = \bigwedge_{i=1}^m g_i'$ and $\overline{g} = \bigvee_{i=1}^n g_i''$,

$$\underline{g} \le f \le \overline{g} \quad \text{and} \quad \overline{g} - \underline{g}|_K < \delta/2.$$

By hypothesis, for all $g_i$'s in question $T_n g_i(x) \to g_i(x)$ a.e. Hence there is a set $B \subset \mathcal{B}$ and an $n_0$ such that $\nu_0(X\setminus B) < \varepsilon/2$ and for all $x \in B$ and $n \ge n_0$

$$T_n g_i'(x) \le T_n f(x) \le T_n g_i''(x), \qquad i=1,2,\ldots,m,$$

and

$$|T_n g_i(x) - g_i(x)| < \delta/2.$$

From this we conclude that for all $x \in B_0 = B \cap K$ and $n \geq n_0$

$$-\delta/2 + g_i'(x) \leq T_n f(x) \leq g_i''(x) + \delta/2$$

$$-\delta + f(x) < -\delta/2 + \underline{g}(x) \leq T_n f(x) \leq \overline{g}(x) + \delta/2 < f(x) + \delta,$$

or for $n \geq n_0$ and all $x \in B_0$, we have $\nu_0(X \setminus B_0) < \varepsilon$ and $|T_n f(x) - f(x)| < \delta$, i.e., $T_n f \to f$ almost uniformly, which implies (13).

## 5. Theorems of Krasnosel'skiĭ and Lifšič

As in Section 4, the Banach lattice E will be a Banach function space $L^\rho$, which consists of equivalence classes of $\nu_0$-measurable functions on a compact metric space X. We assume that

(14) $L^\infty \subset L_\rho \subset M$,

with normal embedding; moreover that $L_\rho$ has the dominated convergence property (10). As a simple corollary of (10) we need here:

(15) If $f_n \to f$ in $L_\rho$, then a subsequence $f_{n'}$ of $f_n$ is bounded: $|f_{n'}| \leq f_0$ for some $f_0 \in L_\rho$.

In [11], Krasnosel'skiĭ and Lifšič define shadows, but their results are about Korovkin sets. Unlike our assumption $\nu_0(\partial_{Ch} G) = 1$ about the Choquet boundary, these authors work with hypotheses about peak points.

A point $x_0 \in X$ is called a peak point for the closed linear hull $\overline{G}$ of G, if there is a function $g_{x_0} \in \overline{G}$ for which

$$g_{x_0}(x_0) = 0, \quad g_{x_0}(x) > 0, \ x \neq x_0.$$

We denote by p(G) the set of all peak points for $\overline{G}$. Obviously, $p(G) \subset \partial_{Ch}(G)$. In general, the inclusion is proper. On the other hand, E. Bishop proved that for compact metrizable X (to which we have restricted ourselves throughout this paper), p(G) is dense in $\partial_{Ch} G$ (see [19, p. 57]). With these definitions, the main theorem (Theorem 1, p. 456) of the paper [11] is as follows:

<u>Theorem 10</u>. Assume that $G = \text{lin } S \subset C(X)$ contains a strictly positive function $g_0$. If $\nu_0(p(G)) = 1$, then $S$ is a Korovkin set for $L_\rho$. In other words, in this case

(16) $\mathcal{S}(S, L_\rho, I) = L_\rho$.

We give here a proof of this, based on the following theorem on lifting. Spaces such as $L^p$ do not have lifting, but principal ideals in these spaces do. A principal ideal is an order ideal in $L_\rho$ generated by a single element $f_C \geq 0$, that is, the set of all $f \in L_\rho$ satisfying $|f| \leq Cf_0$ for some $C > 0$. The following is an easy corollary of [10, Theorem 3, p. 46]:

<u>Theorem L</u>. Let $L_\rho$ be a Banach function space satisfying (14), and let $F$ be a principal ideal of $L_\rho$. Let $F^*$ consist of all a.e. finite functions which belong to the equivalence classes forming the ideal $F$. Then there exists a map (a linear lifting) $U$: $F \to F^*$ that is linear and preserves inequalities.

<u>Proof of Theorem 10</u>: Let $T_n \geq 0$ be a sequence of linear operators mapping $C(X)$ into $L_\rho$, and let $T_n g \to g$, $g \in G$. For a fixed $f_1 \in C(X)$, we wish to prove that $T_n f_1 \to f_1$. It is sufficient to show this for a subsequence $T_{n'}$ of the sequence $T_n$. In what follows, we keep the notation $T_n$ for all possible subsequences.

We first note that for a properly chosen subsequence of the $T_n$, the set $F = \{T_n f : n=1,2,\dots,\ f \in C(X)\}$ is a principal ideal in $L_\rho$. For the strictly positive function $g_0$ we have $T_n g_0 \to g_0$. From (15) we see that an $f_0 \in L_\rho$ exists with the property $T_n g_0 \leq f_0$, at least for a subsequence. Then we have, if $f \in C(X)$ and $C \geq 0$ is such that $|f| \leq C g_0$,

(17) $|T_n f| \leq C T_n g_0 \leq C f_0, \quad n=1,2,\dots$

For the subsequence in question we can replace the operators $T_n$ by the lifted operators $T_n{*}f = UT_nf$.

Returning to the function $f_1$, we see from the dominated convergence property of $L_\rho$ and (17) that it is sufficient to construct a set e, $\nu_0(e) = 1$, such that for a subsequence, $T_n{*}f_1(x) \to f_1(x)$, $x \in e$. Let $e_0 = p(G)$.

Because of (14) we have, for each $g \in G$ and passing to a subsequence if necessary, that $T_n{*}g(x) \to g(x)$ a.e. Let $g_1,\dots,g_p,\dots$ be a dense sequence in G, then we still can find a subsequence and a set $e_1$, $\nu_0(e_1) = 1$, for which $T_n{*}g_p(x) \to g_p(x)$ for all p and all $x \in e_1$. In addition we can assume that the majorant $f_0$ of the $T_n{*}g_0$ as well as all functions $T_n{*}g_p$ are finite on $e_1$. We claim that $T_n{*}g(x) \to g(x)$, $x \in e_1$ for each $g \in G$, and that moreover all these functions are finite.

This follows from the following inequality. For each $\varepsilon > 0$ we can find a $g_p$ for which $|g - g_p| \le \varepsilon\, g_0$. Then

$$|(T_n{*}g(x) - g(x)) - (T_n{*}g_p(x) - g_p(x))|$$
$$\le |T_n(g - g_p)(x)| + |(g - g_p)(x)|$$
$$\le \varepsilon\, f_0(x) + \varepsilon\, g(x).$$

This proves our assertion.

Let $e_2$ be a set, $\nu_0(e_2) = 1$, for which all $T_n{*}f_1(x)$ are finite. We claim that $T_n{*}f_1(x) \to f_1(x)$ for $x \in e = e_0 \cap e_1 \cap e_2$. Let $x_0$ be a fixed point of e. It is sufficient to show that if $T_n{*}f_1(x) \to \lambda$, then $\lambda = f(x_0)$.

On the subspace $G + \mathbb{R}f$ we define the functional

$$L(g + \lambda f) = g(x_0) + a\lambda.$$

(This is the main idea of the proof in [11]). Then L is positive, since

$$L(g + af) = \lim_{n\to\infty} T_n{*}(g + af)(x_0) \ge 0 \quad \text{if } g + af \ge 0.$$

By a theorem of M. G. Kreĭn, L can be extended to a positive linear functional (again denoted by L) onto the whole of C(X). Since $x_0 \in p(G)$, we have the function $g_{x_0}$; and $L(g_{x_0}) = 0$ gives that the whole mass of L is concentrated at $x_0$, that is, $L = a\varepsilon_{x_0}$, where $\varepsilon_{x_0}$ is the evaluation functional. Substituting $g_0$, we get $a = 1$. Then, substituting f, we get $\lambda = f(x_0)$, as required. This completes the proof.

Analyzing this proof (as well as the proof in [11]) we see that it is not simple. It is our opinion that the approach via Lemma 3 of Section 2 is more natural and leads to better results. Thus, our Theorem 6, which is valid under the sole assumption $g_0 \in S$, see Remark 3, contains Theorem 10.

## Appendix

### Shadows for sequences of positive contractions

For sequences of positive linear contractions on the space C(X), the theory of Korovkin sets has been outlined in the lecture notes [15]. This theory is similar to the theory of Korovkin sets for positive operators and for contractions, although Korovkin sets are different in the three cases.

For $L^p$-spaces, $1 \leq p < \infty$, however, the situation is quite different.

Let E be a Banach lattice with norm $\|\cdot\|$. For a sequence $\mathcal{T}$ of (positive) contractions $T_n$, n=1,2,..., on E we define the convergence set $C_{\mathcal{T}}$ to be the set of all $f \in E$ such that $\lim_n T_n f = f$ in E. For a set S in E we define its shadow $\mathcal{S}(S)$ with respect to (positive) contractions to be

$$(18) \quad \mathcal{S}(S) = \cap_{\mathcal{T}} C_{\mathcal{T}}, \qquad \text{where } S \subset C_{\mathcal{T}}.$$

If $\mathcal{S}(S)$ equals E for some S, then we say S is a Korovkin set with respect to (positive) contractions on E.

One can estimate the shadow of a set S from above by means of its definition (18). For example, let T be a fixed (positive) contraction, and let $C_T = \{f \in E\colon Tf = f\}$. Then

(19) $S \subset C_T$ implies $\mathcal{S}(S) \subset C_T$.

Shadows are monotone, i.e., $S_1 \subset S_2$ implies $\mathcal{S}(S_1) \subset \mathcal{S}(S_2)$. Moreover, $\mathcal{S}(\mathcal{S}(S)) = \mathcal{S}(S)$.

In the following theorem we shall characterize the convergence set for a given sequence $\mathcal{T}$ of positive contractions on E for Banach lattices with uniformly strict norm. This means that the following condition is satisfied.

(20) For each $0 < \varepsilon < 1$ there is a $\delta(\varepsilon) > 0$ such that for every pair $f,g \in E$ satisfying $0 \leq g \leq f$, $\varepsilon \leq \|f - g\|$ and $\|f\| = 1$ we have $\|g\| \leq 1 - \delta(\varepsilon)$.

One easily proves that (20) is equivalent to

(21) If $f_n, g_n$, $n=1,2,\ldots$, are two sequences in E such that $0 \leq g_n \leq f_n$, $\|f_n\| = 1$ and $\|g_n\| \to 1$ as $n \to \infty$, then $\|f_n - g_n\| \to 0$ as $n \to \infty$.

Theorem 11. Let $\mathcal{T}$ be a sequence of positive contractions on a Banach space E having a uniformly strict norm. Then its convergence set $C_{\mathcal{T}}$ is a closed vector sublattice of E.

Proof: We only have to prove that $f \in C_{\mathcal{T}}$ implies $f^+ \in C_{\mathcal{T}}$, where $f^+$ is the positive part of f. We have

$$0 \leq (T_n f)^+ \leq T_n f^+,$$

also, since $T_n$ are contractive,

$$\|T_n f^+\| \leq \|f^+\|.$$

For any two elements $g,h \in E$, $|g^+ - h^+| \leq |g - h|$, hence $|(T_n f)^+ - f^+| \leq |T_n f - f|$. This shows that

$$\lim_n (T_n f)^+ = f^+ \text{ in E},$$

in particular,

$$\lim_n \|(T_n f)^+\| = \|f^+\|.$$

Hence by property (21)

$$\lim_n \|(T_n f)^+ - T_n f^+\| = 0.$$

Finally,

$$\|T_n f^+ - f^+\| \leq \|T_n f^+ - (T_n f)^+\| + \|(T_n f)^+ - f^+\| \to 0 \text{ as } n \to \infty.$$

Theorem 1 generalizes a result of the authors [5] for $L^1$-spaces. In this paper, convergence sets for arbitrary contractions on $L^1$-spaces are also studied.

As a consequence of Theorem 11 we have

Theorem 12. Let S be a subset in E, which satisfies (20). The closed vector sublattice generated by S is contained in its shadow $\mathcal{S}(S)$. In particular, if S generates E, then S is a Korovkin set.

The question now arises whether there are "interesting" Banach lattices E the norm of which is uniformly strict. Another question is whether Theorem 12 admits an inverse for certain lattices E.

Property (20) is closely related to uniform convexity. Indeed, every uniformly convex Banach lattice has property (20). The converse is wrong as the example of the $L^1$-space shows. We now discuss some special spaces.

1. Spaces $L^p$. Let $(X,\Sigma,\mu)$ be a measure space, and let $L^p = L^p(X,\Sigma,\mu)$, $1 \leq p \leq \infty$, be the usual Lebesque space. One easily verifies that for $1 \leq p < \infty$ the $L^p$-norm is uniformly strict with $\delta(\varepsilon) = \varepsilon^p/p$ in (20). Since by a result of T. Ando [1] every closed vector sublattice of $L^p$ is the range of a positive contractive projection, we also have: For each set S in $L^p$, $\mathcal{S}(S)$ equals the closed vector sublattice in $L^p$ generated by S. For further information we refer the reader to the useful survey article of H. E. Lacey and S. J. Bernau [12] on the characterization and classification of $L^p$-spaces. In another paper using special properties of the $L^p$-norm S. J. Bernau [6] describes the convergence set for sequences of (not necessarily

positive) contractions in $L^p$, $p \neq 1, 2, \infty$.

2. Spaces $\Lambda(\phi,p)$. Let $\phi(t) > 0$ be a decreasing integrable function on (0,1), and let $\Phi(x) = \int_0^x \phi(t)dt$. We assume that $\Phi(x)$ is normalized by $\Phi(1) = 1$. A Lebesque measurable function f on (0,1) belongs to the Lorentz space $\Lambda(\phi,p)$, $1 \leq p \leq \infty$, if

$$\|f\|_{\phi,p} = \{\int_0^1 \phi(t)f^*(t)^p dt\}^{1/p}$$

(with the obvious modification for $p = \infty$) is finite. Here $f^*$ is the decreasing rearrangement of f. See [14, Chapter III]. Identifying functions which are equal a.e., $\Lambda(\phi,p)$ is a Banach function space in the sense of Luxemberg, see Section 4. If, in particular, $\phi(t) \equiv 1$, then $\Lambda(\phi,p) = L^p$.

For these spaces, $\Lambda(\phi,p)$, $1 \leq p < \infty$, has property (20) exactly when

$$\sup_{0<x<1/2} \frac{\Phi(x)}{\Phi(2x)} < 1. \tag{22}$$

I. Halperin [9] proved that $\Lambda(\phi,p)$, $1 < p < \infty$, is uniformly convex exactly when (22) holds. His proof can be modified to yield the above result.

Uniform strictness of the norm of a Banach lattice E implies that the norm is order continuous. The spaces $\Lambda(\phi,p)$, $1 \leq p < \infty$, for instance, have an order continuous norm. However there are functions $\phi$ which do not satisfy (22). This shows that property (20) is not equivalent to order continuity of the norm. Also, the spaces $\Lambda(\phi,1)$ are not uniformly convex, because their norm is additive for positive decreasing functions.

3. Orlicz spaces. W. A. J. Luxemburg [16] and H. W. Milnes [18] have characterized Orlicz spaces which are uniformly convex. All these spaces satisfy condition (20).

---

[1]Supported, in part, by the NSF Grants GP-34417 and GP-23566.

## References


[1] Ando, T., Banachverbände und positive Operatoren. Math. Z. 109 (1969), 121-130.

[2] Arveson, W. B., An approximation theorem for function algebras. Unpublished manuscript, 9 pages, 1970.

[3] Bauer, H., Šilovscher Rand und Dirichletsches Problem. Ann. Inst. Fourier 11 (1961), 89-136.

[4] Bauer, H., Theorems of Korovkin type for adapted spaces. Ann. Inst. Fourier. In print.

[5] Berens, H. and G. G. Lorentz, Sequences of contractions of $L^1$-spaces. J. Functional Analysis. To appear.

[6] Bernau, S. J., Sequences of contractions of $L_p (1 < p < \infty, p \neq 2)$, an outline. Preprint, 5 pages.

[7] Dzjadyk, V. K., Approximation of functions by positive linear operators and singular integrals (Russian). Math. Sbornik (N.S.) 70 (112) (1966), 508-517, MR 34, #8053.

[8] Franchetti, C., Convergenza di operatori in sottospazi dello spacio C(Q). Boll. d. Un. Matem. Ital., Ser IV, 3 (1970), 668-678.

[9] Halperin, I., Uniform convexity in function spaces. Duke Math. J. 21 (1954), 195-204.

[10] Ionescu Tulcea, A. and C. Ionescu Tulcea, Topics in the Theory of Liftings. Springer-Verlag, Berlin 1969.

[11] Krasnosel'skiĭ, M. A. and E. A. Lifšič, A principle of convergence of sequences of positive linear operators (Russian). Studia Math. 31 (1968), 455-468, MR 38, #6372.

[12] Lacey, H. E. and S. J. Bernau, Characterizations and classifications of some classical Banach spaces. Advances in Math. In print.

[13] Lorentz, G. G., A contribution to the theory of divergent sequences, Acta Math. 80 (1948), 167-190.

[14] Lorentz, G. G., Bernstein Polynomials. University of Toronto Press, Toronto, 1953.

[15] Lorentz, G. G., Korovkin sets (Sets of convergence). Regional Conference at the University of California, Riverside, June 15-19, 1972.

[16] Luxemburg, W. A. J., Banach Function Spaces, Thesis, Delft Technological University, Holland, 1955.

[17] Luxemburg, W. A. J., Rearrangement-invariant Banach function spaces. Queen's Papers in Pure and Applied

Mathematics 10 (1967), 87-144. Queen's University, Canada.

[18] Milnes, H. W., Convexity of Orlicz spaces. Pacific J. Math. 7 (1957), 1451-1483.

[19] Phelps, R. R., Lectures on Choquet's Theorem. Van Nostrand Math. Studies #7, London 1966.

[20] Šaškin, Yu. A., Korovkin systems in spaces of continuous functions. Amer. Math. Soc. Transl. (2) 54 (1966), 125-144 (= Izv. Akad Nauk. S.S.S.R., Ser. Mat. 26 (1962), 495-512).

[21] Šaškin, Yu. A., The Milman-Choquet boundary and approximation theory, Funct. Anal. y Appl. 1 (1967), 170-171.

[22] Šaškin, Yu. A., On convergence of contraction operators, Math. Cluj. 11 (34), 1969, 355-360 (Russian).

[23] Stone, M. H., A generalized Weierstrass approximation theorem. In Studies in Modern Analysis, R. C. Buck ed., MAA Studies in Mathematics Vol. 1, 1962, pp. 30-87.

[24] Vulikh, B. Z., Introduction to the Theory of Partially Ordered Spaces. Jordon and Breach, New York, 1967.

[25] Wolff, M., Darstellung von Banachverbänden und Sätze vom Korovkin-Typ. Math. Ann. 200 (1973), 47-68.

[26] Wolff, M., A general theorem of Korovkin type for vector lattices. These Proceedings.

[27] Wulbert, D. E., Convergence of operators and Korovkin's theorem. J. Approx. Theory 1 (1968), 381-390.

Department of Mathematics
The University of Texas
Austin, Texas 78712

Department of Mathematics
The University of Texas
Austin, Texas 78712
and
California Institute of Technology
Pasadena, California 91109

# A SURVEY OF WORK ON APPROXIMATION AT AACHEN, 1968-1972

P. L. Butzer

In Spring 1971 Professor G. G. Lorentz invited me to prepare a lecture on "Work of the Aachen School in Approximation" to be delivered at the present Symposium on Approximation Theory here at Austin. At first I had great doubts about accepting such an honourful but also difficult task--difficult because reporting on the activities of oneself and ones students and former students is almost impossible to do with the necessary perspective and objectivity. Another reason is that it is actually too big a task to cover some 110 papers (including about a dozen doctoral dissertations) written in Aachen since 1968, even if I restrict myself to the essentials.

Now, however, I feel grateful to G. G. Lorentz for having urged me into accepting this invitation, since it has provided me with a unique opportunity to cast a glance at the work in toto of my students and collaborators who are in some ways the intellectual grandsons and granddaughters of Professor Lorentz. Apart from these personal reasons, it seemed to be worthwhile to attempt such a survey hoping that it would prove of some value not only to approximation theorists but also to other mathematical colleagues, serving as a report on what I consider to be one of the most prospering and thrilling fields of mathematics. It is a field ranging from operator theory, approximation on locally convex or Banach spaces or on compact manifolds, to rather concrete problems such as error estimates for quadrature formulae or a new Walsh-calculus for computer analysis. At least, this is what this survey covers. I have endeavored to present the material in a form understandable by students in the field. This is one of the reasons why the space available for the article was overdrawn, for which I beg indulgence.

This report covers the years 1968-1972, the year 1968 being chosen as a dividing line because the research monographs by Butzer-Berens [A] on semigroups and approximation appeared in 1967, and that of Butzer-Nessel [E] in 1971. However, a major part of the latter book, dealing with Fourier analysis on the circle and the line group in connection with approximation, was essentially completed as of 1968. (See also the shorter monographs Butzer-Scherer [C], Butzer-Trebels [D] of 1968.)

Thus the material covered reflects primarily on the research carried out by one generation of students, those that began their first year analysis with me in 1961, completed their Diploma or Masters theses in 1965/66, and their doctoral dissertations three years or so later. Their names are Hans Johnen, Walter Köhnen, Karl Scherer, Eberhard Stark, Walter Trebels, Ursula Westphal. Also reflected, however, are the contributions of my former students and collaborators, Hubert Berens, Rolf Nessel, Ernst Görlich, who were integral in helping me build up my research group, as well as of my younger students, Werner Kolbe, Heinrich Josef Wagner, Guido Bragard, Josef Junggeburth, Franziska Fehér. Those who joined other institutes and universities should also be mentioned: H. Schulte, D. Ernst, H. G. Neuheuser, S. and A. Pawelke, A. Fetzer, H. Haf, J. Kemper, H. Hövel, J. Kioustelides.

Altogether, it must be emphasized that the contributions of the individuals of our group can only be seen in the background of teamwork; it is this spirit that made possible many of our most exciting advances in the field.

Finally it should be mentioned that part of our work was inspired directly or indirectly by those well-known specialists from all over the world that participated in the conferences on approximation and related topics conducted by myself or in association with Jacob Korevaar, Bela Sz.-Nagy and Jean-Pierre Kahane at Oberwolfach, Black Forest, in 1963, 1965, 1969, and 1971; three of these conferences have appeared as Proceedings (see [F], [G], [H]).

Nine different directions of research in approximation theory are covered in the following presentation, and the table of contents offers a thematic overview. The references listed at the end of the paper do not cover all of the papers written by my group since 1963.

## Table of Contents

## 1. Fundamental Approximation Theorems for Best Approximation and Families of Bounded Linear Operators

In short, this field of research may be regarded as a development parallel to the semigroup theory presented in the research monograph Butzer-Berens [A], with the purpose of building up a theory in an arbitrary Banach space (in the setting of the theory of intermediate spaces) that would now contain a good part of the major results on the classical theory of best approximation as well as analogous results for families of linear operators approximating the identity, the crucial semigroup property here being replaced by commutativity.

### 1.1 Theorems of Best Approximation in Banach Spaces

Denoting by $E_n(C_{2\pi};f)$ the best approximation of degree n of $f \in C_{2\pi}$ by trigonometric polynomials $t_n(x)$ of degree n, the theorem of Weierstrass states simply that $E_n(C_{2\pi};f)\to 0$ as $n\to\infty$ for each $f \in C_{2\pi}$. Moreover, there exists a unique $t_n^*(f;x) \in T_n$ (= the class of all $t_n(x)$) such that $E_n(C_{2\pi};f) = \|f - t_n^*\|_{C_{2\pi}}$. The direct theorem of D. Jackson asserts that if the $r^{th}$ derivative (*) $f^{(r)} \in \mathrm{Lip}^*\alpha$, i.e., the second modulus of continuity $\omega_2(f^{(r)},t;C_{2\pi}) = O(t^\alpha)$, $0<\alpha\leq 2$, then (**) $E_n(C_{2\pi};f) = O(n^{-r-\alpha})$. Conversely, the inverse theorem of S.N.Bernstein (in the Zygmund version of 1944) states that (**) implies (*) for $0<\alpha<2$. Related to the latter is an assertion on simultaneous approximation stating that $E_n(C_{2\pi};f) = O(n^{-\beta})$ for $f \in C_{2\pi}$ with $\beta > 0$ implies $f^{(k)} \in C_{2\pi}$ and $\|f^{(k)} - t_n^{*(k)}\|_{C_{2\pi}} = O(n^{k-\beta})$ for $k<\beta$, $k \in P$ [1] (see S. B. Stečkin, J. Czipszer-G. Freud, A. L. Garkavi in [21]). Finally a result of M. Zamansky of 1949 states that if

$\|f-t_n\|_{C_{2\pi}} = O(n^{-\beta})$ for $\beta>0$, then $\|t_n^{(k)}\|_{C_{2\pi}} = O(n^{k-\beta})$ for $k>\beta$. The converse to this result is true provided $t_n$ is the polynomial $t_n^*$ of best approximation. This was conjectured and established for the Hilbert space $L^2_{2\pi}$ by Butzer-Pawelke [18] in 1967 and then for $C_{2\pi}$ (and $L^p_{2\pi}$, $1\leq p<\infty$) by G. Sunouchi [[25]].

The primary goal of Butzer-Scherer [20, 21, 23, 24] is to present the foregoing results in a Banach space setting in a form stating that the assertions of Jackson, Bernstein, Zamansky, and Stečkin are equivalent to each other for polynomials of best approximation. Let us just state a few representative theorems and comment on them.

Let X be any Banach space and $P_0 = \{0\} \subset P_1 \subset \ldots \subset P_n \subset \ldots \subset X$ [2] a sequence of (closed) linear subspaces of X, the best approximation of degree n to $f \in X$ being defined by

$$E_n(f;X) = \inf_{p_n \in P_n} \|f - p_n\|_X \qquad (n \in N).$$

Assume (as a generalization of the Weierstrass theorem) that

(W) $\lim_{n\to\infty} E_n(f;X) = 0 \qquad (f \in X)$

(or equivalently that $\cup_0^\infty P_n := P$ is dense in X), and the existence of some element $p_n^*(f) \in P_n$ (not necessarily unique) such that

(E) $E_n(f;X) = \|f - p_n^*(f)\|_X$.

The essentials for the proofs of the following theorems are the existence of a Jackson and Bernstein-type inequality. To formulate them, assume the existence of a Banach subspace Y of X with respect to the norm $\|.\|_Y = \|.\|_X + |.|_Y$, $|.|_Y$ being a semi-norm defined on Y. Moreover, let $P \subset Y$. The Jackson-type inequality of order $\sigma \geq 0$ with respect to $Y \subset X$ now states that

(J) $E_n(f;X) \leq C\, n^{-\sigma}|f|_Y \qquad (f \in Y;\ n \in N)$

and the Bernstein-type inequality of order $\sigma$

(B) $\|p_n\|_Y \le D\, n^{\sigma}\|p_n\|_X \qquad (p_n \in P_n;\ n \in \mathbb{N})$,

C and D being positive constants depending only upon Y and $\sigma$. As a generalization of the (higher) modulus of continuity consider the (modified) K-functional

$$K(f,t;X,Y) := \inf_{g\in Y} (\|f-g\|_X + t|g|_Y) \qquad (f \in X;\ t>0).$$

$K(f,t)$ is continuous and monotone increasing in t with $\lim_{t\to 0} K(f,t)=0\ \forall\ f \in X$ provided Y is dense in X.

Denoting by $\ell^q_*, 1\le q\le\infty$, the space of all sequences of numbers $\{a_n\}_1^\infty$ with $\sum_{n=1}^\infty |a_n|^q(1/n)<\infty$ ($\sup_{1\le n<\infty}|a_n| < \infty$ for $q=\infty$), a typical theorem reads (see [20, Satz 3], [21, Thm. 3.2], [56, Kor. 1]):

<u>Theorem 1</u>. Let X, $\{P_n\}_0^\infty$, (W), (E) be given as above. Assume that (J) and (B) are satisfied for two B-subspaces $Y_1$ and $Y_2$ of X with orders $\sigma_1$ and $\sigma_2$, respectively. The following four assertions are equivalent for $0\le\sigma_1<\theta<\sigma_2$, $1\le q\le\infty$ and $f \in X$:

(a) $\{n^\theta E_n(f;X)\} \in \ell^q_*$,

(b) $f \in Y_1$, $\{n^{\theta-\sigma_1}|f - p_n^*(f)|_{Y_1}\} \in \ell^q_*$,

(c) $\{n^{\theta-\sigma_2}|p_n^*(f)|_{Y_2}\} \in \ell^q_*$,

(d) $\{n^\theta K(f, n^{-\sigma_2}; X, Y_2)\} \in \ell^q_*$ .

Consider the simplest case $q=\infty$, $X=C_{2\pi}$, $Y_1=C^k_{2\pi}$, $Y_2=C^\ell_{2\pi}$ (with $|f|_{Y_2} = \|f^{(\ell)}\|_{C_{2\pi}}$) and $P_n = T_n$. It covers the approximation theorems mentioned in the introduction since (J) and (B) turn out to be the classical inequalities of Jackson and Bernstein for $t_n(x)$ with $\sigma_1=k<\ell=\sigma_2$. Assertion (d) is equivalent to $\omega_1(f,t;C_{2\pi}) = O(t^\theta)$, or to $\omega_2(f^{(r)},t;C_{2\pi}) = O(t^\alpha)$ with

$f^{(r)} \in C_{2\pi}$ in case $\theta = r+\alpha$, $0<\alpha<2$, $r \in$ P (see [20], [A, p.192]).

Theorem 1 can be formulated in a more general framework (see [24, 56]), revealing the full power of the results. This is achieved by introducing a continuous parameter t, $t\to 0+$ (instead of n), i.e., a family $\{P_t;\ 0<t\leq 1\}$ of subspaces of X with corresponding $E_t(f;X)$, Jackson and Bernstein-type inequalities of order y(t) (instead of $n^\sigma$), as well as the notion of a function norm $\Phi$(i.e., a functional defined on the class $M_+$ of nonnegative measurable functions $\phi(t)$, $t \in (0,1]$) to describe the order of approximation more broadly. In this setting Theorem 1 gives the equivalence of the assertions (see [56, Satz 4]) $\Phi[E_t(f;X)]<\infty$, $\Phi[y_1(t)|f-p_t^*(f)|_{Y_1}]<\infty$, $\Phi[y_2(t)|p_t^*(f)|_{Y_2}]<\infty$, $\Phi[K(f,y_2(t);X,Y_2)]<\infty$ for an $f \in X$. Of course $y_1(t)$, $y_2(t) \in M_+$ and $\Phi$ have to satisfy certain conditions, for instance, upper and lower boundedness of $\Phi$ by $y_1(t)$ and $y_2(t)$, corresponding to restriction $0\leq\sigma_1<\theta<\sigma_2$. These conditions generalize those of S. M. Losinskiĭ and N. K. Bari-S. B. Stečkin, to which they reduce in the case $P_t = P_{[1/t]}$ and $\Phi[\phi(t)]=\sup_{t>0}(\Omega([1/t]))^{-1}\cdot\phi([1/t])$ (i.e., discrete parameter $n=[1/t]$, $n \in$ N) when $\Phi[E_t(f;X)]<\infty$ means $E_n(f;X) = O(\Omega(n))$, $n\to\infty$, $\Omega \in M_+$ being increasing.

There are applications to best approximation by entire functions of exponential type [56], to the n-dim. torus [48], to best approximation by almost periodic functions [11], to best approximation by algebraic polynomials [79].

## 1.2 Approximation Theorems for Linear Processes

A programme corresponding to the preceding was also carried out for a general class of linear approximation processes, thus for a sequence of operators $T = \{T_n\}_{n=0}^{\infty} \subset E(X)$ (=Banach algebra of endomorphisms of X) having the approximation property

$(W_T)$ $\quad \lim_{n\to\infty} \|T_n f - f\|_X = 0 \qquad (f \in X)$

and satisfying a Jackson-type inequality of order $\sigma$ with respect to some (B)-subspace Y of X, namely

$(J_T)$ $\quad \|T_n f - f\|_X \leq C_T\, n^{-\sigma} |f|_Y \qquad (f \in Y; n \in \mathbb{N}),$

$C_T$ being independent of f and n. There are two possible definitions of the corresponding Bernstein-type inequality: either, let $P_n$ be the span of the ranges of $T_k$, $0\leq k\leq n$ (here again $P_n \subset P_{n+1} \subset \ldots \subset X$) and assume (strong Bernstein-type of order $\sigma$)

$(B_T^S)$ $\quad P_n \subset Y,\ |p_n|_Y \leq D_T^S\, n^{\sigma} \|p_n\|_X \qquad (p_n \in P_n;\ n \in \mathbb{N}),$

or the weak Bernstein-type inequality, namely

$(B_T^W)$ $\quad T_n f \in Y,\ |T_n f|_Y \leq D_T^W\, n^{\sigma} \|f\|_X \quad (f \in X;\ n \in \mathbb{N}).$

In the first instance a typical result reads (see [23, Thm. 1] for $q=\infty$):

Theorem 2. Let $T$ satisfy assumptions $(W_T)$ and $(J_T)$, $(B_T^S)$ for $Y_2 \subset X$ with order $\sigma_2 > 0$. If $0<\theta<\sigma_2$, $1\leq q\leq\infty$, the following three assertions are equivalent for an $f \in X$:

(i) $\{n^\theta E_n(f;X)\} \in \ell_*^q,$

(ii) $\{n^\theta \|T_n f - f\|_X\} \in \ell_*^q,$

(iii) $\{n^\theta K(f, n^{-\sigma_2}; X, Y_2)\} \in \ell_*^q.$

Since (i) $\Leftrightarrow$ (ii), such $T$ may be called "optimal" with respect to their approximation order. To obtain further assertions of type (b), (c) of Theorem 1 which are equivalent to (i) - (iii) we have to assume the commutativity of the operators $T_n$, i.e., $T_n T_m = T_m T_n$. But then the weaker inequality suffices. Indeed

Theorem 3. Let $T$ be a sequence of commutative operators satisfying $(W_T)$ and $(J_T)$, $(B_T^W)$ for $Y_2 \subset X$ with order $\sigma_2 > 0$, as well as $(J_T)$, $(B_T^W)$ for a second subspace $Y_1 \subset X$ with order $\sigma_1$, $0<\sigma_1<\theta<\sigma_2$. Then assertions (ii), (iii) of Theorem 2 are equivalent for

$1 \leq q \leq \infty$ to:

$$\text{(iv)}\quad \{n^{\theta-\sigma_1} |T_n f - f|_{Y_1}\} \in \ell^q_*,$$

$$\text{(v)}\quad \{n^{\theta-\sigma_2} |T_n f|_{Y_2}\} \in \ell^q_*.$$

The equivalence of (ii) with (iii) and (v) is already to be found in [C], that with (iv) under assumption ($B^S_T$) in [21], under the weaker ($B^W_T$) only in [24, Cor. 3]. Note that (i) of Theorem 2 is not equivalent to (ii) - (v) under condition ($B^W_T$). A (counter)-example is the singular integral of de La Vallée Poussin having order of approximation half as good as order of best approximation.

Theorems 2,3 present a major innovation in the sense that such results could previously only be established for a class of summation processes satisfying certain identities (see [18]) or for semigroups of operators (see [A]). Since an application of the above theorems reduces simply to checking whether Jackson and Bernstein-type inequalities of highest possible or saturation order (see Chapter 2.1) are satisfied, they cover the (non-optimal) approximation behavior of practically all known summation methods of Fourier series (see [C, Kap. 4], [21, 23, 24]). For further applications see Chapters 2.3, 3, 5, 8.

Just as Theorem 1, Theorems 2,3 may also be formulated in the general setting of function norms $\Phi$, $T = \{T(t);\ 0<t\leq 1\}$ now being a family in E(X) depending on a continuous parameter t, $t \to 0+$ (see [24]). Such a theory subsumes many further applications, in particular, approximation theory of Butzer-Berens [A] for holomorphic semigroups of operators, for resolvent operators of the infinitesimal generator of a semigroup, for the Riesz means of the Fourier-inversion integral for $L^p(R^1)$, $1 \leq p \leq 2$, etc. (see [24]).

## 1.3 Intermediate Spaces; Duality Theory

The results of Sec. 1,2 can be formulated in the setting of the theory of intermediate spaces to obtain connections with the interpolation theory for Banach spaces. Thus we define the discrete K-interpolation spaces of X and Y [C, p. 42], namely (the seminorm $|.|_Y$ in definition of K-functional now being replaced by the norm $\|.\|_Y$:

$$[X,Y]_{\theta,q;K} = \{f \in X; \|f\|_{\theta,q;K} = (\sum_{n=1}^{\infty} [n^{\theta} K(f,n^{-1};X,Y)]^q \frac{1}{n})^{1/q} < \infty\}$$

$$(0<\theta<1,\ 1\leq q<\infty;\ 0\leq\theta\leq 1,\ q=\infty),$$

the so-called K-approximation spaces [C, p. 53]

$$X^K_{\theta,q} = \{f \in X; \|f\|_{\theta,q;X} = \|f\|_X + (\sum_{n=1}^{\infty} [n^{\theta} E_n(f;X)]^q \frac{1}{n})^{1/q} < \infty\}$$

as well as the spaces associated with the approximation process [C, p. 74], i.e.,

$$[X]_{\theta,q}(T_n) = \{f \in X; \|f\|_{\theta,q;T} = \|f\|_X + (\sum_{n=1}^{\infty} [n^{\theta}\|T_n f - f\|_X]^q \frac{1}{n})^{1/q} < \infty\}$$

(with obvious modifications for $q=\infty$). These spaces are all (B)-spaces. In this framework, an equivalence for approximation assertions is expressed simply by the equivalence of the corresponding spaces. For example, the relations

$$(1.1)\qquad X^K_{\theta,q} \cong [X]_{\theta,q}(T_n) \cong [X,Y_2]_{\theta/\sigma_2,q;K} \qquad (0<\theta<\sigma_2)$$

express nothing but Theorem 2. Note that the limiting cases $\theta=0$ and $\theta=\sigma_2$ correspond to the spaces X and Y in view of $(W_T)$ and $(J_T)$. Thus (1.1) may be interpreted as an interpolation theorem between Jackson and Bernstein-type inequalities, and $X^K_{\theta,q}$, $[X]_{\theta,q}(T_n)$ and $[X,Y_2]_{\theta/\sigma_2,q;K}$ as intermediate spaces between X and $Y_2$, representing the "intermediate" order of approximation $\theta$. Moreover, the $X^K_{\theta,q}$ are a family of subspaces of X which, for fixed $\theta$, increase monotonely for q increasing from 1 to $\infty$. Thus

$$X^K_{\theta',q} \subset X^K_{\theta,q} \subset X^K_{\theta,p} \qquad (0\le\theta\le\theta',\ 1\le q\le p\le\infty).$$

Since $(\int_0^\infty [t^{-\theta/\sigma_2} K(f,t;X,Y_2)]^q dt/t)^{1/q}$ defines an equivalent norm for the space $[X,Y_2]_{\theta/\sigma_2,q;K}$, one has the connection with the "continuous" interpolation theory for Banach spaces, in particular in the form presented in [A] (see also Ch. 2.1).

A further complex is concerned with best approximation in a dual setting, i.e., determination and characterization of the duals of the intermediate spaces introduced above. This was carried out in [72, 22, 73]. To mention a few results, one has e.g., for $0<\theta<1$, $1\le q<\infty$ [71, p. 17].

$$(1.2) \qquad [X,Y]'_{\theta,q;K} \cong [Y',X']_{1-\theta,q';K} \qquad (q^{-1}+(q')^{-1}=1),$$

the dash indicating the dual space, i.e., the space of continuous linear functionals defined on the corresponding (B)-spaces. Under the hypotheses of Jackson and Bernstein-type inequalities $(J_T)$, $(J_{T'})$, $(B_T)$, $(B_{T'})$,

$$([X]_{\theta,q}(T_n))' \cong [Y']_{\sigma-\theta,q'}(T'_n) \cong \{f\in Y';\{n^{-\theta}\|T'_n f'\|_{X'}\}\in \ell^q_*\}$$

is a characterization [72, p. 25] for commutative approximation processes (where $T' = \{T'_n\}_{n=0}^\infty$, $T'_n$ denoting the operator conjugate to $T_n$).

Such results are carried out in the general setting of function norms $\Phi$ in [73], the concept of a dual function norm $\Phi'$ being of importance. This theory yields not only known characterizations of Banach spaces of periodic distributions (see Taibleson [[27]]) but also many new ones. Also assertion (1.2) can be formulated [73] (more generally) in terms of function norms $\Phi$ as well as for any pair of (B)-spaces X,Y (without restriction $Y\subset X$); this yields an extension together with a simple proof of an unpublished result of J. Peetre.

To characterize the conjugates of the K-approximation spaces there is introduced [72, p.49; 22] the dual $D'_X$ of the locally convex space $D_X = \cap_{r=1}^{\infty} X^K_{r,\infty}$. A main result yields the characterization

$$(X^K_{\theta,q})' = \{f' \in D'_X;\ \{n^{-\theta} \|f'\|_{(P_n,X)}\} \in \ell_*^{q'}\},$$

$\|f'\|_{(P_n,X)}$ denoting the (finite) norm of the restriction of $f' \in X'$ to $P_n$ with respect to X. The quantity $\|f'\|_{(P_n,X)}$ is brought in connection [22,72] with the "dual best approximation" $E_r'(f',X')$, the approximating subspaces now being the annihilators $(P_n,X)^{\perp}$. It is shown that Jackson and Bernstein-type inequalities are satisfied in the original space if and only if there hold "dual" Bernstein and Jackson inequalities, respectively, in the dual spaces. In particular, the dual version of Theorem 1 reads (see [22, Thms. III1, III2]):

<u>Theorem 4</u>. Let X, $Y_1$, $Y_2$ be defined as in Theorem 1. The following are equivalent for $f' \in D'_X$:

(a) $\{n^{-\theta}\|f'\|_{(P_n,X)}\} \in \ell_*^q$,

(b) $\{n^{\sigma_1-\theta} \|f'_{X,n}\|_{Y'_1}\} \in \ell_*^q$,

(c) $f' \in Y'_2$, $\{n^{\sigma_2-\theta} \|f' - f'_{X,n}\|_{Y'_2}\} \in \ell_*^q$,

(d) $f' \in Y'_2$, $\{n^{\sigma_2-\theta} K(f', n^{-\sigma_2}; Y',X')\} \in \ell_*^q$.

Here $f'_{X,n}$ is a functional in X' satisfying, for each $p_n \in P_n$,

$$\langle f'_{X,n}, p_n\rangle = \langle f', p_n\rangle,\ \|f'_{X,n}\|_{X'} = \|f'\|_{(P_n,X)}.$$

It is clear by the above that Jackson and Bernstein-type inequalities are the essential ingredients for approximation theory in regard to degree of convergence. This gave one main orientation for our further research, in particular displaying the situation when there is given an orthogonal system in a Banach space; see Ch. 3,4.

In connection with Bernstein's inequality Görlich [41, 42] has supplied a number of variants of this inequality for the case of trigonometric polynomial approximation in the spaces $C_{2\pi}$, $L^p_{2\pi}$, $1 \le p < \infty$. In particular, he considered logarithmic and exponential "limiting cases" of Bernstein's inequality, namely

$$\left\| \sum_{k=-n}^{n} \phi(|k|)\, c_k e^{ikx} \right\| \le 2\phi(n) \left\| \sum_{k=-n}^{n} c_k e^{ikx} \right\| ,$$

where e.g. $\phi(x)=\log(1+x)$ or $\phi(x) = e^{\beta x}$ and $\Sigma_{k=-n}^{n} c_k e^{ikx}$ is any trigonometric polynomial of degree $\le n$. Moreover, on this basis he constructed an approximation theory parallel to the foregoing, using suitable semigroup operators (instead of the ordinary translation) in order to form a modulus of continuity.

## 1.4 Banach-Steinhaus Theorem; Error Estimates for Quadrature Formulae

The fundamental theorem of Banach-Steinhaus essentially states that a family of bounded operators is convergent on a whole space if and only if the operators are uniformly bounded as well as convergent on a dense subspace. This theorem has been equipped with an order of approximation by Butzer-Scherer-Westphal [25].

Let X,Y and A be locally convex Hausdorff spaces with topologies generated by the families of filtrating seminorms $\{p\}$, $\{q\}$, and $\{\bar{p}\}$, respectively, A being continuously embedded in X (i.e., to each $p \in \{p\}$ there is $\bar{p} \in \{\bar{p}\}$ and $c>0$ with $p(f) \le c\bar{p}(f)\ \forall\ f \in A$). Define a modification of the K-functional by

$$K(f,t;X,A)_{p,\bar{p}} = \inf_{g \in A}\{p(f-g) + t\bar{p}(g)\} \qquad (t>0;\ f \in X).$$

<u>Theorem 5.</u> If $T_n$, $n \in \mathbb{N}$, $T$ are bounded linear mappings defined on $X$ into $Y$, $X$ and $A$ being barralled, then the following two assertions are equivalent:

To each $q \in \{q\}$ there is $p \in \{p\}$ and $\bar{p} \in \{\bar{p}\}$ such that for $\delta>0$

$$(1.3) \quad q[(T_n-T)f] = \mathcal{O}(K(f,n^{-\delta};X,A)_{p,\bar{p}}) \qquad (\forall f \in X),$$

to each $q \in \{q\}$ there is $p \in \{p\}$ and $M>0$ such that for $\delta>0$

$$(1.4) \quad \text{(i)} \quad \sup_{n\geq 0} q[(T_n-T)f] \leq Mp(f) \qquad (\forall f \in X)$$

and

$$\text{(ii)} \quad q[(T_n-T)f] = \mathcal{O}(n^{-\delta}) \qquad (\forall f \in A).$$

In comparison with the classical Banach-Steinhaus theorem, condition (1.3) and the Jackson-type (1.4,ii) involving an order of approximation replace two assertions on pure convergence of that theorem, namely that $q[(T_n-T)f]\to 0$ as $n\to\infty$ for all $f \in X$, respectively for all $f \in A$, $A$ being a total set in $X$. Condition (1.4,i) is the same for both theorems. Note that if $A$ is total in $X$, then $\lim_{t\to 0+}K(f,t;X,A)_{p,\bar{p}} = 0$.

In numerical integration it is of interest to estimate the error of approximation of $Qf := \int_a^b f(x)dx$ by

$$Q_n^{\mu}f := \sum_{i=1}^{n} A_{i,n}f(x_{i,n}) + \sum_{\nu=1}^{\mu}\sum_{i=1}^{n} B_{i,n}^{\nu} f^{(\nu)}(x_{i,n}^{\nu})$$

with given nodes $x_{i,n}, x_{i,n}^{\nu} \in [a,b]$ and weights $A_{i,n}$, $B_{i,n}^{\nu}$, provided $f \in C^{\mu}[a,b]$, $\mu \in P$ fixed. For this purpose one assumes that the quadrature formula $Q_n^{\mu}f \approx Qf$, $n$ large, is exact for polynomials $p_m$ of fixed degree $m$ $(\geq\mu)$, i.e., $Q_n^{\mu}\, p_n \equiv Qp_m$. An application of a weaker version of Theorem 5 together with Peano's theorem yields that

$$(1.5) \quad \left|\int_a^b f(x)dx - Q_n^{\mu}f\right| \leq c_{\mu,m}\frac{1}{n^{\mu}}\,\omega_{m+1-\mu}(f^{(\mu)};\tfrac{1}{n}) \qquad (\forall f \in C^{\mu}[a,b])$$

if and only if

$$\left.\begin{aligned} &(i)\quad (\mu\geq 1): \int_a^b |\chi^{\mu}_{n,\mu-1}(u)|du \\ &\qquad (\mu=0): \sum_{i=1}^{n} |A_{i,n}| \end{aligned}\right\} = O(n^{-\mu}),$$

$$(ii)\quad \int_c^b |\chi^{\mu}_{n,m}(u)|du = O(n^{-m-1}) .$$

Here $\chi_{n,\ell}(u) = (1/\ell!)(Q^{\mu}_n - Q)_x (x-u)^{\ell}_+$, the index x meaning that the functional $Q^{\mu}_n - Q$ is applied to the truncated power $(x-u)^{\ell}_+$ considered as a function of x, and $\omega_{\ell}(f,\delta)$ stands for the modulus of continuity of f of order $\ell$.

Note that (i), (ii) may be verified for many examples, for instance in case $\mu=0$ for the composite Newton-Cotes formulae (cf. P. J. Davis - P. Rabinowitz [[C]]). In such cases (1.5) gives error estimates which are free of derivatives.

## 2. Topics Related to Semigroup Theory

### 2.1 A General Saturation Theorem

Whereas Chapter 1 was essentially concerned with non-optimal approximation, one object here is to investigate the optimal or saturation case. In continuation of the results in [A] on the approximation by semigroup operators, Berens in his lecture notes (Habilitationsschrift) [B] studies approximation processes of the more general type: Let $\{T_\rho : \rho>0\}$ be a family of uniformly bounded, linear, commutative operators in E(X) for which

$$\|T_\rho f\| \leq M\|f\|, \qquad \lim_{\rho\to\infty}\|T_\rho f - f\| = 0 \qquad (\forall\ f \in X).$$

Whereas the $T_\rho$ need not necessarily satisfy a functional equation with respect to the parameter $\rho$ (comparable to the semigroup property), it is here assumed, however, that there exists a closed linear operator B with domain D(B) dense in X and range in X (corresponding to the generator for semigroups) which is associated with $\{T_\rho\}$ such that the Voronovskaja-type relation

$$(2.1) \qquad \lim_{\rho\to\infty}\|\rho^{\tau}[T_\rho f - f] - Bf\| = 0 \qquad (\forall\, f \in D(B))$$

is satisfied for some $\tau > 0$. Under these hypotheses Berens considers--the second major innovation of his paper--the saturation behavior of $\{T_\rho\}$ by making use of E. Gagliardo's concept of relative completion: the completion of D(B) relative to X, denoted by $\widetilde{D(B)}^X$, is the space of all $f \in X$ for which there exists

$$(2.2) \qquad \{f_n\} \subset D(B) : \lim_{n\to\infty} \|f_n - f\| = 0, \ \sup_n\|Bf_n\| < \infty$$

and which is normable under $\inf\{\sup_n(\|f_n\|+\|Bf_n\|)\}$, the infimum taken over all sequences $\{f_n\}$ satisfying (2.2).

<u>Saturation Theorem</u> ([B, Satz 3.2, Bem. 3.4], [E, p. 502]). With $\{T_\rho;\rho>0\}$ and B given as above, let there be associated a <u>regularization process</u> $\{J_n;\, n \in \mathbb{N}\}$, i.e., a family of bounded linear operators $J_n \in \mathcal{E}(X)$ which commute with $T_\rho$ and such that $J_n[X] \subset D(B)$, $n \in \mathbb{N}$, $\lim_{n\to\infty}\|J_n f - f\| = 0 \ \forall\, f \in X$. Then the family $\{T_\rho\}$ is <u>saturated</u> in X with order $O(\rho^{-\tau})$, $\rho\to\infty$, and the <u>saturation class</u> is characterized by $\widetilde{D(B)}^X$; i.e.,

$$(2.3) \qquad \|T_\rho f - f\| = \begin{cases} o(\rho^{-\tau}) \iff f \in D(B) \text{ and } Bf = 0, \\ O(\rho^{-\tau}) \iff f \in \widetilde{D(B)}^X. \end{cases}$$

In case X is reflexive, this class is simply D(B), normed under $\|f\|+\|Bf\|$.

The approximation behavior is described most comprehensively by the spaces (compare Ch. 1.3)

$$(2.4) \qquad (X)_{\theta,q}\,(T_\rho) = \{f \in X;\, \rho^{\theta}\|T_\rho f - f\|\} \in L^q_*(0,\infty)\},$$

$0<\theta<\tau$, $1\le q\le\infty$ and $\theta=\tau$, $q=\infty$, associated with the process $\{T_\rho : \rho>0\}$. Here $L^q_*(a,b)$, $0\le a<b\le\infty$, is the space of all numerically valued

functions $h(t)$ defined on $(a,b)$ which are $q^{th}$ power integrable with respect to Haar's measure $dt/(t-a)$. One has

(2.5) $$(X)_{\theta,q}(T_\rho) \cong (X,D(B))_{\theta/\tau,q;K} \qquad (0<\theta<\tau,\ 1\leq q\leq\infty)$$

and

(2.6) $$(X)_{\tau,\infty}(T_\rho) \cong (X,D(B))_{1,\infty;K} \cong \widetilde{D(B)}^X,$$

$(X,D(B))_{\theta/\tau,q;K}$ $(0<\theta<\tau,\ 1\leq q\leq\infty;\ \theta=\tau,\ q=\infty)$ being the continuous K-interpolation space of $X$ and $D(B)$ consisting of all $f \in X$ for which $t^{-\theta/\tau}K(f,t;X,D(B)) \in L^q_*(0,\infty)$. (2.5) is a result on non-optimal approximation (valid under additional assumption of a Bernstein-type inequality); see also the more general approach in Ch. 1.2. (2.6) concerns the saturation case treated above. In the foregoing we only surveyed those contributions of Professor Berens which he obtained before his joining the University of California at Santa Barbara in Autumn 1968.

Let us finally note that Köhnen [60] generalized part of the results in [A] to n-parameter semigroups of operators, particularly the saturation theorem. He also proved approximation theorems for Taylor and Riemann operators of n-parameter semigroups, as well as theorems about the adjoint n-parameter semigroup. A very modern direction in functional analysis is the non-linear semigroup theory as developed by T. Kato, Y. Komura, F. Browder, R. Dorroh, H. Brezis, M. G. Crandall, A. Pazy, T. Liggett, I. Miyadera, G. F. Webb, and others. For a non-linear semigroup generated by A, A a not necessarily single-valued accretive operator (see [[5]]), Westphal [101] showed that the saturation class also turns out to be a kind of relative completion, namely the family of all $f \in \overline{D(A)}$ for which there exists a sequence $\{f_n\} \subset D(A)$ such that $\lim_{n\to\infty}\|f_n - f\| = 0$ and $\sup_n \inf\{\|g_n\|;\ [f_n,g_n] \in A\}<\infty$.

## 2.2 Differential Equations in Banach Spaces and Perturbations

Consider the initial value problem for the time-dependent

inhomogeneous evolution equation in the (B)-space X, namely

$$(2.7)\qquad \frac{du}{dt} - A(t)u(t) = F(t) \qquad (0<t\leq t_0), \quad u(0) = f \qquad (f \in X),$$

where the operator A(t), depending on t, generates a holomorphic semigroup $\{T(s;A(t));\ s\geq 0\}$ for each fixed $t \in [0,t_o]$ such that its domain D(A(t)), however, is independent of t, and the inhomogeneous term F(t) is a vector-valued function of t. Under suitable conditions upon A(t) (i.e., $A(t)A(s)^{-1}$ being Hölder continuous in the uniform operator topology on $[0,t_0]$) and F(t), H. Tanabe and P. E. Sobolevskiĭ (see R. W. Carroll [[B]] and literature cited there) have shown that (2.7) has the unique solution

$$u(t;f) = U(t,0)f + \int_0^t U(t,v)F(v)dv.$$

Here U(t,s), $0\leq s\leq t\leq t_0$, is the evolution operator associated with (2.7) which may be written as a <u>perturbation</u> of the semigroup $\{T(t;A(s));\ t\geq 0\}$ generated by A(s) in the form

$$U(t,s) = T((t-s);A(s)) - S(t,s),$$

where S(t,s) is a certain linear operator for each pair (s,t), $0\leq s\leq t\leq t_0$. In [58] Köhnen shows that for fixed s the evolution operators U(t,s) have the same approximation behavior for $t\to s+$ as do the semigroup operators T(t; A(s)) for $t\to 0+$. A typical result in [58, II, Theorems 4.2, 5.1] giving the approximation invariance of u(t;f), U(t,s)f, and T(t;A(s))f, reads (u'(t;f) being the strong derivative of u(t;f)):

<u>Theorem.</u> The following assertions are equivalent for $0<\theta<1$, $1\leq q<\infty$ or $0<\theta\leq 1$, $q=\infty$ and an $f \in X$:

(i) $\{t^{-\theta}\|u(t;f) - f\|\} \in L^q_*(0,t_0)$,

(ii) $\{(t-s)^{-\theta}\|U(t,s)f - f\|\} \in L^q_*(s,t_0)$ (any fixed $s \in [0,t_0]$),

(iii) $\{t^{-\theta}\|T(t;A(s'))f - f\|\} \in L^q_*(0,t_0)$ (any fixed $s' \in [0,t_0]$),

(iv) $\{t^{1-\theta}\|u'(t;f)\|\} \in L^q_*(0,t_o)$.

Corresponding results also hold if in equation (2.7) there is added to A(t) a linear perturbing operator B(t) which satisfies assumptions of H. Tanabe and M. Z. Solomjak. Köhnen [58] also treats several examples, A(t) either being an ordinary or partial differential operator, including transient problems in physics as heat conduction, electrical conductivity and diffusion.

The investigations have their origin in Butzer-Köhnen [12] concerned with the stability of the approximation behavior of semigroups under perturbations. It was shown that two arbitrary semigroups of operators of class $(C_o)$ possess the same approximation behavior, optimal as well as non-optimal, provided only that the domains of their infinitesimal generators coincide. A more elementary proof of this fact can be found in Köhnen [59].

## 2.3 Ergodic Theorems and Approximation

The mean ergodic theorem is concerned with the convergence of certain averages of the iterates $T^n$ of a power-bounded operator $T \in E(X)$. Whereas all investigations connected with this matter seem to have been restricted to mere convergence, Butzer-Westphal [33,34,35] first dealt with the rate of convergence of the averages in question.

Consider the (C,k)-means of the sequence $\{T^n; n \geq 0\}$

$$\sigma_n^k(T) = (C_n^{k+1})^{-1} \sum_{i=0}^{n} C_{n-i}^k T^i,$$

where $C_n^k = k(k+1)\ldots(k+n-1)/n!$, $k \in N$, $n \in P$, $T \in E(X)$ being power-bounded, i.e., $\|T^n\| \leq M$, $n \in N$. The classical mean ergodic theorem states that $\lim_{n\to\infty}\|\sigma_n^k(T)f - Pf\| = 0$ for each $f \in X_o$, where $X_o = N(I-T) \oplus \overline{R(I-T)}$ is the direct sum of the null-space and closure of the range of (I-T), and P is the bounded linear projection of $X$ onto $N(I-T)$ along $\overline{R(I-T)}$.

The theorem on optimal approximation now reads that the process $\sigma_n^k(T)$ is saturated for each fixed $k \in \mathbb{N}$ with order $O(n^{-1})$, and the Favard class is equal to $\tilde{Y}^{X_0}$ (or even to $Y$ provided $X$ is in addition reflexive), where $Y = N(I-T) \oplus R(I-T_0)$, $T_0 = T/X_0$. The counterpart on non-optimal approximation states that for $f \in X_0$, each $k \in \mathbb{N}$,

$$[X_0]_{\theta,q}(\sigma_n^k(T)) \cong (X_0,Y)_{\theta,q;K} \qquad (0<\theta<1,\ 1\leq q\leq\infty)$$

(recall definitions in Chapters 1.3, 2.1).

Of interest is that the proof of the latter result follows as an application of the general Butzer-Scherer theorem (see Theorem 3, Ch. 1.3) observing that $\{J_n^k(T) := P+I-\sigma_n^k(T);\ n\geq 0\}$ is a sequence of commutative operators in $E(X_0)$ satisfying the Jackson and Bernstein-type inequalities

$$\|J_n^k(T)f-f\|_{X_0} \leq Cn^{-1}\|f\|_Y, \quad \|J_n^k(T)f\|_Y \leq Dn\|f\|_{X_0}$$

for $f \in Y$, $f \in X_0$, respectively.

Note that the above results are also valid (see [64]) for the Cesàro-means $(C,\beta)$ of fractional order $\beta\geq 1$. Corresponding results are also valid for the Abel ergodic theorem in the discrete as well as in the continuous case.

## 2.4 Fractional Powers of Closed Operators

The study of fractional differentiation has a long history connected with the names G. W. Leibniz, N. H. Abel, J. Liouville, B. Riemann, H. J. Holmgren, H. Weyl, A. Marchaud and more recently with M. Riesz, S. Bochner and W. Feller. However, it is only since about 1950 that the general theory of fractional powers of operators was developed, particularly for powers $(-A)^\gamma$, $\gamma>0$, $A$ being the infinitesimal generator of a uniformly bounded semigroup $\{T(t);\ t\geq 0\}$ of class $(C_0)$ on a real or complex (B)-space $X$. Names are e.g., R. S. Phillips,

T. Kato and A. V. Balakrishnan. The latter author defines $(-A)^{\gamma}$, $n-1\leq\gamma<n$, $n\in\mathbb{N}$, as the smallest closed extension of the operator $B^{\gamma}$, given on $D(A^n)$ by

$$(2.8)\qquad B^{\gamma}f = -\frac{\sin \pi\gamma}{\pi}\int_0^{\infty}\lambda^{\gamma-n}(\lambda I-A)^{-1}A^n f\,d\lambda;$$

see [[1]], compare also H. Komatsu [[12]].

This work was carried on at Aachen in a series of papers [5, 6, 3, 99, 100, 47]. Westphal [100,I] defines the fractional power $(-A)^{\gamma}$ or order $\gamma$, $\gamma$ real $>0$, by

$$(2.9)\qquad \begin{aligned} (-A)^{\gamma}f &= \text{s-lim}_{\varepsilon\to 0+}(-A)_{\gamma,\varepsilon}f \\ (-A)_{\gamma,\varepsilon}f &= [C_{\gamma,n}]^{-1}\int_{\varepsilon}^{\infty}u^{-\gamma}[I-T(u)]^n f\,du/u, \end{aligned}$$

whenever this strong limit exists for some $f\in X$, the domain being denoted by $D((-A)^{\gamma})$. Here $n\in\mathbb{N}$, $n>\gamma$, the limit being independent of this n and

$$C_{\gamma,n} = \int_0^{\infty}u^{-1-\gamma}(1-e^{-u})^n du.$$

For the sake of simplicity let us restrict the exponents $\gamma$ to $0<\gamma<1$. In distinction to the definition via (2.8), $(-A)_{\gamma,\varepsilon}$ may actually be interpreted as a difference quotient, namely

$$(-A)_{\gamma,\varepsilon}f = \frac{1}{\gamma C_{\gamma,1}}\quad\frac{f - S_{\gamma,\varepsilon}f}{\varepsilon^{\gamma}}$$

$$S_{\gamma,\varepsilon}f = \gamma\int_1^{\infty}u^{-\gamma}\,T(\varepsilon u)f\,du/u \qquad (f\in X),$$

just as $(-A)$ is defined as a limit of a difference quotient. This quotient may, for its part, be written as a product of $(-A)$ and the regularization process $\{\int_0^1 T(\varepsilon u)du;\ \varepsilon>0\}$; i.e.,

$$\varepsilon^{-1}[I-T(\varepsilon)]f = \begin{cases} -A\int_0^1 T(\varepsilon u)f\,du & (f\in X) \\ \int_0^1 T(\varepsilon u)(-A)f\,du & (f\in D(A)), \end{cases}$$

an elementary representation of fundamental importance in classical semigroup theory, yielding e.g. that the operator $(-A)$ is closed. In the fractional power case there are two counterparts of this equation (see [10], I, Theorems 2.5, 3.3]).

Proposition. For each $f \in X$ and $\varepsilon > 0$ $J_{\gamma,\varepsilon}f := [C_{\gamma,1}]^{-1} \int_0^\infty \cdot q_{\gamma,1}(u)T(\varepsilon u)f\,du$ $(0<\gamma<1)$ and $I_{\gamma,\varepsilon}f := \int_0^\infty p_{\gamma,1}(u)T(\varepsilon u)f\,du$ $(0<\gamma\leq 1)$ belong to $D((-A)^\gamma)$, $q_{\gamma,1}$ and $p_{\gamma,1}$ being certain $L^1(0,\infty)$-functions, and there holds

$$(2.10)\quad (-A)_{\gamma,\varepsilon}f = \begin{cases} (-A)^\gamma J_{\gamma,\varepsilon}f & (f \in X) \\ J_{\gamma,\varepsilon}(-A)^\gamma f & (f \in D((-A)^\gamma)) \end{cases}$$

as well as

$$(2.11)\quad \varepsilon^{-\gamma}[I-T(\varepsilon)]f = \begin{cases} (-A)^\gamma I_{\gamma,\varepsilon}f & (f \in X) \\ I_{\gamma,\varepsilon}(-A)^\gamma f & (f \in D((-A)^\gamma)). \end{cases}$$

This proposition is central for the Westphal approach. Note that the family $\{J_{\gamma,\varepsilon};\varepsilon>0\}$ is a regularization process associated with $(-A)^\gamma$ and $S_{\gamma,\varepsilon}$. (2.10) and (2.11) may be regarded as inversion formulae for (2.9). While (2.10) gives a converse to the limiting process in (2.9), (2.11), in addition, takes into account the integral operation in (2.9). From (2.10) it follows readily e.g. that $(-A)^\gamma$ is a closed operator. (2.11) provides the estimate $\|[I-T(t)]f\| = O(t^\gamma)$ if $f \in D((-A)^\gamma)$, which in turn yields the inclusion $D((-A)^{\gamma_1}) \subset D((-A)^{\gamma_2})$ for $\gamma_1 > \gamma_2$, and furthermore the basic additivity property $(-A)^{\gamma_1} \cdot (-A)^{\gamma_2} = (-A)^{\gamma_1+\gamma_2}$ by proceeding to the limit in the equation

$$(2.12)\quad (-A)_{\gamma_1+\gamma_2,\varepsilon}f = (-A)^{\gamma_1}(-A)^{\gamma_2} J_{\gamma_1+\gamma_2,\varepsilon}f \qquad (f \in X),$$

the structure of which is similar to that of (2.10). One also has the multiplicativity property $((-A)^{\gamma_1})^{\gamma_2} = (-A)^{\gamma_1\gamma_2}$ and the coincidence $(-A)^{\gamma} = -A$ for $\gamma=1$ (for the latter see also J. L. Lions - J. Peetre [[16]]). Definition (2.9) is actually equivalent to those of Balakrishnan [[1]] and Komatsu [[12]]. However, it has the advantage that it is defined by a simple limiting process of a difference quotient and not via an indirect smallest closed extension process. Moreover, there is the immediate parallelism with the theory for the natural powers $(-A)^n$.

The proofs of the theory sketched above are rather simple; the essential tool is the Laplace transformation as well as several basic identities. For example, the function $q_{\gamma,1}$ occurring in the integral operator $J_{\gamma,\varepsilon}$ is defined by its Laplace transform

$$\hat{q}_{\gamma,1}(s) = s^{-\gamma} \int_0^{\infty} u^{-\gamma}(1-e^{-su})du/u \qquad (\mathrm{Re}\ s>0).$$

The proofs of (2.10) - (2.12) reduce to the verification of identities, the proofs of which depend only upon the classical convolution and uniqueness theorems for Laplace transforms. In the case of (2.10) the identity reads

$$(-A)_{\gamma,\eta}\, J_{\gamma,\varepsilon} f = (-A)_{\gamma,\varepsilon}\, J_{\gamma,\eta} f \qquad (f \in X;\ \varepsilon,\eta>0).$$

## 2.5 Generalizations and Logarithmic Derivatives

In Part II of [100] the foregoing theory was also developed in case A is the infinitesimal generator of a uniformly bounded group $\{G(t);\ -\infty<t<\infty\}$ of operators of class $(C_0)$. Here the limit for $\varepsilon\to 0+$ of the integral ($0<\gamma<2$, $\bar{C}_{\gamma,2}$ a constant)

$$(2.13) \qquad [\bar{C}_{\gamma,2}]^{-1} \int_{\varepsilon}^{\infty} u^{-\gamma}[G(u/2) - G(-u/2)]^2 f\, du/u$$

is studied, again giving a closed operator which turns out to be $(-A^2)^{\gamma/2}$.

In the instance that A is not the generator of some semi-group but a closed linear operator whose resolvent $R(\lambda;A)$ exists for each $\lambda>0$ such that $\sup_{\lambda>0}\|\lambda R(\lambda;A)\| < \infty$, then it is also possible to introduce fractional powers. Indeed, $(-A)^{\gamma}$, $0<\gamma<1$, may be defined (see [47]) by the limit for $N\to\infty$ of ($C'_{\gamma,1}$ a constant)

$$(2.14)\quad [C'_{\gamma,1}]^{-1}\int_0^N \lambda^{\gamma}[I-\lambda R(\lambda;A)]f\,d\lambda/\lambda.$$

This is actually the most general class of operators in a (B)-space for which such powers may be constructed.

Whereas the Fourier transform is the major tool in the development of (2.13), it is the Stieltjes transform with (2.14). In the investigations mentioned so far, $\gamma$ was restricted to the positive reals. Komatsu [[12]] also built up his calculus in case of complex powers; for purely imaginary powers see M. J. Fisher [[6]]. In this context there is a logarithmic analog established by Görlich in his Habilitationsschrift [41] (see also [42]) for the special case of $2\pi$-periodic functions of one variable; in some sense a limiting case of the above. A function $f\in C_{2\pi}$ is said to have a logarithmic derivative $D_Lf$ in $C_{2\pi}$ if $f^{\wedge}(k)\log(1+k^2)^{1/2}$ are the Fourier coefficients of some function $g\in C_{2\pi}$, and in this case $g=D_Lf$. Denoting the set of all such f by $D(D_L)$, one has the characterization

$$f\in D(D_L)\iff \exists\ \text{s-}\lim_{\varepsilon\to 0+}\frac{1}{2}\int_{\varepsilon}^{\infty}\frac{f(x+t)-2f(x)+f(x-t)}{te^t}\,dt,$$

and the strong limit of the integral equals $(D_Lf)(x)$. The relative completion of $D(D_L)$ with respect to $C_{2\pi}$ may be interpreted as the saturation class of a certain fractional integral of order $\gamma$ (namely the inverse of the Bessel derivative) when considered as an approximation process for $\gamma\to 0+$.

2.6 Landau-Kallman-Rota-Hille-Inequality

This inequality, which has received rather wide attention

in the past three years (see E. Hille [[8]]), is a generalization of the Landau inequality for functions on $(0,\infty)$ which are twice differentiable, namely

$$\|\frac{d}{dx} f\|^2 \leq 4\|f\| \, \|\frac{d^2}{dx^2} f\|,$$

with the differential operators $(d/dx)$, $(d/dx)^2$ replaced by semigroup generators A and their powers of arbitrary integral order. Trebels - Westphal [97] have extended this inequality to fractional powers of $(-A)$. By arguments making essential use of the inversion formula (2.11) they showed that

$$\|(-A)^{\alpha} f\|^{\gamma} \leq K_{\gamma,\alpha}^{\gamma} \|f\|^{\gamma-\alpha} \|(-A)^{\gamma} f\|^{\alpha}$$

for $f \in D((-A)^{\gamma})$, $0<\alpha<\gamma$, $K_{\gamma,\alpha}$ being a constant. In case $X=L_{2\pi}^{p}$, $1\leq p<\infty$, or $= C_{2\pi}$, this inequality may be applied to deduce the basic Bernstein inequality for fractional derivatives [41], namely

$$\|t_n^{(\alpha)}\| \leq K_{j,\alpha} n^{\alpha} \|t_n\| \qquad (0<\alpha<j, \ j \in \mathbb{N}),$$

where $t_n(x) = \Sigma_{k=-n}^{n} c_k e^{ikx}$.

## 2.7 Fractional Powers and Intermediate Spaces

Before turning to the application cited, note that saturation classes of singular integrals of Laplace transform type

$$\rho \int_0^x f(x-u)\chi(\rho u)du \qquad (f \in L^p(0,\infty), \ 1<p<\infty)$$

may be characterized via fractional powers under suitable conditions upon the kernel $\chi \in L^1(0,\infty)$. In Berens-Butzer [1] it was shown that this class is equal to the set of all $f \in L^p(0,\infty)$ for which there exists $g \in L^p(0,\infty)$ such that

$$s^{\gamma} f^{\wedge}(s) = g^{\wedge}(s) \qquad (\text{Re } s>0),$$

$f^{\wedge}(s)$ denoting the Laplace transform of f. An application of the calculus for fractional powers in the particular case of the semigroup of translations, namely $[T(t)f](x) = f(x-t)$, yields that the saturation class may be characterized as the set of all f for which

$$[-\Gamma(-\gamma)]^{-1} \int_{\varepsilon}^{\infty} u^{-\gamma}[f(x) - f(x-u)]du/u \qquad (0<\gamma<1)$$

converges in the $L^p$-norm for $\varepsilon\to 0+$ (cf. [5]).

Now to the characterization of the approximation behavior of the semigroup $T(t)$ on $D((-A)^{\gamma})$ for $t\to 0+$. This is, indeed, connected with the intermediate spaces of X and $D((-A)^{\gamma})$ constructed via the K-interpolation method of J. Peetre.

If $(X)_{\theta,q}(T(t))$ is the approximation space (2.4), now associated with the process $\{T(t);\ t\geq 0\}$, $\rho$ being replaced by $t^{-1}$ and $\tau = 1$, the major result which extends [A, Theorems 3.4.2, 3.4.3] for $\gamma=1$ to $0<\gamma<1$ reads [100,B,24]:

Theorem. (i) If $0<\theta<\gamma<1$, $1\leq q\leq\infty$, then

$$(X,D((-A)^{\gamma}))_{\theta/\gamma,q;K} \cong (X)_{\theta,q}(T(t)).$$

(ii) If $\theta=\gamma$ , $q=\infty$, then

(a) $(X,D((-A)^{\gamma}))_{1,\infty;K} \cong \widetilde{D((-A)^{\gamma})}^{X}$,

(b) $(X,D((-A)^{\gamma}))_{1,\infty;K}$ is equal to the space $\{f \in X;\ \sup_{\varepsilon>0}\|(-A)_{\gamma,\varepsilon}f\|<\infty\}$ with equivalent norm $\|f\|+\sup_{\varepsilon>0}\|(-A)_{\gamma,\varepsilon}f\|$.

Part (ii) follows immediately from (2.6) by taking $\{S_{\gamma,\varepsilon};\varepsilon>0\}$ as the approximation process with $\rho=1/\varepsilon$ $(\varepsilon\to 0+)$, $\tau=\gamma$, $B=-\gamma C_{\gamma,1}(-A)^{\gamma}$. For (i) cf. [B].

## 3. Fourier Analysis on $R^n$ and Other Transform Methods

A further main field of research of our group is the treatment of approximation theoretical problems via integral transform methods, in particular via Fourier transforms. The initial factor here is probably the saturation problem posed by Jean Favard. To give a representative account of the Fourier analytical aspect, the comments to follow are restricted to the n-dim. theory essentially developed after 1968--a major part of the results solved by one-dim. Fourier transform methods being found in the monograph of Butzer-Nessel [E].

The underlying aspect of all integral transform methods is the well-known property that a difficult "convolution product" is transformed into a pointwise product.

The convolution product in the instance of Fourier analysis on $R^n$ is defined by

$$(3.1) \qquad f * d\mu(x) = (2\pi)^{-n/2} \int_{R^n} f(x-u)\,d\mu(u) \qquad (f \in L^p(R^n)),$$

where $\mu \in M(R^n)$, the class of all bounded Borel measures. The associated Fourier (-Stieltjes) transform of $\mu \in M(R^n)$ is the (continuous, bounded) function

$$[d\mu]^\wedge(v) = (2\pi)^{-n/2} \int_{R^n} e^{-iv\cdot x}\,d\mu(x),$$

$v\cdot x = \Sigma_{k=1}^{n} v_k x_k$ being the inner product of $x, v \in R^n$. If $\mu$ is absolutely continuous, i.e., $\mu' = g \in L^1(R^n)$, the Fourier transform of $g$ is defined by

$$g^\wedge(v) = (2\pi)^{-n/2} \int_{R^n} e^{-iv\cdot x} g(x)\,dx.$$

Since for $f \in L^1(R^n)$ one has $[f * d\mu]^\wedge(v) = f^\wedge(v)\cdot[d\mu]^\wedge(v)$, this suggests that methods of Fourier analysis may enable one to handle approximation processes of the convolution type

$$(3.2) \qquad T_\rho f = f * d\mu_\rho,$$

where the kernel $\{\mu_\rho\} \subset M(R^n)$ satisfies the properties

$$(3.3) \qquad \int_{R^n} d\mu_\rho = 1, \quad \int_{R^n} |d\mu_\rho| \le C, \quad \lim_{\rho\to\infty} \int_{|u|\ge\delta} |d\mu_\rho(u)| = 0 \qquad (\delta > 0).$$

## 3.1 The Saturation Problem

The first systematic contribution here is the doctoral dissertation of Nessel [65]. Observing that approximation processes of this type enable one to separate function and kernel in the transformed space, crucial for solution of the saturation problem is a precise discussion of the behavior of $[d\mu_\rho]^\wedge - 1$ for $\rho\to\infty$, already noted by Favard. Motivated by the one-dimensional theory one poses the following hypotheses (see [14], [65]):

Suppose there exists a continuous function $\psi(v)$ on $R^n$

(F) with isolated zeros such that

$$\lim_{\rho\to\infty} \frac{1}{\phi(\rho)} \{[d\mu_\rho]^\wedge(v) - 1\} = \psi(v) \qquad (v \in R^n),$$

where $\phi(\rho)$ is a positive function tending monotonely to zero as $\rho\to\infty$.

In addition to (F), suppose there exists a family $\{\nu_\rho\}$

(F*) of uniformly bounded measures such that the representation

$$[d\mu_\rho]^\wedge(v) - 1 = \phi(\rho)\psi(v)[d\nu_\rho]^\wedge(v)$$

holds for any $v \in R^n$ and $\rho>0$.

Under these multiplier conditions it was essentially possible to solve the saturation problem, to begin with for $1\le p\le 2$ in [14,66], [65], then for $2<p<\infty$ in [4] using a functional or dual method (since the classical Fourier transform is bounded only for $1\le p\le 2$).

The next step was to find a unified approach to saturation and characterization theorems valid for all $L^p$-spaces. Inspired by a short note by H. Buchwalter (1960), E. Görlich in his dissertation [36] and [9,37,38] built up a distributional method in the instance $\psi(v) = |v|^\alpha$, $\alpha>0$, for $1\le p<\infty$. Using a lemma of E. M. Stein [[E, p. 133]], he was the first to introduce the space of Bessel potentials

$$L_\alpha^p = \left\{ f \in, L^p;\ f = \begin{cases} G_\alpha * d\mu, \ \mu \in M(R^n), \ p=1, \\ G_\alpha * h, \ \ h \in L^p(R^n), \ 1<p\leq\infty \end{cases} \right\},$$

into saturation theory ($G_\alpha \in L^1(R^n)$ being defined via $G_\alpha^\wedge(v) = (1+|v|^2)^{-\alpha/2}$). For $\alpha \in N$ compare also Kojima-Sunouchi [[15]], whereas J. Boman [[3]] developed a method for positive homogeneous functions $\psi(v)$, i.e., $\psi(\tau v) = \tau^\alpha \psi(v)$, $\tau,\alpha>0$, in a pure distributional way without using the spaces $L_\alpha^p$, noting that $\psi(v)f^\wedge$ is a tempered distribution.

A different type of generalization is due to Trebels [95]; here a generalized version of the cited Stein result is stated as a hypothesis. It means that to a closed operator there is associated an equivalent closed operator with continuous inverse. A typical result (for not necessarily radial kernels), whose proof depends on classical Fourier analysis and density arguments reads (see [95]):

Theorem 1. Let $X(R^n)$ be one of the spaces $L^p(R^n)$, $1\leq p<\infty$, or $C(R^n)$; let the operators $\{T_\rho; \rho>0\}$ be defined on $X(R^n)$ by (3.2).

(a) If (F) is satisfied, then $\|T_\rho f - f\| = o(\phi(\rho))$, $\rho\to\infty$, implies $f(x)=0$ a.e. or $f$ is a constant in the continuous case.

(b) If (F*) holds and to $\psi(v)$ there exist $g \in L^1(R^n)$ and $\chi_1,\chi_2,\chi_3 \in M(R^n)$ such that

(3.4) $\psi(v) = [d\chi_1]^\wedge(v)/g^\wedge(v)$,

(3.5) $1/g^\wedge(v) = [d\chi_2]^\wedge(v) + \psi(v)[d\chi_3]^\wedge(v)$,

then the Favard class $F[X(R^n); T_\rho]$ is given by

$$L_g^p = \left\{ f \in X(R^n);\ f = \begin{cases} g * d\mu, \ \mu \in M(R^n), \ p=1, \\ g * h, \ \ h \in L^p(R^n), \ 1<p\leq\infty \end{cases} \right\}.$$

Note that in case $\psi(v) = |v|^{\alpha}$ Stein's result gives a possible choice for g, namely $g(x) = G_{\alpha}(x)$; for a further relation of type (3.4), (3.5) compare Butzer-Kolbe-Nessel [13]. For an extension of this theory to n-parameter approximation with product kernels see Berens-Nessel [4]. For a mixed approach via Fourier analysis and semigroup theory see [[17]].

In the course of this development the following particular approximation processes were studied: Gauss-Weierstrass, Abel-Poisson, Picard, Bochner-Riesz, spherical means, Riesz means, generalized singular integral of Weierstrass, etc.; see [4], [36, 37, 38], [65, 66], [95], [[3]], [[15]], [[17]].

The verification of condition (F*) was of essential importance in this respect. The following methods were applied: (i) functional equations in [2] (for spherical harmonics) and in [38], [65, 66], and a direct generalization in [[15]], (ii) moment conditions in [69], [[3]], (iii) direct application of multiplier theory in [[3]], [38], [[15]], [66], [[17]], [95], etc.

## 3.2 Characterizations of Favard Classes

In connection with Theorem 1 one wishes to characterize the space $L_g^p(R^n)$. This has been carried out especially in the case of the space of Bessel potentials $L_{\alpha}^p(R^n)$, for $1<p<\infty$ the multiplier theorem of Marcinkiewicz-Mikhlin-Hörmander being used. A typical result reads:

Theorem 2. If $\alpha=2$ the following assertions are equivalent:

(i) $f \in L_2^p$ $(1\leq p\leq\infty)$,

(ii) $\sum_{k,j=1}^{n} \|\frac{\partial^2}{\partial x_k \partial x_j} f\|_p = O(1)$ (=Sobolev space of order 2) ($1<p<\infty$; A. P. Calderón 1961)

(iii) $\|f(x+2u) - 2f(x+u) + f(x)\|_p = O(|u|^2)$ $(u \in R^n; 1<p<\infty,$ [36, 37]),

(iv) $\|f(x+\tau e^k) - f(x) - \tau \frac{\partial f(x)}{\partial x_k}\|_p = O(\tau^2)$ $(\tau>0;\ 1\leq k\leq n;$ $1<p<\infty;$ [37,38]),

$e^k \in R^n$ being the unit coordinate vector along the $k$-axis,

(v) $\|\sum_{k=1}^{n} \{f(x+2\tau e^k) - 2f(x+\tau e^k)+f(x)\}\|_p = O(\tau^2)$

$(\tau>0;\ 1\leq p\leq\infty;$ [29], [88,93]),

(vi) $\|\int_{|u|\geq\varepsilon} \{f(x+3u) - 3f(x+2u)+3f(x+u) - f(x)\}|u|^{-n-2}du\|_p$

$= O(1)$ $(\varepsilon>0;\ 1\leq p\leq\infty;$ [[15]], [92]).

For further characterizations see Görlich [37] and the literature cited, as well as the survey paper by V. I. Burenkov of 1968. Analogous results are also known for $\alpha\neq 2$ (see [36,37], [92,93], [[29]]). In case $\alpha$ is odd, an important role is played by the Riesz transform

$$R_k f = \lim_{\varepsilon\to 0} c_k \int_{|u|\geq\varepsilon} (u_k/|u|^{n+1}) f(x-u)du \qquad (1\leq k\leq n).$$

For fractional $\alpha$, particularly for $0<\alpha<2$, $L_\alpha^p$ may be characterized by hypersingular integrals of type (vi) Theorem 2, namely

$$(3.6) \qquad \| \frac{2^\alpha}{\pi^{n/2}} \frac{\Gamma((n+\alpha)/2)}{\Gamma(-\alpha/2)} \int_{|u|\geq\varepsilon} \frac{f(x+u)-f(x)}{|u|^{n+\alpha}} du\|_p = O(1) \quad (\varepsilon>0).$$

Note that this integral is Hadamard's finite part of the fractional Riesz integral of negative order for $\varepsilon\to 0+$ (see fine treatment in Horváth [[10]]). For $1<p<\infty$ (3.6) is due to Stein [[E, p. 162]]; for $1\leq p\leq\infty$ there are several proofs, see [[15]], [92], [[29]]. Wheeden [[29]] generalizes (3.6) for

$1<p<\infty$ by multiplying $|u|^{-n-\alpha}$ by $\Omega(u/|u|)$ ($\Omega$ integrable over the unit sphere); Trebels [92] showed that this generalization reduces to an application of the Marcinkiewicz-Mikhlin-Hörmander multiplier theorem provided $\Omega$ satisfies the conditions of the Calderón-Zygmund theory.

Naturally there exist many saturation spaces other than $L^p_\alpha$ (see [14], [4]). In particular, the saturation classes for n-parameter methods do principally not possess radial structure; for their characterization see, e.g., [90].

### 3.3 Other Types of Integral Transforms

In a similar fashion other types of integral transforms may be used to treat corresponding approximation problems (compare [E, p. 479 ). Thus, e.g., in case of the n-dim. torus $T^n$ the Fourier coefficient method may be applied for similar problems. This is carried out in Görlich [39] and Nessel-Pawelke [69] (indeed, performed earlier than the $R^n$-case). Note however, that the Fourier multiplier approaches on $X(R^n)$ and $X(T^n)$ essentially coincide (cf. [[F; Ch. 7]]). In case of the n-dim. unit sphere the same problems were discussed by Berens-Butzer-Pawelke [2] in the framework of expansions into spherical harmonics. (For the latter two transform methods see the general approach in Ch. 4).

Finally, in connection with the approximation of boundary values by solutions of harmonic problems in a wedge Mellin transform methods have proved to be very useful, as is demonstrated in [63, 62] (cf. also [61]). In particular, W. Kolbe [61] in his dissertation discusses biharmonic problems arising from elasticity theory. In all these cases more than one boundary condition is involved so that one has to extend the previous notions to systems of coupled approximations. In particular, simultaneous saturation has to be considered. In these papers this has been achieved simply by an extension of the classical concepts to product spaces. On account of the linear structure involved matrix methods are employed.

### 3.4 Concluding Remarks

Summarizing, integral transform methods are especially useful in solving saturation problems; hereby it turns out that only a discussion of the transformed kernel (as e.g., expressed in (F*)) is essential. Once this was noted it was also possible, by posing suitable multiplier conditions upon the kernel, to solve another problem of Favard, namely to give comparison estimates for two different methods of approximation $\{T_\rho;\rho>0\}$, $\{S_\rho;\rho>0\}$ in form of

$$\|S_\rho f-f\| \leq C\|T_\rho f-f\|,$$

see J. Löfstrom [[17]] and H. S. Shapiro [[24]] on $X(R^n)$. Analogously Trebels [95] provided simple multiplier conditions (upon the kernel) for the validity of Jackson, Bernstein (and Zamansky) type inequalities on $X(R^n)$, thus for the validity of the two basic hypotheses of the general theorems of Butzer-Scherer on non-optimal approximation (see Theorems 1-4 in Ch. 1) for the particular instance $X(R^n)$.

The key point in the investigations of this chapter is a suitable, easy to apply multiplier theory, as built up e.g., in Fourier analysis in [[3]], [[17]], [[20]], [[E]], [[F]], [94]. It is our opinion that a further development of multiplier theory (see Ch. 4), or, equivalently, of the theory of singular convolution operators, would lead to sharper results in approximation theory.

## 4. Summation Processes of Fourier Expansions in Banach Spaces

In the previous Chapter we saw that one can treat a series of approximation problems in the particular space $L^p(R^n)$ using

Fourier transform methods. In view of the general approximation theorems of Ch. 1 the question now arises whether it would be possible to find relatively simple sufficient conditions such that general approximation processes in arbitrary Banach spaces satisfy Jackson, Bernstein, and Zamansky type inequalities or that saturation or comparison theorems follow for such processes. It is indeed possible to solve these questions for summation processes of Fourier expansions in Banach spaces, in particular for most of the classical orthogonal expansions.

For this purpose, let X be an arbitrary (real or complex) (B)-space with norm $\|.\|$. Let $\{P_k\}_{k=0}^{\infty} \subset E(X)$ be a total, fundamental system of mutually orthogonal projections on X, i.e., (i) $P_k f = 0$ for all $k \in P$ implies $f=0$ (total), (ii) the linear span of $\cup_{k=0}^{\infty} P_k[X]$ is dense in X (fundamental), (iii) $P_k P_j = \delta_{kj} P_k$ for all $k,j \in P$ (orthogonal).

Then with each $f \in X$ one may associate its (formal) Fourier series

$$(4.1) \qquad f \sim \sum_{k=0} P_k f \qquad\qquad (f \in X).$$

With s the set of all sequences $\gamma = \{\gamma_k\}_{k=0}^{\infty}$ of scalars, $\gamma \in s$ is called a <u>multiplier</u> for X (corresponding to $\{P_k\}$), in notation $\gamma \in M$, if for each $f \in X$ there exists an element $f^{\gamma} \in X$ such that $P_k f^{\gamma} = \gamma_k P_k f$ for all $k \in P$. Defining $Gf = f^{\gamma}$ one clearly has $G \in E(X)$ and $Gf \sim \Sigma \gamma_k P_k f$. Operators admitting such an expansion are called multiplier operators, in symbols $G \subset E(X)_M$. M is a commutative (B)-algebra, isometric and isomorphic to $E(X)_M$ with respect to the natural vector operations, coordinatewise multiplication and norm

$$\|\gamma\|_M = \sup_{f \in X, \|f\| \leq 1} \{\|f^{\gamma}\|\} = \|G\|_{E(X)} \, .$$

For basic results on bases and decompositions in (B)-spaces compare the books by J. T. Marti (1969) and I. Singer (1970).

Jackson and Bernstein-type inequalities suggest that one

introduces a subspace $X^{\phi}$ of X defined for some given $\phi \in s$ by

$$X^{\phi} = \{f \in X; \exists\, f^{\phi} \in X \ni P_k f^{\phi} = \phi_k P_k f,\ k \in \mathbb{P}\}$$

as well as the closed linear operator $B^{\phi}$: $f \to f^{\phi}$ on $X^{\phi}$.

These concepts suffice to solve the problems raised above for approximation processes of multiplier-type. For the following theorem compare Butzer-Nessel-Trebels [16, 17], Görlich-Nessel-Trebels [43, 44], Trebels [96]:

Theorem 1. Let $\{T_\rho; \rho>0\} \subset E(X)_M$ be a strong approximation process with associated multiplier family $\{\tau(\rho); \rho>0\}$.

(a) If there exists a non-negative, monotonely increasing function $\chi(\rho)$ with $\lim_{\rho\to\infty} \chi(\rho) = \infty$, $\phi \in s$, and a uniformly bounded multiplier family $\{\gamma(\rho)\} \subset M$, i.e., $\|\gamma(\rho)\|_M \leq C$, with

$$(4.2) \qquad \chi(\rho)\,\{\tau_k(\rho) - 1\} = \phi_k \gamma_k(\rho) \qquad (k \in \mathbb{P};\ \rho>0),$$

then one has the Jackson-type inequality

$$(4.2^*) \qquad \chi(\rho)\|T_\rho f - f\| \leq \sup_{\rho>0}\|\gamma(\rho)\|_M \|B^{\phi} f\| \qquad (f \in X^{\phi}).$$

If, in particular, $\lim_{\rho\to\infty} \gamma_k(\rho) = 1\ \forall\, k \in \mathbb{P}$ and $\phi_k = 0$ iff $\tau_k(\rho)=1$, then there holds the saturation theorem

(i) $\chi(\rho)\|T_\rho f - f\| = o(1)$ for $\rho\to\infty$ implies $T_\rho f = f\ \forall\, \rho>0$, i.e., f is an invariant element;

(ii) the Favard class of the process $\{T_\rho; \rho>0\}$ is $\widetilde{X^{\phi}}^X$; the following semi-norms are equivalent on $\widetilde{X^{\phi}}^X$:

(α) $\sup_{\rho>0} \chi(\rho)\|T_\rho f - f\|$, (β) $\sup_{\rho>0}\|B^{\phi} G_\rho f\|$.

(b) Let $\chi$, $\phi$, $\gamma(\rho)$ be given as in (a), with (4.2) replaced by

$$(4.3) \qquad \phi_k \tau_k(\rho) = \chi(\rho)\gamma_k(\rho) \qquad (k \in \mathbb{P};\ \rho>0).$$

Then there holds the Bernstein-type inequality

(4.3*) $\|B^{\phi}T_{\rho}f\| \leq \chi(\rho) \sup_{\rho>0}\|\gamma(\rho)\|_{M}\|f\|$ (f ∈ X)

(c) Let $\chi$, $\phi$, $\gamma(\rho)$ be given as in (a) with (4.2) replaced by

(4.4) $\phi_k\tau_k(\rho) = \chi(\rho)\gamma_k(\rho)\ \{\tau_k(\rho) - 1\}$ (k ∈ P; ρ>0).

Then one has the Zamansky-type inequality

(4.4*) $\|B^{\phi}T_{\rho}f\| \leq \chi(\rho)\ [\sup_{\rho>0}\|\gamma(\rho)\|_{M}]\ \|T_{\rho}f-f\|$ (f ∈ X).

(d) Let $\{S_{\rho};\rho>0\} \in E(X)_V$ be a second strong approximation process with associated multiplier family $\{\sigma(\rho);\rho>0\}$, and let $\{\gamma(\rho)\}$ be as in (a) with

(4.5) $\tau_k(\rho) - 1 = \gamma_k(\rho)\ \{\sigma_k(\rho) - 1\}$ (k ∈ P; ρ>0),

then there holds the comparison theorem

(4.5*) $\|T_{\rho}f - f\| \leq \{\sup_{\rho>0}\|\gamma(\rho)\|_{M}\}\ \|S_{\rho}f - f\|$ (f ∈ X).

This easy to prove theorem reveals the versatility of the multiplier approach (see Ch. 3) provided one has a suitable multiplier theory at one's disposal. Whereas such a theory was available in the Fourier transform case, it must first be built up for the present setting. But this does not seem possible under the hypotheses[3] upon the pair X, $\{P_k\}$ hitherto existing. The applications suggest that an additional assumption be the uniform boundedness of the (C,α)-means of (4.1). Recalling Chapter 2.3, these are defined by

(4.6) $(C,\alpha)_n f := \sum_{k=0}^{n} \frac{A_{n-k}^{\alpha}}{A_n^{\alpha}} P_k f,\qquad A_n^{\alpha} = \frac{\Gamma(n+\alpha+1)}{\Gamma(n+1)\Gamma(\alpha+1)}$ (α ∈ R)

and one assumes[4] that for some α≥0

(4.7) $\|(C,\alpha)_n f\| \leq C_{\alpha}\|f\|$ (f ∈ X).

This assumption has been the starting point for many investigations in classical $(C,\alpha)$-summability theory for numerical series. In this connection, in order to describe the multiplier class $M$ in the present frame, let us consider

$$(4.8)\qquad bv_{\alpha+1} := \{\gamma \in \ell^{\infty};\ \|\gamma\|_{bv_{\alpha+1}} := \sum_{k=0}^{\infty} A_k^{\alpha}|\Delta^{\alpha+1}\gamma_k| + \lim_{k\to\infty}|\gamma_k| < \infty\}$$

$$(\alpha \geq 0).$$

Here the fractional difference operator is defined by

$$\Delta^{\alpha+1}\gamma_k = \sum_{i=0}^{\infty} \gamma_{k+i}\ A_i^{-\alpha-2} \quad .$$

However, $\gamma \in \ell^{\infty}$ together with the boundedness of the sum in (4.8) implies the existence of $\lim_{k\to\infty} \gamma_k = \gamma_{\infty}$. Note that for $0\leq\beta\leq\alpha$ one has $bv_{\alpha+1} \subset bv_{\beta+1}$ in the sense of continuous embedding.

Theorem 2. Let $X$, $\{P_k\}$ be given as above, and let (4.7) be satisfied for some $\alpha\geq 0$. Then $bv_{\alpha+1} \subset M$ in sense of continuous embedding.

This result follows by a suitable modification of the well-known paper by A. F. Andersen (1928), considering the element

$$f^{\gamma} = \sum_{k=0}^{\infty} A_k^{\alpha}\ \Delta^{\alpha+1}\gamma_k (C,\alpha)_k f + \gamma_{\infty} f$$

for $\gamma \in bv_{\alpha+1}$ and $f \in X$. In the case of 1-dimensional trig. series (in various function spaces) this criterion is already to be found in C. N. Moore (1933) and G. Goes (1960).

But in general it is difficult to check whether $\gamma \in bv_{\alpha+1}$. Therefore one estimates the sum $\sum_{k=0}^{\infty} A_k^{\alpha}|\Delta^{\alpha+1}\gamma_k|$ by a suitable integral. For this purpose we extend $\gamma$, defined on $P$, to the positive real axis, i.e., we look for a suitable function $g(x)$, $x\geq 0$, with $g(k)=\gamma_k$ and satisfying certain regularity properties to be stated in the definition of $BV_{\alpha+1}$ below. First consider the fractional integral of order $(1-\xi)$, $0<\xi<1$,

$$I_\omega^{1-\xi}[g](x) := \frac{1}{\Gamma(1-\xi)} \int_x^\omega (u-x)^{-\xi} g(u)du,$$

and, following J. Cossar (1941), define the fractional derivative of order $\xi$ by

$$g^{(\xi)}(x) := \lim_{\omega\to\infty} - \frac{d}{dx} I_\omega^{1-\xi}[g](x).$$

The usual differentiation of $g^{(\xi)}$ yields pure fractional derivatives of order $\alpha = [\alpha] + \xi$, i.e.,

$$g^{(\alpha)}(x) = (\frac{d}{dx})^{[\alpha]} g^{(\xi)}(x).$$

As an analog for $bv_{\alpha+1}$, $\alpha>0$, consider now the class

$$BV_{\alpha+1} = \{g \in C_0;\ g^{(\xi)},\dots,g^{(\alpha-1)} \in AC_{loc}(0,\infty),\ g^{(\alpha)} \in BV_{loc}(0,\infty),\ \|g\|_{BV_{\alpha+1}} := \sup_{x\geq 0} |g(x)| + \frac{1}{\Gamma(\alpha+1)} \int_0^\infty x^\alpha |dg^{(\alpha)}(x)| < \infty\}.$$

Here $C_0$ is the set of all $f \in C[0,\infty)$ vanishing at infinity (this is no restriction; otherwise consider $\{\gamma_k - \gamma_\infty\}_{k=0}^\infty$). If $\alpha=0$ then $BV_1$ is the usual class of functions of bounded variation on $[0,\infty)$ which vanish at infinity. One again has $BV_{\alpha+1} \subset BV_{\beta+1}$, $0\leq\beta\leq\alpha$, in the sense of continuous embedding; moreover, one has for $\gamma_k = g(k)$

$$\sum_{k=0}^\infty A_k^\alpha |\Delta^{\alpha+1}\gamma_k| \leq D_\alpha \int_0^\infty x^\alpha |dg^{(\alpha)}(x)| \qquad (\alpha\geq 0).$$

This implies (see [17] for $\alpha \in \mathbb{N}$, [96] for $\alpha\geq 0$):

<u>Theorem 3</u>. Let X, $\{P_k\}$ be given as above, (4.7) being satisfied for some $\alpha\geq 0$. For $\gamma(\rho) \in s$ let there exist some $g \in BV_{\alpha+1}$ such that $\gamma_k(\rho) = g(k/\rho)$. Then

$$\|\gamma(\rho)\|_M \leq D_\alpha' \|g\|_{BV_{\alpha+1}} \qquad \text{(uniformly in } \rho>0).$$

The assumption that $\gamma(\rho)$ be of Fejér's type (i.e.,

$\gamma_k(\rho) = g(k/\rho))$, may be weakened considerably, namely to (*) $\gamma_k(\rho) = g(\Phi(k)/\psi(\rho))$, where $\psi(\rho)>0$, $\Phi(x)$ is strictly monotone increasing for $x\geq 0$ such that $\lim_{x\to 0+}\Phi(x) = 0$, $\lim_{x\to\infty}\Phi(x) = \infty$ and $\Phi$ is $([\alpha]+2)$-times differentiable on $(0,\infty)$ with

$$|x^k\Phi^{(k+1)}(x)| \leq c|\Phi'(x)| \qquad (0\leq k\leq[\alpha]+1).$$

(In case $\alpha$ is purely fractional, one must also assume $\Phi'(x)$ to be monotone). Under these assumptions one again has (see [96]) $\|\gamma(\rho)\| \leq D^*_\alpha\|g\|_{BV_{\alpha+1}}$ uniformly in $\rho>0$ in case $\gamma(\rho)$ satisfies (*). This is to be seen as an analog to Hardy's "Second Theorem of Consistency" for numerical series. Finally it should be mentioned that one can set up a Lipschitz condition for $g^{([\alpha]+1)}$ which implies $g \in BV_{\alpha+1}$ (cf. [96]); this is a suitable modification of a theorem of H. Weyl (1917) concerning Lipschitz conditions and the existence of fractional derivatives.

Theorem 3 implies that the Riesz means of order $\lambda$, namely

$$R_{\lambda;\rho}f = \sum_{k<\rho} (1-\frac{k}{\rho})^\lambda P_k f \qquad (\rho>0),$$

are uniformly bounded operators on X for $\lambda\geq\alpha$. In particular

$$\|(C,\alpha)_n f\| \leq C_\alpha\|f\| \qquad (n \in P,\ f \in X)$$

implies

$$\|R_{\alpha;\rho}f\| \leq C^*_\alpha\|f\| \qquad (\rho>0,\ f \in X). \tag{4.9}$$

But the converse implication can be shown by an argument analogous to one by J. J. Gergen (1937) for numerical series. This shows that Theorems 2 and 3 are in this sense best possible under hypothesis (4.7).

In connection with Theorem 1, applications of the above multiplier theory to Cesàro, Riesz, de La Vallée Poussin, Abel-Cartwright and Picard means are to be found in [17], [43], [96].

In the foregoing general results the choice of the X, $\{P_k\}$

is still free. Particular instances for one and n-dim. Fourier series, for Jacobi series (which of course embrace expansions according to Legendre, Chebyshev and ultraspherical polynomials), for Laguerre, Hermite, Walsh, and Haar series, as well as for expansions with respect to Bessel functions and spherical harmonics are to be found in [17], [43], [96]. The abstract theory sketched above yields many new multiplier theorems for most of these classical series.

Consider the case for expansions with respect to ultraspherical polynomials: Askey-Hirschman (1963) established the uniform boundedness of their $(C,\alpha)$-means, but the analog of the Marcinkiewicz multiplier theorem due to Muckenhoupt-Stein (1965) does not seem to fit in so well with $(C,\alpha)$-summability theory. But in our frame $(C,\alpha)$-summability and multiplier theory correspond to another precisely. Thus the problem arises whether this correspondence can be established for concrete orthogonal systems. To our mind this interconnection would be worthwhile examining.

## 5. Approximation Theorems on Compact Manifolds

The goal of this field of our research is to generalize the main approximation theorems valid on the one-dimensional torus $T^1$ to compact $C^\infty$-manifolds (without boundary). For the first aim, to establish the analogs of the fundamental theorems on best approximation (see Theorems 1,2, Ch. 1), basic concepts such as that of a Lipschitz class or a modulus of continuity must first be defined.

Let G be a Lie group with invariant Riemannian metric $\rho$, and let G act on the manifold M by $p \mapsto s^{-1}p$, $s \in G$, $p \in M$, as a Lie transformation group (S. Helgason [[D, p. 112]]). For a function f on M and $s \in G$ define $T(s)f$ by $T(s)f(p) = f(s^{-1}p)$, $p \in M$. On each Banach space X of functions on M such that $f \in X$ implies $T(s)f \in X$ for all $s \in G$ one has

$$(5.1) \qquad T(s_1 s_2) = T(s_1)T(s_2) \qquad (s_1, s_2 \in G).$$

If, furthermore, $s \to T(s)f$ is continuous from G to X for every $f \in X$ and $T(s) \in E(X)$, then the mapping T of G into E(X) possesses all properties of a continuous representation of G on X. For each f of the representation space X the $r^{th}$ modulus of continuity is now defined by

$$(5.2) \qquad \omega_r(f;\delta) = \sup_{\rho(s,e) \leq \delta} \{\| \sum_{k=0}^{r} (-1)^{r-k} \binom{r}{k} T(s^k) f \| \} \qquad (\delta > 0),$$

e being the unit element of G. Particularly for the continuous representations of a compact connected Lie group G these moduli have practically the same nice properties as those of the classical moduli of continuity on $T^1$ (see, Butzer-Johnen [10], Johnen [51, 52, 53, 54]). Moreover, since Riemannian metrics are equivalent on compact manifolds, the moduli of continuity are then equivalent for different Riemann metrics on G.

If G is a compact connected Lie group and H a compact (Lie) subgroup of G, then, by the preceding observations, moduli of continuity can be defined on $L^p(G/H)$, $1 \leq p < \infty$, and $C(G/H)$, the $L^p$-spaces being understood with respect to the unique normalized left invariant measure on G/H. Special instances of these homogeneous spaces G/H are G itself, the matrix groups SO(m), U(m), SU(m), Sp(n), $T^m$, the m-sphere $S^m = SO(m+1)/SO(m)$, and the real projective space $P^m$.

To formulate the Jackson and Bernstein theorems on G/H under the hypotheses stated above we still need the concept of a trigonometric polynomial of a definite degree. This was first given by D. L. Ragozin [[21]] as follows: Take a faithful finite dimensional unitary representation D of G on some $C^m$ (C the complex field) with its usual inner product, and form the m'×m' matrix (m'= 3m)

$$[D \oplus \overline{D} \oplus D(e)](s) = \begin{pmatrix} D(s) & 0 & 0 \\ 0 & \overline{D}(s) & 0 \\ 0 & 0 & D(e) \end{pmatrix} \qquad (s \in G),$$

where $\overline{D}(s) = (\overline{d}_{ij}(s))$ is defined by $D(s) = (d_{ij}(s))$, the bar denoting complex conjugation. Then set

$$D_n(s) = \otimes^n [D \oplus \overline{D} \oplus D(e)](s),$$

where $\otimes^n$, the $n^{th}$ Kronecker power of a matrix $v$ in $U(m')$, is defined inductively by $\otimes^1 v = v$,

$$\otimes^{n+1} v = \begin{pmatrix} v_{11} \otimes^n v \cdots\cdots\cdots\cdots v_{1m'} \otimes^n v \\ \vdots \\ v_{m'1} \otimes^n v \cdots\cdots\cdots\cdots v_{m'm'} \otimes^n v \end{pmatrix} .$$

Then one calls a function a trigonometric polynomial of degree n on G if it is representable as a linear combination of entries of $D_n$. If a function on G/H arises by projection from $L^2(G)$ into $L^2(G/H)$ of a trig. polynomial of degree n on G (see Johnen [55]), it is called a <u>trig. polynomial of degree n on G/H</u>. This definition of degree is practically independent of the representation D. Indeed, if D, D' are two faithful finite dim. unitary representations with $P_n(D)$, $P_n(D')$ (= set of trig. poly. of degree n on G/H with respect to D,D'), then there exists $d = d(D,D') \in \mathbb{N}$ such that $P_n(D) \subset P_{dn}(D')$.

Having fixed a suitable representation D of G and so the space $P_n$ on G/H, then the best approximation for f in $L^p(G/H)$ or $C(G/H)$ is defined by

$$E_n(f) = \inf_{t_n \in P_n} \|f - t_n\|.$$

<u>Theorem.</u> Let $\ell_1,\ldots,\ell_m$ be an orthonormal basis for the Lie algebra $G$ of G with respect to the positive-definite bilinear form B, and let $\rho$ be the Riemannian metric on G induced by B. Then for all $f \in L^{p,k}(G/H)$, $1 \leq p < \infty$, $f \in C^k(G/H)$ (see [55]) there exists a constant $c_k$ such that

$$(5.3) \qquad E_n(f) \leq c_k n^{-k} \sum_{i_1,\ldots i_k=1}^{m} \omega_2(\ell_{i_1} \cdots \ell_{i_k} f; n^{-1}).$$

This Jackson-type theorem was (implicitly) proved in its full generality in Johnen [53] by a constructive method (see also [55]). For the particular compact symmetric spaces of rank 1 there exist earlier constructive proofs by Ragozin [[22, 23]], and for U(m), SO(m), Sp(m) by Johnen [50, 51]. In case the second modulus is replaced by the first, see Ragozin [[22]]; for the unitary group and k=0 see Gong-Sheng. For further conclusions to be drawn from (5.3), for estimates even stronger than (5.3) derived from the "weaker" inequalities

$$E_n(f) \leq c_k n^{-k} \sum_{i_1,\dots,i_k=1}^{m} \|\ell_{i_1} \cdots \ell_{i_k} f\| \qquad (k \in \mathbb{N}),$$

see Johnen [52, 53]. There it is also shown how the non-commutativity of G brings in complications.

Already Ragozin [[21]] proved a Bernstein-type inequality for arbitrary left invariant vector fields $\ell$ on G/H and $t_n \in \mathcal{P}_n$, namely

$$\text{(5.4)} \qquad \sup_{p \in G/H} |\ell t_n(p)| \leq c_\ell\, n \sup_{p \in G/H} |t_n(p)|,$$

which Johnen [55] extended to $L^p(G/H)$, $1 \leq p < \infty$. On the compact homogeneous spaces of the foregoing type there therefore exist those inequalities which play the fundamental role in the general theorem of Butzer-Scherer (see Ch. 1.1). For the connection between the K-functional involved there and the moduli of continuity here see Johnen [52], which also treats approximation processes on representation spaces of Lie groups.

A further problem on compact homogeneous spaces is the saturation problem. For the general theory see Bragard-Johnen [7], which extends results of H. Buchwalter [[4]] on commutative groups. The saturation classes for particular processes had up to then been determined explicitly only for compact symmetric spaces of rank 1 such as $S^m$ (see [2]), SU(2), $P^m$ (see [7], H. Bavinck [[2]]), as well as for one-parameter processes.

So far we have only treated approximation on Lie transformation spaces. On an ordinary manifold, however, there exist only local transformation groups. If the m-dim. manifold M is compact, an $r^{th}$ modulus of continuity of $f \in C(M)$ can be introduced as follows (see Butzer-Johnen [10]): Take a finite atlas $\mathcal{U} = (U_i, \psi_i)_{i \in I}$ of M such that

$$Q := \{x \in R^m;\ |x| \leq 1\} \subset \psi_i(U_i), \qquad M = \bigcup_{i \in I} \psi_i^{-1}(Q^o),$$

$Q^o$ being the interior of Q. Define

$$\omega_r^{\mathcal{U}}(f;\delta) = \sum_{i \in I} \sup_{\substack{x, x+rh \in Q \\ |h| \leq \delta}} \left| \sum_{k=0}^{r} (-1)^{r-k} \binom{r}{k} f \circ \psi_i^{-1}(x+kh) \right|.$$

For two different such finite atlasses $\mathcal{U}, \mathcal{V}$ there exist two positive constants $c_1 = c_1(\mathcal{U}, \mathcal{V})$, $c_2$ such that for all $f \in C(M)$ and $0 < t \leq 1$

$$c_1 [t^r \|f\|_\infty + \omega_r^{\mathcal{U}}(f;t)] \leq t^r \|f\|_\infty + \omega_r^{\mathcal{V}}(f;t) \leq c_2 [t^r \|f\|_\infty + \omega_r^{\mathcal{U}}(f;t)].$$

This was shown by Ragozin [[22]] for r=1, by Johnen [49] for r=2, in Butzer-Johnen [10] for general r. Thus one may also introduce Lipschitz classes in this case. In particular, the assertions of the theorems of Jackson, Bernstein, Zamansky and Stečkin (Theorem 1, Ch. 1) are also equivalent on compact submanifolds of $R^r (r > m)$ for best approximation by restrictions of algebraic polynomials of degree n (see [10, Theorem 6]) provided a Bernstein-type inequality exists. A necessary condition for its existence is that our compact manifold be contained in an algebraic manifold (see [[22]]). Sufficiency conditions are still open.

## 6. Nikolskiĭ (Best Asymptotic) Constants

In 1940 S. M. Nikolskiĭ [[19]] initiated the study of the measure of approximation with respect to Lipschitz classes, beginning with the example of the Fejér-means. It was followed up by B. Sz.-Nagy, I. P. Natanson, A. F. Timan, L. Lorch, V. A. Baskakov, L. I. Bausov, P. P. Korovkin, R. Taberski, Y. Matsuoka (references in [45, 67, 85] make up $\geq 100$ papers). For periodic singular integrals of convolution type with positive even kernel, i.e., for $f \in C_{2\pi}$ and

$$(f * \chi_\rho)(x) = \frac{1}{\pi} \int_{-\pi}^{\pi} f(x-u)\chi_\rho(u)du \qquad (\rho>0, \rho\to\infty)$$

$$\chi_\rho(x) = \frac{1}{2} + \sum_{k=1}^{\infty} \chi_\rho^{\wedge}(k) \cos kx,$$

the measure of approximation with respect to $\mathrm{Lip}_2^*\alpha$ is defined by (see [E, p. 82 ff])

$$\Delta_\rho(\chi;\alpha) := \sup_{f \in \mathrm{Lip}_2^*\alpha} \|f*\chi_\rho - f\|_{C_{2\pi}} \qquad (0<\alpha\leq 2).$$

The problem is to establish an asymptotic expansion (if it exists) for this quantity having as many terms as possible; thus, for some $m=1,2,\ldots$ let there exist constants $N_k(\chi;\alpha)$, $k=1,2,\ldots,m$ and positive functions $\psi_k(\rho)$, $k=1,2,\ldots,m+1$ satisfying $\psi_1(\rho)=o(1)$, $\psi_{k+1}(\rho)=o(\psi_k(\rho))$, $\rho\to\infty$, such that

$$(6.1) \qquad \Delta_\rho(\chi;\alpha) = \sum_{k=1}^{m} N_k(\chi;\alpha)\psi_k(\rho) + O(\psi_{m+1}(\rho)) \qquad (\rho\to\infty)$$

with $\psi_{m+1}(\rho)$ being so that $[R_{m+1}(\rho)]^{-1}$ is bounded, where

$$R_{m+1}(\rho) := [\psi_{m+1}(\rho)]^{-1}\{\Delta_\rho(\chi;\alpha) - \sum_{k=1}^{m} N_k(\chi;\alpha)\psi_k(\rho)\} \qquad (\rho\to\infty),$$

i.e., the order of the remainder term in (6.1) cannot be increased. (In case $m=1$ the existence of (6.1) with $N_1(\chi;\alpha)>0$ is a direct consequence of $\chi_\rho(x)$ being an approximate identity). $N_k(\chi;\alpha)$ is called <u>Nikolskiĭ constant</u> of order k (of $f*\chi_\rho$ with respect to $\mathrm{Lip}_2^*\alpha$). If $m=\infty$ in (6.1) one speaks of the <u>complete asymptotic</u> (c.a.) expansion of $\Delta_\rho(\chi;\alpha)$. If there exists m such that $R_{m+1}(\rho)$ is bounded but

(6.2) $\lim_{\rho\to\infty} R_{m+1}(\rho) \nexists$,

then (6.1) is called the (m-term) essential asymptotic (e.a.) expansion of $\Delta_\rho(\chi;\alpha)$. Nevertheless, the limit in (6.2) may exist for each m provided $\rho$ is restricted to some subrange. In this instance one speaks of restricted c.a. expansions.

Essential a. expansions for particular poly. kernels (Fejér, Jackson, Jackson-de La Vallée Poussin) were considered by Butzer-Stark [26], Stark [80] in 1969. For the integral of Fejér with kernel of degree (n-1) the result reads (m=2, $\alpha$=1)

$$(6.3)\qquad \Delta_n(F;1) = \frac{2}{\pi}\frac{\log n}{n} + \frac{2}{\pi}(1+\gamma+\log 2)\frac{1}{n} + \mathcal{O}\left(\frac{1}{n^3}\right) \quad (n\to\infty),$$

$\gamma$ being Euler's constant. The method of proof, using the coefficients $\chi_\rho^\wedge(k)$, is restricted to (the non-fractional) $\alpha$=1. Independently in 1969 S. A. Teljakovskiĭ [[28]] obtained the restricted c.a. expansion (n even or odd in view of factor $(-1)^n$ below) for Fejér's integral for all $0<\alpha\leq 1$. Here the term $\mathcal{O}(n^{-3})$ is precisely

$$\frac{2}{\pi}\sum_{k=1}^{\infty}\frac{2k-1}{2k}\{1+(-1)^n(1-2^{-2k})\}\, B_{2k}\,\frac{1}{n^{2k+1}},$$

$B_{2k}$ being the Bernoulli numbers. His proof is based upon the closed representation of Fejér's kernel. This result was reestablished by Stark [33] for $\alpha$=1 with a simple proof based on [26, 80].

An actual c.a. expansion was derived by Stark [86] for the singular integral of Abel-Poisson, which has a continuous approximation parameter $r\to 1-$, namely[5]

$$\Delta_r(A;1) = \frac{2}{\pi}\sum_{k=1}^{\infty}\{\alpha_k(1-r)^k\log\frac{1}{1-r} + \beta_k(1-r)^k\},$$

$$\alpha_k = \frac{1}{k},\ \beta_k = \frac{1}{k}\left(\log 2 + \frac{1}{k} - \sum_{j=1}^{k-1}\frac{1}{j2^j}\right) \qquad (k\in\mathbb{N}).$$

Essential or complete asymptotic expansions are, so far,

only known for particular examples. On the other hand, first Nikolskiĭ constants may be determined for general classes of singular integrals. For instance, if the kernel has a certain generating function or satisfies the Korovkin condition ([E, p. 85f, p. 450])

$$(6.4) \qquad \lim_{\rho\to\infty} \frac{1 - \chi_\rho^\wedge(k)}{1 - \chi_1^\wedge(1)} = k^2 \qquad (k=2,3,\ldots).$$

Another approach was presented by Nessel [67] in his Habilitationsschrift (1970). One of his general results for singular integrals having kernels of Fejér's type (i.e., $\Phi_\rho^*(x) := \sum_{k=-\infty}^{\infty} \cdot \rho\Phi(\rho[x+2k\pi])$ with $\Phi \in L^1(R)$, $\Phi(x) \geq 0$ (a.e.), $\int_{-\infty}^{\infty} \Phi(x)dx = \pi$) states:

If $\Phi_\rho^*(x)$ is a kernel of Fejér's type with an even generating function $\Phi$, and (the $\alpha^{th}$ moment)

$$m(\Phi^*;\alpha) := \frac{2}{\pi} \int_0^\infty u^\alpha \Phi(u)du < \infty \quad (0<\alpha\leq 2),$$

then

$$\Delta_\rho(\Phi^*;\alpha) = m(\Phi^*;\alpha)\rho^{-\alpha} + o(\rho^{-\alpha}) \qquad (\rho\to\infty).$$

Thus $N_1(\Phi^*;\alpha) = m(\Phi^*;\alpha)$ for all $0<\alpha\leq 2$. One of the great advantages of the Nessel results is that they are also considered in the n-dim. situation not treated before.

Though the latter theorem covers a series of well-known examples, even some classical kernels are not of Fejér's type. For this reason Stark [85] introduced kernels of perturbed Fejér's type. A kernel $K_\rho(x)$ is said to be of this type provided there exists a decomposition

$$\Phi_\rho^*(x) = K_\rho(x)\{1 + \varepsilon_\rho(x)\}$$

such that $K_\rho(x)$ and the Fejér-type (comparison) kernel $\Phi_\rho^*(x)$ both satisfy condition (6.4) together with

$$\lim_{\rho\to\infty} (1-\Phi^{*\wedge}_\rho(k))/(1-K_\rho^\wedge(k)) = 1$$

i.e., $N_1(\Phi^*;2) = N_1(K;2)$, the perturbation term $\varepsilon_\rho(x)$ satisfying

$$r_\rho(K;\alpha) := \frac{2}{\pi}\int_0^\pi u^\alpha |\varepsilon_\rho(u)|\, K_\rho(u)du = o(\rho^{-2}) \qquad (0<\alpha\leq 2;\ \rho\to\infty).$$

Under these hypotheses it is shown that $N_1(K;\alpha) = N_1(\Phi^*;\alpha)$ $[= m(\Phi^*;\alpha)]$ for all $0<\alpha\leq 2$. Particular perturbed Fejér-type kernels are those of Jackson (in comparison with the Fejér-type kernel of (Jackson) - de La Vallée Poussin) as well as Fejér-Korovkin (in comparison with that of Bohman-Zheng Wei-xing of Fejér-type).

A third problem in this connection is to find relations between Nikolskiĭ constants and other important constants occurring in approximation theory. Görlich-Stark [45,46] showed that in the saturation case for kernels satisfying (6.4) one has

$$N_1(\chi;2) = 2\, c_S(\chi) = 2\, c_V(\chi)$$

where $c_S(\chi)$ is the constant in the saturation limit

$$\lim_{\rho\to\infty} \rho^\tau(1 - \chi_\rho^\wedge(k)) = c_S(\chi)\, k^2 \qquad (\tau=1,2)$$

and $c_V(\chi)$ is the leading constant of the Voronovskaja-type expansion

$$f * \chi_\rho(x) = f(x) + c_V(\chi)\, D^2 f(x)\, \rho^{-\tau} + o(\rho^{-\tau}) \qquad (\rho\to\infty).$$

In case of non-optimal approximation such connections are still an open problem.

## 7. Periodic Convolution Integrals Having Kernels of Finite Oscillation

### 7.1 Approximation Improvement

If $f \in C_{2\pi}$ and $\chi_n(x)$ is a positive even trig. poly. kernel of degree n, then P. P. Korovkin showed (in particular) that the approximation of f by $f * \chi_n$ cannot exceed $O(n^{-2})$, $n\to\infty$, i.e., there exist functions f such that $\lim_{n\to\infty} n^2 \|f*\chi_n - f\|_{C_{2\pi}} \neq 0$. The question arises whether it is possible to improve the order of approximation if the kernel is allowed to have a <u>finite</u> number of changes of sign; see [E, p. 93].

First attempts at formulating and solving this question are due to P. P. Korovkin[6] (three papers in 1962) and independently to J. R. Rice who discussed this problem with the author at the Gatlinburg conference of 1963.

Writing $\chi_n \in S_{2m}$, $m \in \mathbb{N}$, if $\chi_n$ has for each $n \in \mathbb{N}$, $n > m$, exactly $m(m\neq m(n))$ changes of sign (zeros of simple multiplicity) at the points $\alpha_{j,n}$ with $0 < \alpha_{1,n} < \dots < \alpha_{m,n} < \pi$, Scherer [71] in his Diplomarbeit (1966) (in the revised form of Butzer-Nessel-Scherer [15]) showed that for $\chi_n \in S_{2m}$, $\chi_n(0) > 0$, at least one of the (m+1) sequences

$$\{n^{2+2m}\| \cos k\circ * \chi_n(x) - \cos kx\|_{C_{2\pi}}\} \qquad (1\leq k\leq m+1)$$

does not tend to zero as $n\to\infty$. Thus the order of approximation for kernels $\chi_n \in S_{2m}$ can be improved to $O(n^{-2-2m})$, provided such kernels exist. Basic here is that any $\chi_{n+m} \in S_{2m}$ can be uniquely decomposed as

$$(7.1) \quad \chi_{n+m}(x) = c_{n,m}\chi_n(x) \cdot \prod_{j=1}^{m} (\cos x - \cos \alpha_{j,n})$$

$$= 1/2 + \sum_{k=1}^{n+m} \chi^{\wedge}_{n+m}(k)\cos kx$$

with (factor) kernel $\chi_n \in S_0$ and another poly. exhibiting the changes of sign. Whereas $\lim_{n\to\infty} \alpha_{m,n} = 0$ is necessary for approximation improvement, nevertheless (*) $\pi/(n+m+2) < \alpha_{1,n}$ (distance condition).

The existence problem for $\chi_n \in S_{2m}$ was solved by A. I. Kovalenko [[14]] in 1966 for a particular class of kernels, namely those generated by $\phi$-functions satisfying numerous constraints: if $f \in C_{2\pi}^{2+2m}$, then

$$\lim_{n\to\infty} n^{2+2m}[f * \chi_{n+m}(x) - f(x)] = c_{V,m} f^{(2+2m)}(x).$$

The first explicit example for $\chi_{n+1} \in S_2$ (with two symmetrical zeros at $\pm\alpha_{1,n}$ cn $(-\pi,\pi)$) was given by Butzer-Stark [27]. It is characterized by the saturation limit

$$(7.2) \qquad \lim_{n\to\infty} n^4(1-\chi_{n+1}^\wedge(k)) = \frac{3}{8}\pi^4(k^4 + k^2) \qquad (k \in \mathbb{N})$$

with zeros at $\pm\pi\sqrt{5}/n$. The idea behind this construction was extended by Stark [81] to a general class of kernels in $S_2$ delivering approximation up to $O(n^{-4})$.

In his dissertation Stark [82] builds up an intricate analysis leading to a one-parameter family (in $\Omega$) of $S_2$-kernels characterized by the limit

$$\lim_{n\to\infty} n^{2\tau}(1-\chi_{n+1}^\wedge(k)) = c_S(\chi)[k^4-(1+12\Omega)k^2]$$

$$(\tau=1,2;\ k \in \mathbb{N};\ -\infty<\Omega\leq 0).$$

In this second approach the higher order trig. moments $M_n^t(2\sigma) := (1/\pi)\int_{-\pi}^{\pi}(2\sin u/2)^{2\sigma}\chi_n(u)du$ together with the existence of a $\mu$-term asymptotic expansion

$$1-\chi_n^\wedge(k) = \sum_{j=1}^{\mu} (-1)^{j+1}\psi_j(k)n^{-\tau j} + o(n^{-\tau\mu})$$

$$(\tau=1,2;\ k \in \mathbb{N};\ \mu \geq 3)$$

play a decisive role. Condition (*) is here sharpened to

$$\frac{1}{2}\frac{M_n^t(4)}{M_n^t(2)} \leq 1 - \cos\alpha_{1,n} \leq \frac{1}{2}\frac{M_n^t(6)}{M_n^t(4)}$$

An application yields a series of concrete optimal ($\tau=2$) kernels of class $S_2$ (including those which are not necessarily of poly. type) as well as the fact that most classical positive kernels cannot be suitably used as factor kernels in (7.1) giving optimal or at least better approximation.

In [87] Stark constructs a family of integrals which originate from the Abel-Poisson kernel yielding kernels which are either positive or of class $S_2$, the saturation order being improved from $O(1-r)$ to $O[(1-r)^2]$, $r\to 1-$. Here a new phenomenon occurs: the zeros may be fixed or vary with r. The results include from a unified point of view some well-known singular integrals[7] of T. Anghelutza (1924) and from biharmonic potential problems (S. Kaniev, P. Pych, 1963 ff), etc.; see [87].

Another different and independent approach was developed by C. J. Hoff [[9]] in his 1968 thesis (under Professor J. R. Rice); it includes approximation of nonperiodic functions continuous on R.

Let us finally remark that at least 40 papers have been written on kernels of finite oscillation since 1962; let us mention G. N. Vinogradova (1965), A. I. Kovalenko, L. M. Zybin (1966), V. I. Bui (1968), R. M. Min'kova, Ju. A. Šaškin (1969), V. I. Dudin, L. G. Labsker (1970), G. Freud, Z. Ditzian, E. Görlich [40]. In the latter note there are considered certain features of saturation classes of positive poly. approximation processes and the problem whether analogous properties are true for processes having kernels of class $S_{2m}$

### 7.2 Equivalence Theorems for Fourier Coefficients

If $\chi_\rho(x)$ is any positive kernel, then condition (6.4) is valid for all k=2,3,... if and only if it is valid for k=2. This well-known result of Korovkin has been extended by several further equivalent assertions in Görlich-Stark [45] (see also [46]), in particular by equivalences involving trig. and algebraic moments such as È. A. Komleva's condition. In this survey (containing 138 references) also a great number of examples has been treated from a unified point of view, in particular with best constants (cf. Ch. 6).

Stark [82, 84] generalized these results to a certain class of kernels of finite or infinite oscillation, namely: for arbitrary kernels satisfying $[\chi_\rho]^\wedge(1) \neq 1$ one has that (6.4) is valid for all $k=2,3,\ldots$, if and only if

$$\lim_{\rho\to\infty} M^t_\rho(2\sigma)/M^t_\rho(2) = 0 \qquad (\sigma=2,3,\ldots).$$

A simple corollary is Korovkin's result which may be enlarged by two further equivalences, namely

$$\{(6.4)\ \forall\, k=2,3,\ldots\} \Longleftrightarrow \lim_{\rho\to\infty} \frac{M^t_\rho(4)}{M^t_\rho(2)} = 0 \Longleftrightarrow \lim_{\rho\to\infty} \frac{M^t_\rho(2\sigma)}{M^t_\rho(2)} = 0$$

$$\forall\ \sigma = 2,\ 3,\ \ldots\ .$$

The proofs do not use any techniques of approximation theory (in contrast to the original Korovkin proof) but depend upon the following formula together with its new inversion

$$M^t_\rho(2\sigma) = 2 \sum_{k=1}^{\sigma} (-1)^{k+1}\binom{2\sigma}{\sigma-k}(1-[\chi_\rho]^\wedge(k)) \qquad (\sigma \in \mathbb{N}),$$

$$1-[\chi_\rho]^\wedge(k) = \sum_{\sigma=1}^{k} (-1)^{\sigma+1} \frac{k}{k+\sigma}\binom{k+\sigma}{2\sigma} M^t_\rho(2\sigma) \qquad (k \in \mathbb{N}).$$

## 8. Spline Approximation

One object here is to apply the general approximation theorem of Chapter 1.1 to linear approximation by splines, i.e., by the linear subspaces $S_p(n,k;\Delta)$ of $L^\infty(a,b)$, $[a,b]$ being finite, defined by

$$(8.1)\qquad S_p(n,k;\Delta)=\{s \in L^{\infty,k}(a,b);\, s(x) \in P_n,\ x \in (x_{i-1},x_i),$$

$$i=1,\ldots,N\},$$

where $0\leq k\leq n$ and $\Delta$ is the (fixed) partition (or mesh) $\Delta = \{a=x_0<x_1<\ldots<x_N=b\}$ of $[a,b]$, $P_n$ denoting the set of all algebraic polynomials of degree n. A spline of class $S_p(n,k;\Delta)$ is said to

have deficiency n+l-k.

K. Scherer [74] first considered best approximation of continuous functions by the spline subspace $S_p(n,n;\Delta_t)$, the meshes $\Delta_t$ being of form $(0<t\leq(b-a)/2)$

$$(8.2)\qquad \Delta_t = \{x_i, b;\ x_i = a+it,\ 0\leq i\leq i_0,\ t/2\leq b-x_{i_0} < 3t/2\}.$$

Then Theorem 1, Chapter 1 can be formulated for $X = C[a,b]$, $Y = L^{1,k}(a,b)$, $1\leq k\leq n$, subspaces $P_t = S_p(n,n;\Delta_t)$ with

$$(8.3)\qquad E_t(f;C[a,b]) = \inf_{s\in S_p(n,n;\Delta_t)} \|f-s\|_C.$$

This yields new characterizations of the classical Lipschitz spaces Lip $(\alpha,r)$ and completes previous error bounds by C. de Boor, I. J. Schoenberg and others into an equivalence theorem. The associated Jackson and Bernstein-type inequalities needed for the proofs are easily derived from known results. An interesting feature here is that a Jackson-type inequality also holds for $Y = L^{\infty,n+1}(a,b)$, whereas an associated Bernstein-type inequality cannot hold since the $n^{th}$ derivative of a spline in $S_p(n,n;\Delta)$ is a discontinuous step function. In this case an inverse theorem must be established separately, estimating the $(n+1)^{th}$ modulus of continuity by $E_t(f)$. This inverse result is sharper than the comparable classical estimate in Bernstein's theorem on best approximation by trig. polynomials. A conseeuqnce is the unexpected occurrence of saturation for $E_t(f)$ in (8.3). On the other hand, this gives a new characterization of Lip(n+1, n+1), $n \in P$, as saturation classes of a best approximation.

These results were carried over to $L^p$-functions, i.e., $X = L^p(a,b)$, $1\leq p\leq\infty$, in [75]. Here a Jackson-type inequality had to be proved. In [77] this result is sharpened further. Firstly, the family $\Delta_t$ is restricted to the sequence of equidistant meshes $\Delta_N = \{x_i;\ x_i = a+i(b-a)/N,\ 0\leq i\leq N\}$, $n \in N$;

secondly, the spaces $S_p(r,k;\Delta)$ of less smooth splines are considered; thirdly, the inverse theorem is sharpened to a form which may be regarded as optimal. This yields a complete characterization of all Besov spaces by best approximation by splines. An important feature is that in the Zamansky-type characterization there occur the jumps of the derivatives of the approximating splines if these are not smooth enough, where this has no influence on reduction type assertions. In particular, results of J. Nitsche, D. Gaier, etc., are improved.

Another complex is to study corresponding approximation theorems for linear spline operators, notably of interpolating type. In contrast to the Lagrange interpolation problem, for natural interpolating splines one has convergence for <u>all</u> continuous functions. This is the main object in [76], giving a new proof for equidistant meshes which simplifies and extends a corresponding result of Swartz-Varga [[26]].

Direct theorems for interpolating polynomial splines of lowest deficiency are studied in greater detail in [78]. Here different boundary conditions necessary to determine a unique interpolating spline have to be compared. The resulting comparison theorems yield error bounds for interpolatory splines of odd degree satisfying the well-known first or second integral relation (see Ahlberg-Nilson-Walsh [[A]]) for functions belonging to the spaces $W^{p,j}(a,b)$, $j=n,2n$, $2\leq p\leq\infty$. This extends well-known results in case $p=2$, however, under the restriction of asymptotically uniform meshes. Also a stability theorem is shown for such meshes and general boundary conditions, extending a corresponding result of Swartz-Varga [[26]]. The case of even degree splines is considered briefly. Finally, these direct theorems are used to yield approximation theorems for linear spline operators in form of an equivalence theorem in of the type of Theorems 2,3, Chapter 1.

## 9. A New Calculus for Walsh Functions

The theory of Walsh functions with their manifold applications to information science, to communication engineering and system theory has become a rather popular field of research in the past five years since these functions are intimately connected with the representation of numerical data in binary notation, and so with the binary digital computer. Although about three hundred papers were written on Walsh functions till 1970 (discovered by J. L. Walsh in 1923), the theoretical ones employing such essential concepts as continuity, Lipschitz conditions, only a handful were concerned with the basic concept of a derivative first introduced by J. E. Gibbs in 1969.

This concept was modified and built up by Butzer-Wagner [30, 31, 32] and Wagner [98] into a basic differential and integral calculus for Walsh functions as follows: Consider the dyadic group G, i.e., the set of all infinite sequences $x=(x_1,x_2,\ldots,x_k,\ldots)$ with $x_k \in \{0,1\}$, $k \in \mathbb{N}$ and termwise addition modulo 2, in symbols $\dot{+}$. Consider the unique dyadic expansion of $n \in \mathbb{P}$

$$n = \sum_{k=0}^{N} n_k 2^k \qquad (n_k \in \{0,1\},\ 2^N \le n < 2^{N+1},\ N \in \mathbb{P}).$$

The $n^{th}$ Walsh-Paley function is then defined by

$$\psi_n(x) = (-1)^{\sum_{k=0}^{N} n_k x_{k+1}} \qquad (n \in \mathbb{P}).$$

These are characters on G, i.e., continuous functions such that $\psi_n(x\dot{+}y) = \psi_n(x)\psi_n(y)$, $x,y \in G$; they form an orthonormal system on G, i.e., $\int_G \psi_n(x)\psi_m(x)dx = \delta_{n,m}$ ($\int_G dx$ being the Haar integral on the LCA group G). The Walsh functions, being step-functions, are not differentiable in the classical sense. The <u>derivative</u> of f on G (in the pointwise or norm sense on $L^1(G)$) is here defined (see [31, 98]) by

$$(Df)(x) = \frac{1}{2} \sum_{j=0}^{\infty} 2^j [f(x) - f(x \dotplus e_{j+1})] \qquad (x \in G),$$

where $e_j = (\delta_{j,1}, \ldots, \delta_{j,k}, \ldots)$, and the integral concept via the convolution

$$(If)(x) = \int_G f(x \dotplus u) W(u) du,$$

$W(u)$ being a certain function in $L^1(G)$. The operators D, I have a number of properties in common with the classical derivative and integration. Although D here is a global and not a local property, it is a closed linear operator, and the $\psi_n$ are arbitrary often differentiable with $D^r \psi_n = n^r \psi_n$. A counterpart of the classical fundamental theorem of the differential and integral calculus is valid for these two concepts. A calculus parallel to the Leibniz-Newton calculus is developed which fits the analysis concerned with Walsh functions.

In this setting, for example, the Walsh functions turn out to be the non-trivial proper solutions of a first order linear "differential equation" with an initial condition. Our notion of derivative enables one to estimate the degree of approximation of a function by the partial sums of the Walsh-Fourier series, to estimate the order of magnitude of the Walsh-Fourier coefficients, as well as to establish an analog of the fundamental theorem on best approximation (Ch. 1.1) by Walsh polynomials. Partial "differential equations" of the wave equation type are set up and solved. This answers a problem originally envisaged by H. F. Harmuth in 1969. Finally it should be mentioned that the above theory may also be carried out for the Walsh-Paley system replaced by the Walsh-Kaczmarz system, which is sequency ordered.

What is lacking is an appropriate meaning for our concept of a derivative in terms of those sciences in which Walsh functions have been employed. In the words of J. E. Gibbs it is to be hoped that "our calculus will serve information science in the universal way that Newton's calculus has served physical science."

---

[1]In the sequel, $P = \{0,1,2,\ldots\}$, $N = \{1,2,\ldots\}$.

[2]$Y \subset X$ means that, algebraically, the space Y is contained in X, topologically, that Y is continuously embedded in X, i.e., $\|f\|_X \le \|f\|_Y$ for all $f \in Y$. $X \cong Y$ means that $Y \subset X$ and $X \subset Y$.

[3]The hypothesis that the projection-family $\{P_k\}$ be fundamental is irrelevant for the following multiplier criteria.

[4]One could also assume the uniform boundedness of the Abel, de La Valleé Poussin, or Euler means, for example.

[5]This result was already obtained by L. V. Malei [[18]] in 1961 with a different proof; it is not cited in the Russian literature in the field and was rediscovered by Stark.

[6]See [[13]]; the two further papers, which appeared only in 1965, are contained in the Proceedings of the Baku Conference (1962); see [82], [E]. Korovkin considered this problem from a more abstract side, the algebraic case being explicitly carried out.

[7]Perhaps the first such integral of class $S_2$ is due to M. Ghermanesco [[7]] in 1933, long before the problem was recognized and formulated as above.

## References

### Books, Monographs, Proceedings of Conferences


[A] Butzer, P.L. and H.Berens, Semi-Groups of Operators and Approximation. Berlin-Heidelberg-New York, 1967, xi + 318 pp.

[B] Berens, H., Interpolationsmethoden zur Behandlung von Approximationsprozessen auf Banachräumen. Lecture Notes in Math. 64, Berlin-Heidelberg-New York, 1968, iv+90 pp.

[C] Butzer, P. L. and K. Scherer, Approximationsprozesse und Interpolationsmethoden. (Hochschulskripten 826/826a) Mannheim 1968, 172 pp.

[D] Butzer, P. L. and W. Trebels, Hilberttransformation, gebrochene Integration und Differentiation. (Forschungsber. des Landes Nordrhein-Westfalen 1889) Köln-Opladen 1968, 82 pp.

[E] Butzer, P. L. and R. J. Nessel, Fourier Analysis and Approximation. Vol. I, One-dimensional Theory. Basel-New York 1971, xvi+553 pp.

[F] Butzer, P. L. and J. Korevaar (Eds.), On Approximation Theory. (Proc. Conf. Math. Res. Inst. Oberwolfach, Black Forest, 4-10 August 1963; ISNM 5) Basel-Stuttgart 1964, xvi+261 pp.

[G] Butzer, P. L. and B. Sz. Nagy (Eds.), Abstract Spaces and Approximation. (Proc. Conf. Math. Res. Inst. Oberwolfach, Black Forest, 18-27 July 1968; ISNM 10) Basel-Stuttgart 1969, 423 pp.

[H] Butzer, P. L., J. P. Kahane and B. Sz. Nagy (Eds.) Linear Operators and Approximation. (Proc. Conf. Math. Res. Inst. Oberwolfach, Black Forest, 14-22 August 1971; ISNM 20) Basel-Stuttgart 1972, 506 pp.

[[A]] Ahlberg, J. H., E. N. Nilson and J. L. Walsh, The Theory of Splines and Their Applications. New York 1967, xi+284 pp.

[[B]] Carroll, R. W., Abstract Methods in Partial Differential Equations. New York-Evanston-London 1969, ix+374 pp.

[[C]] Davis, P. J. and P. Rabinowitz, Numerical Integration. Waltham, Mass. 1967, ix+225 pp.

[[D]] Helgason, S., Differential Geometry and Symmetric Spaces. New York 1963, xiv+486 pp.

[[E]] Stein, E. M., Singular Integrals and Differentiability Properties of Functions. Princeton 1970, xiv+290 pp.

[[F]] Stein, E. M. and G. Weiss, Introduction to Fourier Analysis on Euclidean Spaces. Princeton 1971, x+297 pp.

[[G]] Alexits, G. and S. B. Stechkin (Eds.), Proceedings of the Conference on Constructive Theory of Functions (Approximation Theory). (Proc. Conf., Budapest, 24 August-3 September 1969) Budapest 1972, 538 pp.

[[H]] Penkov, B. and D. Vačov (Eds.), Constructive Function Theory (Proc. Internat. Conf. Constructive Function Theory, Varna/Bulgaria, 19-25 May 1970) Sofia 1972, 363 pp.

[[I]] Sz.-Nagy, B. (Ed.), Hilbert Space Operators and Operator Algebras. (Coll. Math. Soc. János Bolyai, 5; Hilbert Space Operators, Tihany/Hungary, 14-18 September 1970) Amsterdam-London 1972, 544 pp.

Original Papers

[1] Berens, H. and P. L. Butzer, On the best approximation for singular integrals by Laplace-transform methods. In: [F]; pp. 24-42.

[2] Berens, H., P. L. Butzer, and S. Pawelke, Limitierungsverfahren von Reihen mehrdimensionaler Kugelfunktionen und deren Saturationsverhalten. Publ. Res. Inst. Math. Sci. Ser. A 4 (1968), 201-268.

[3] Berens, H., P. L. Butzer, and U. Westphal, Representation of fractional powers of infinitesemal generators of semi-groups. Bull. Amer. Math. Soc. 74 (1968), 191-196.

[4] Berens, H. and R. J. Nessel, Contributions to the theory of saturation for singular integrals in several variables, IV: Product kernels and n-parameter approximation; V: Saturation in $L_p(E^n)$, $2<p<\infty$. Nederl. Akad. Wetensch. Proc. Ser. A 71 = Indag. Math. 30 (1968), 325-335; A 72 = 31 (1969), 71-76.

[5] Berens, H. and U. Westphal, Zur Charakterisierung von Ableitungen nichtganzer Ordnung im Rahmen der Laplace-Transformation. Math. Nachr. 38 (1968), 115-129.

[6] Berens, H. and U. Westphal, A Cauchy problem for a generalized wave equation. Acta Sci. Math. (Szeged) 29 (1968), 93-106.

[7] Bragard, G. K. and H. Johnen, Saturation auf kompakten Gruppen und kompakten symmetrischen Räumen. J. Reine Angew. Math.; to appear.

[8] Butzer, P. L. and E. Görlich, Zur Charakterisierung von Saturationsklassen in der Theorie der Fourierreihen. Tôhoku Math. J. (2) 17 (1965), 29-54.

[9] Butzer, P. L. and E. Görlich, Characterizations of Favard classes for functions of several variables. Bull. Amer Math. Soc. 74 (1968), 149-152.

[10] Butzer, P. L. and H. Johnen, Lipschitz spaces on compact manifolds. J. Functional Analysis 7 (1971), 242-266.

[11] Butzer, P. L. and J. Kemper, Operatorenkalkül von Approximationsverfahren fastperiodischer Funktionen. In: Forschungsber. des Landes Nordrhein-Westfalen 2157; Köln-Opladen 1970, 53 pp.; pp. 23-53.

[12] Butzer, P. L. and W. Köhnen, Approximation invariance of semi-group operators under perturbations. J. Approximation Theory 2 (1969), 389-393.

[13] Butzer, P. L., W. Kolbe and R. J. Nessel, Approximation by functions harmonic in a strip. Arch. Rational Mech. Anal. 44 (1972), 329-336.

[14] Butzer, P. L. and R. J. Nessel, Contributions to the theory of saturation for singular integrals in several variables; I: General theory. Nederl. Akad. Wetensch. Proc. Ser. A 69 = Indag. Math. 28 (1966), 515-531.

[15] Butzer, P. L., R. J. Nessel and K. Scherer, Trigonometric convolution operators with kernels having alternating signs and their degree of convergence. Jber. Deutsch. Math.-Verein. 70 (1967), 86-99.

[16] Butzer, P. L., R. J. Nessel and W. Trebels, On the comparison of approximation processes in Hilbert spaces. In: [H]; pp. 234-253.

[17] Butzer, P. L., R. J. Nessel and W. Trebels, On summation processes of Fourier expansions in Banach spaces; I: Comparison theorems; II: Saturation theorems. Tôhoku Math. J. (2) 24 (1972), 127-140; 25 (1973), 551-569.

[18] Butzer, P. L. and S. Pawelke, Ableitungen von trigonometrischen Approximationsprozessen. Acta Sci. Math. (Szeged) 28 (1967), 173-183.

[19] Butzer, P. L. and S. Pawelke, Semi-groups and resolvent operators. Arch. Rational Mech. Anal. 30 (1968), 127-147.

[20] Butzer, P. L. and K. Scherer, Über die Fundamentalsätze der klassischen Approximationstheorie in abstrakten Räumen. In: [G]; pp. 113-125.

[21] Butzer, P. L. and K. Scherer, On the fundamental approximation theorems of D. Jackson, S. N. Bernstein and theorems of M. Zamansky and S. B. Stečkin. Aequationes Math. 3 (1969), 170-185.

[22] Butzer, P. L. and K. Scherer, On fundamental theorems of approximation theory and their dual versions. J. Approximation Theory 3 (1970), 87-100.

[23] Butzer, P. L. and K. Scherer, Approximation theorems for sequence of commutative operators in Banach spaces. In: [[H]]; pp. 137-145.

[24] Butzer, P. L. and K. Scherer, Jackson and Bernstein-type inequalities for families of commutative operators in Banach spaces. J. Approximation Theory 5 (1972), 308-342.

[25] Butzer, P. L., K. Scherer and U. Westphal, On the Banach-Steinhaus theorem and approximation in locally convex spaces. Acta Sci. Math. (Szeged); to appear.

[26] Butzer, P. L. and E. L. Stark, Wesentliche asymptotische Entwicklungen für Approximationsmaße trigonometrische singulärer Integrale. Math. Nachr. 39 (1969), 223-237.

[27] Butzer, P. L. and E. L. Stark, On a trigonometric convolution operator with kernel having two zeros of simple multiplicity. Acta Math. Acad. Sci. Hungar. 20 (1969), 451-461.

[28] Butzer, P. L. and W. Trebels, Hilbert transforms, fractional integration and differentiation. Bull. Amer. Math. Soc. 74 (1968), 106-110.

[29] Butzer, P. L. and W. Trebels, Opérateurs de Gauß-Weierstraß et de Cauchy-Poisson et conditions lipschitziennes dans $L^1(E_n)$. C. R. Acad. Sci. Paris Sér. A 268 (1969), 700-703.

[30] Butzer, P. L. and H. J. Wagner, Approximation by Walsh polynomials and the concept of a derivative. In Applications of Walsh functions (Proc. Sympos. Naval Res. Lab., Washington, D. C., 27-29 March 1972; Ed. R. W. Zeek-A. E. Showalter) Washington, D. C. 1972, xi+401 pp.; pp. 388-392.

[31] Butzer, P. L. and H. J. Wagner, On a Gibbs-type derivative in Walsh-Fourier analysis with applications. In: Proceedings of the 1972 National Electronics Conference. Chicago, 9-10 October 1972; pp. 393-398.

[32] Butzer, P. L. and H. J. Wagner, Walsh-Fourier series and the concept of a derivative. Applicable Anal. 3 (1973), 29-46.

[33] Butzer, P. L. and U. Westphal, On the Cayley transform and semigroup operators. In: [[I]]; pp. 89-97.

[34] Butzer, P. L. and U. Westphal, The mean ergodic theorem and saturation. Indiana Univ. Math. J. 20 (1971), 1163-1174.

[35] Butzer, P. L. and U. Westphal, Ein Operatorenkalkül für das approximationstheoretische Verhalten des Ergodensatzes im Mittel. In: [H]; pp. 102-114.

[36] Görlich, E., Distributionentheoretische Methoden in der Saturationstheorie. Dissertation, Rheinisch-Westfälische Technische Hochschule Aachen, i+69 pp.; 1967.

[37] Görlich, E., Distributional methods in saturation theory. J. Approx. Theory 1 (1968), 111-136.

[38] Görlich, E., Saturation theorems and distributional methods. In: [G]; pp. 218-232.

[39] Görlich, E., Zur Saturation von Summationsverfahren mehrdimensionaler Fourierreihen. Österreich. Akad. Wiss. Math.-Natur. Kl. S.-B. II 117 (1968), 171-202.

[40] Görlich, E., Über optimale Approximationsoperatoren. In: [[H]]; pp. 187-191.

[41] Görlich, E., Logarithmische und exponentielle Ungleichungen vom Bernstein-Typ und verallgemeinerte Ableitungen. Habilitationsschrift, Rheinisch-Westfälische Technische Hochschule Aachen, 65 pp.; 1971.

[42] Görlich, E. Logarithmic and exponential variants of Bernstein's inequality and generalized derivatives. In: [H]; pp. 325-337.

[43] Görlich, E., R. J. Nessel and W. Trebels, Bernstein-type inequalities for families of multiplier-operators in Banach spaces with Cesàro decompositions; I: General theory, II: Applications. Acta Sci. Math. (Szeged); to appear.

[44] Görlich, E., R. J. Nessel and W. Trebels, Zur Approximationstheorie für Summationsprozesse von Fourier-Entwicklungen in Banach-Räumen: Vergleichssätze und Ungleichungen vom Bernstein-Typ. In:Proc. Conf. "Theory of Approximation"; Poznan, 22-26 August 1972; to appear.

[45] Görlich, E. and E. L. Stark, Über beste Konstanten und asymptotische Entwicklungen positiver Faltungsintegrale und deren Zusammenhang mit dem Saturationsproblem. Jber. Deutsch. Math.-Verein. 72 (1970) 18-61.

[46] Görlich, E. and E. L. Stark, A unified approach to three problems on approximation by positive linear operators. In:[[G]]; pp. 201-208.

[47] Hövel, H. W. and U. Westphal, Fractional powers of closed operators. Studia Math. 42 (1972),177-194.

[48] Johnen, H., Über Sätze von M. Zamansky und S. B. Stečkin und ihre Umkehrungen auf dem n-dimensionalen Torus. J. Approx. Theory 2 (1969) 97-110.

[49] Johnen, H., Klassen stetiger Funktionen und Approximation auf kompakten Mannigfaltigkeiten. In:Forschungsber. des Landes Nordrhein-Westfalen 2078; Köln-Opladen 1970, 60 pp.; pp. 3-25.

[50] Johnen, H., Best approximation on the unitary group. In: [[I]]; pp. 295-303.

[51] Johnen, H., Stetigkeitsmoduli und Approximationstheorie auf kompakten Liegruppen. Dissertation, Rheinisch-Westfälische Technische Hochschule Aachen; ii-61 pp.;1970.

[52] Johnen, H., Darstellungen von Liegruppen und Approximationsprozesse auf Banachraumen. J. Reine Angew. Math. 254 (1972), 160-187.

[53] Johnen, H., Sätze vom Jackson-Typ auf Darstellungsräumen kompakter, zusammenhängender Liegruppen. In: [H]; pp.254-272.

[54] Johnen, H., Inequalities connected with moduli of smoothness. Mat Vesnik 9 (24) (1972), 289-303.

[55] Johnen, H., Best approximation on compact homogeneous spaces. In: Proc. Conf. "Functional Analysis and its Applications"; Madras, 1-7 January 1973; to appear.

[56] Junggeburth, J., K. Scherer and W. Trebels, Zur besten Approximation auf Banachräumen mit Anwendungen auf ganze Funktionen. (Forschungsber. des Landes Nordrhein-Westfalen 2311) Köln-Opladen; to appear.

[57] Kemper, J. and R. J. Nessel, Gewichtete Approximation durch variationsvermindernde Operatoren vom Faltungstyp. (Forschungsber. des Landes Nordrhein-Westfalen 2311) Köln-Opladen; to appear.

[58] Köhnen, W., Das Anfangswertverhalten von Evolutionsgleichungen in Banachräumen, Teil I: Approximationseigenschaften von Evolutionsoperatoren; Teil II: Anwendungen. Dissertation, Rheinisch-Westfälische Technische Hochschule Aachen, ii+76 pp.; 1969 = (revised) Tôhoku Math. J. (2) 22 (1970), 566-596; 23 (1971), 621-639.

[59] Köhnen, W., Approximationsprobleme bei der Störung von Halbgruppenoperatoren. In: [[H]]; pp. 201-207.

[60] Köhnen, W., n-parametrige Halgruppen von Operatoren in der Approximationstheorie. In: Proc. Conf. "Theory of Approximation; Poznań, 22-26 August 1972; to appear.

[61] Kolbe, W., Simultane Approximationsprozesse im Zusammenhang mit biharmonischen Randwertproblemen. Dissertation, Rheinisch-Westfälische Technische Hochschule Aachen, iii+100 pp.; 1972.

[62] Kolbe, W. and R. J. Nessel, Simultaneous saturation in connection with Dirichlet's problem for a wedge. Applicable Anal. 1 (1971), 231-240.

[63] Kolbe, W. and R. J. Nessel, Saturation theory in connection with Mellin transform methods. SIAM J. Math. Anal. 3 (1972), 246-262.

[64] Leviatan, D. and U. Westphal, On the mean ergodic theorem and approximation. Mathematica (Cluj); to appear.

[65] Nessel, R. J., Das Saturationsproblem für mehrdimensionale singuläre Integrale und seine Lösung mit Hilfe der Fouriertransformation. Dissertation, Rheinisch-Westfälische Technische Hochschule Aachen, iii+145 pp.; 1965.

[66] Nessel, R. J., Contributions to the theory of saturation for singular integrals in several variables; II: Applications; III: Radial kernels. Nederl. Akad. Wetensch. Proc. Ser. A 70 = Indag. Math. 29 (1967), 52-64; 65-73.

[67] Nessel, R. J., Über Nikolskii-Konstanten von positiven Approximationsverfahren bezüglich Lipschitzklassen. Habilitationsschrift, Rheinisch-Westfälische Technische Hochschule Aachen, 51 pp.; 1970 = (revised) Jber. Deutsch. Math.-Verein. 73 (1971), 6-47.

[68] Nessel, R. J., Nikolskii constants of positive operators. In:[[H]]; pp. 239-244.

[69] Nessel, R. J. and A. Pawelke, Über Favardklassen von Summationsprozessen mehrdimensionaler Fourierreihen. Compositio Math. 19 (1968), 196-212.

[70] Nessel, R. J. and W. Trebels, Gebrochene Differentiation und Integration und Charakterisierungen von Favard-Klassen. In:[[G]]; pp. 331-341.

[71] Scherer, K., Die Abhängigkeit der optimalen Approximationsordnung trigonometrischer Faltungsoperatoren von der Anzahl der Vorzeichenwechsel ihrer Kerne. Diplomarbeit, Rheinisch-Westfälische Technische Hochschule Aachen, July 1966, iii+31 pp. (unpublished).

[72] Scherer, K., Dualität bei Interpolations- und Approximationsräumen. Dissertation, Rheinisch-Westfälische Technische Hochschule Aachen, 76 pp.; 1969.

[73] Scherer, K., Über die Dualen von Banachräumen, die durch lineare Approximationsprozesse erzeugt werden, und Anwendungen für periodische Distributionen. Acta Math. Acad. Sci. Hungar. 23 (1973), 343-365.

[74] Scherer, K., On the best approximation of continuous functions by splines. SIAM J. Numer. Anal. 7 (1970), 418-423.

[75] Scherer, K., Über die beste Approximation von $L^p$-Funktionen durch Splines. In:[[H]]; pp. 277-286.

[76] Scherer, K., Über die Konvergenz von natürlichen interpolierenden Splines. In:[H]; pp. 487-492.

[77] Scherer, K., Characterization of generalized Lipschitz classes by best approximation with splines. To appear.

[78] Scherer, K., A comparison approach to direct theorems for polynomial spline approximation. In:Proc. Conf. "Theory of Approximation"; Poznań, 22-26 August 1972; to appear.

[79] Scherer, K. and H. J. Wagner, An equivalence theorem on best approximation of continuous functions by algebraic polynomials. Applicable Anal. 1 (1972), 343-354.

[80] Stark, E. L., Über die Approximationsmaße spezieller singulärer Integrale. Computing 4 (1969), 153-159.

[81] Stark, E. L., On approximation improvent for trigonometric singular integrals by means of finite oscillation kernels with separated zeros. In:[[H]]; pp. 337-344.

[82] Stark, E. L., Über trigonometrische singuläre Faltungsintegrale mit Kernen endlicher Oszillation. Dissertation, Rheinisch-Westfälische Technische Hochschule Aachen, iv+85 pp.; 1970.

[83] Stark, E. L., The complete expansion of the measure of approximation for the Fejér means (Russ.) Ukrain. Mat. Ž. 24 (1972), 562-566 = Transl.: Ukrainian Math. J. 24 (1972); in print.

[84] Stark. E. L., An extension of a theorem of P. P. Korovkin to singular integrals with not necessarily positive kernels. Nederl. Akad. Wetensch. Proc. Ser. A 75 = Indag. Math. 34 (1972), 227-235.

[85] Stark, E. L., Nikolskiĭ constants for positive singular integrals of perturbed Fejér-type. In:[H]; pp. 348-363.

[86] Stark, E. L., The complete asymptotic expansion for the measure of approximation of Abel-Poisson's singular integral for Lip 1 (Russ.). Mat. Zametki 13 (1973), 21-28 = Transl.: Math. Notes; in print.

[87] Stark, E. L., On a generalization of Abel-Poisson's singular integral having kernels of finite oscillation. Studia Sci. Math. Hungar.; to appear.

[88] Trebels, W., Charakterisierungen von Saturationsklassen in $L^1(E_n)$. Dissertation, Rheinisch-Westfälische Technische Hochschule Aachen, ii+93 pp., 1969.

[89] Trebels, W., Über absolute Summierbarkeit von n-dimensionalen Fourierreihen und Fourierintegralen. J. Approx. Theory 2 (1969), 394-399.

[90] Trebels, W., Einige n-parametrige Approximationsverfahren und Charakterisierung ihrer Favardklassen. In:Forschungsber. des Landes Nordrhein-Westfalen 2078, Köln-Opladen 1970, 60 pp.; pp. 27-60.

[91] Trebels, W., Besselpotentiale gerader Ordnung und äquivalente Lipschitzräume. In:Forschber. des Landes Nordrhein-Westfalen 2157, Köln-Opladen 1970, 53 pp.; pp. 5-22.

[92] Trebels, W., Imbedding theorems of spaces of hypersingular integrals and Bessel potentials. J. Approx. Theory 6 (1972), 202-214.

[93] Trebels, W., Generalized Lipschitz conditions and Riesz derivatives on the space of Bessel potentials $L_\alpha^p$; I: Conditions for elements of $L_\alpha^p$ and their Riesz transforms for $0<\alpha\leq 2$; II: The case $\alpha>1$; Riesz derivatives. Applicable Anal. 1 (1971) 75-99; 3 (1973); in print.

[94] Trebels, W., On a Fourier - $L^1(E_n)$ - multiplier criterium. Acta Sci. Math. (Szeged) 34 (1973); in print.

[95] Trebels, W., Über Jackson- and Bernstein-Typ Ungleichungen für lineare Approximationsverfahren in $L^p(E_n)$. Manuscripta Math. 8 (1973), 287-318.

[96] Trebels, W., Multipliers for $(C,\alpha)$-bounded Fourier expansions in Banach spaces and approximation theory. 103 pp.; to appear.

[97] Trebels, W. and U. Westphal, A note on the Landau-Kallman-Rota-Hille inequality. In:[H]; pp. 115-119.

[98] Wagner, H. J., Ein Differential- und Integralkalkul in der Walsh-Fourier-Analysis mit Anwendungen. Dissertation, Rheinisch-Westfalische Technische Hochschule Aachen, 71 pp.; 1972 = (Forschungsber. des Landes Nordrhein-Westfalen) Köln-Opladen; in print.

[99] Westphal, U., Über Potenzen von Erzeugern von Halbgruppenoperatoren. In:[G]; pp. 82-91.

[100] Westphal, U., Ein Kalkül für gebrochene Potenzen infinitesimaler Erzeuger von Halbgruppen und Gruppen von Operatoren; Teil I: Halbgruppenerzeuger; Teil II: Gruppenerzeuger. Dissertation, Rheinisch-Westfälische Technische Hochschule Aachen, ii+93 pp. 1969 = (revised) Compositio Math. 22 (1970), 67-103, 104-136.

[101] Westphal, U., Sur la saturation pour des semi-groupes non linéaires. C. R. Acad. Sci. Paris Sér. A 274 (1972), 1351-1353.

[[1]] Balakrishnan, A. V., Fractional powers of closed operators and the semigroups generated by them. Pacific J. Math. 10 (1960), 419-437.

[[2]] Bavinck, H., Jacobi Series and Approximation. Dissertation, Mathematisch Centrum-Amsterdam ix+97 pp.; 1972.

[[3]] Bohman, J., Saturation problems and distribution theory. In Lecture Notes in Math. 187 (1971), 249-266.

[[4]] Buchwalter, H., Saturation sur un groupe abélien localement compact. C. R. Acad. Sci. Paris 250 (1960), 808-810.

[[5]] Crandall, M. G. and T. M. Liggett, Generation of semi-groups of nonlinear transformations on general Banach spaces. Amer. J. Math. 93 (1971), 265-298.

[[6]] Fisher, M. J., Purely imaginary powers of certain differential operators; I. Amer. J. Math. 93 (1971), 452-478.

[[7]] Ghermanesco, M., Sur l'intégrale de Poisson. Sur l'intégrale de Poisson (suite). Bull. Sci. École Polytechnique de Timisoara 4 (1932), 159-184; 5 (1933), 41-74.

[[8]] Hille, E., Generalizations of Landau's inequality to linear operators. In: [H]; pp. 20-32.

[[9]] Hoff, C. J., Approximation with kernels of finite oscillations. Thesis, Purdue University, June 1968, 116 pp. = id., Part I, Convergence. J. Approx. Theory 3 (1970), 213-228; id., Part II, Degree of approximation. To appear ibid.

[[10]] Horváth, J., Finite parts of distributions. In: [H]; pp. 142-158.

[[11]] Jerome, J. W. and R. S. Varga, Generalizations of spline functions and applications to nonlinear boundary value and eigenvalue problems. In: Theory and Applications of Spline Functions (Proc. Sem. Math. Res. Center, Univ. Wisconsin, October 1968; Ed. T.N.E. Greville) New York 1969, xi+212 pp.; pp. 103-155.

[[12]] Komatsu, H., Fractional powers of operators. Pacific J. Math. 19 (1966), 285-346; II. Interpolation spaces. Ibid. 21 (1967), 89-111; III. Negative powers. J. Math. Soc. Japan 21 (1969), 205-220.

[[13]] Korovkin, P. P., Convergent sequences of linear operators (Russ.). Uspehi Mat. Nauk 17, no. 4 (106) (1962), 147-152.

[[14]] Kovalenko, A. I., Certain summability methods for Fourier series (Russ.). Mat. Sb. (N.S.) 71 (113) (1966), 598-616.

[[15]] Kozima, M. and G. Sunouchi, On the approximation and saturation by general singular integrals. Tôhoku Math. J. (2) 20 (1968), 146-169.

[[16]] Lions, J. L. and J. Peetre, Sur une classe d'espaces d'interpolation. Inst. Hautes Études Sci. Publ. Math. 19 (1964), 5-68.

[[17]] Löfström, J., Some theorems on interpolation spaces with applications to approximation in $L_p$. Math. Ann. 172 (1967), 176-196.

[[18]] Maleĭ, L. V., Precise estimate for approximation of quasismooth functions by Poisson integrals (Russ.). Vesci Akad. Navuk BSSR Ser. Fiz.-Tèhn. Navuk 1961, no. 3 (1961), 25-32.

[[19]] Nikolskiĭ, S. M., Sur l'allure asymptotique du reste dans l'approximation au moyen des sommes de Fejér des fonctions vérifiant la condition de Lipschitz (Russ.; French sum.). Izv. Akad. Nauk SSSR Ser. Mat. 4 (1940), 501-508.

[[20]] Peetre, J., Applications de la théorie des espaces d'interpolation dans l'analyse harmonique. Ricerche Mat. 15 (1966), 3-36.

[[21]] Ragozin, D. L., Approximation theory on SU(2). J. Approx. Theory 1 (1969), 464-475.

[[22]] Ragozin, D. L., Polynomial approximation on compact manifolds and homogeneous spaces. Trans. Amer. Math. Soc. 150 (1970), 41-53.

[[23]] Ragozin, D. L., Constructive polynomial approximation on spheres and projective spaces. Trans. Amer. Math. Soc. 162 (1971), 157-170.

[[24]] Shapiro, H. S., Some Tauberian theorems with applications to approximation theory. Bull. Amer. Math. Soc.. 74 (1968), 500-504.

[[25]] Sunouchi, G., Derivatives of a polynomial of best approximation. Jber. Deutsch. Math.-Verein. 70 (1968), 165-166.

[[26]] Swartz, B. K. and R. S. Varga, Error bounds for spline and L-spline interpolation. J. Approx. Theory 6 (1972), 6-49.

[[27]] Taibleson, M. H., On the theory of Lipschitz spaces of distributions on Euclidean n-space; I: Principal properties; II: Translation invariant operators, duality, and interpolation. J. Math. Mech. 13 (1964), 407-479; 14 (1965) 821-839.

[[28]] Teljakovskiĭ, S. A , On approximation by Fejér means to functions satisfying Lipschitz conditions (Russ.).

Ukrain. Mat Ž. 21 (1969) 334-342 = Transl.: Ukrainian Math. J. 21 (1969), 279-286 (1970).

[[29]] Wheeden, R. L., On hypersingular integrals and Lebesgue spaces of differentiable functions I, II. Trans. Amer. Math Soc. 134 (1968), 421-435; 139 (1969), 37-53.

Lehrstuhl A für Mathematik
Technological University of Aachen
5100 Aachen
West Germany

# UNIFORM ALGEBRAS ON PLANE SETS

T. W. Gamelin[1]

## Introduction

The object of this report is to give a sketch of an area in which the abstract theory of uniform algebras has contributed to the solution of concrete problems in approximation theory. Let K be a compact subset of the complex plane C. Some typical problems we shall treat are the following.

Problem I. Find conditions on a continuous complex-valued function f on K which guarantee that f be uniformly approximable by functions analytic in a neighborhood of K. In particular, what conditions on K guarantee that every function continuous on K and analytic on $K^0$ be approximable?

The classical result here is the Weierstrass approximation theorem. One extension of the Weierstrass Theorem is the Hartogs-Rosenthal Theorem, which asserts that $R(K) = C(K)$ whenever K has zero area. A more recent result, already classical, is Mergelyan's Theorem, which asserts that every $f \in A(K)$ can be approximated uniformly on K by analytic polynomials providing $C \setminus K$ is connected.

Problem II. What conditions on K guarantee that every continuous real-valued function on $\partial K$ can be approximated uniformly by the real parts of functions which are analytic on a neighborhood of K?

The classical result bearing on this problem is the Walsh-Lebesgue Theorem, which asserts that every function in $C_R(\partial K)$ can be approximated uniformly on $\partial K$ by harmonic polynomials in x and y providing that $C\setminus K$ is connected.

Problem III. Find conditions on a bounded analytic function f on $K^0$ which guarantee that f be a pointwise limit on $K^0$ of a sequence of functions which are analytic in a neighborhood of K and uniformly bounded on K. In particular, what conditions on K guarantee that every such f is approximable?

Here the prototypal result is the Farrell-Rubel-Shields Theorem, asserting that if $C\setminus K$ is connected, then every $f \in H^\infty(K^0)$ can be approximated pointwise on $K^0$ be a sequence $\{f_n\}$ of analytic polynomials satisfying $\|f_n\|_K \leq \|f\|_{K^0}$.

A great deal of progress has been made on these problems in the past two decades. On the purely constructive side, there is the work of the Soviet school led by A. G. Vitushkin. On the functional-analytic side, there is the seminal work of E. Bishop, the abstract proof of Mergelyan's Theorem due to I. Glicksberg and J. Wermer, and a series of recent contributions, the most prominent of which is the theorem of A. M. Davie on pointwise bounded convergence.

Our purpose is to describe the current state of affairs in this area by laying out a rough outline of the logical (as opposed to chronological) development of the theory. The report is organized as follows.

Sections 1, 2 and 3 cover the constructive approach, highlighting work of S. N. Mergelyan, A. G. Vitushkin and M. S. Melnikov. Here a solution to Problem I is presented.

Sections 4, 5 and 6 cover the earlier developments of the abstract approach. The central notions involve representing measures, orthogonal measures, peak points and Gleason parts.

Sections 7, 8 and 9 are concerned with Davie's Theorem and some of its applications. This includes in Section 9 a solution to Problem III, and it leads to a sharpened form of Vitushkin's solution to Problem I.

Section 10 is directed towards Dirichlet algebras and a solution to Problem II. Section 11 is devoted to D. Sarason's solution of the problem of determining the weak-star closure of the analytic polynomials in $L^{\infty}(\nu)$, where $\nu$ is a compactly supported measure on C.

Section 12 contains a discussion of the work of B. Cole and of O. Bekken dealing with approximation on polysets in $C^{n}$.

In recent years, much progress has also been made on several related approximation problems, which will not be covered in this report. These include the problem of approximation in $L^{p}$ norm by analytic functions, and more generally the problem of approximation in $L^{p}$ norm by solutions of some fixed elliptic differential equation. Recent work in this area, and further references, are found in the papers of J. Brennan [5], L. Hedberg [27,28], and J. Polking [36].

## Notations and Conventions

Throughout this report, X is a compact space, K is a compact subset of C, and U is a bounded open subset of C. Our notation will be generally that of [20], except that $\Delta(z;\delta)$ will denote the <u>open</u> disc centered at z with radius $\delta$, and the extended complex plane will be denoted by $C^{*}$. All measures will be regular finite complex Borel measures.

The space of continuous complex-valued functions on X is denoted by C(X). The space of real-valued continuous functions on X is denoted by $C_{R}(X)$. The norm of uniform convergence on X is given by

$$\|f\|_{X} = \sup\{|f(x)|: x \in X\}.$$

The dual of C(X) is the space M(X) of measures on X. A uniform algebra on X is a closed subalgebra of C(X) which contains the constants and separates the points of X. If A is a uniform algebra on X, and $\nu$ is a measure on X, then the weak-star closure of A in $L^\infty(\nu)$ is denoted by $H^\infty(\nu)$.

The uniform closure in C(K) of the functions analytic in a neighborhood of K is denoted by R(K). The algebra of bounded analytic functions on U is denoted by $H^\infty(U)$. We define $A(U) = C(\overline{U}) \cap H^\infty(U)$ and $A(K) = C(K) \cap H^\infty(K^0)$.

## 1. A Scheme for Approximation

This section is built around the basic scheme of approximation employed by Mergelyan and Vitushkin. First we sketch the constructive proof of Mergelyan's Theorem, then we state Vitushkin's Theorem and discuss related results.

Let $f \in A(K)$. We seek conditions which will place f in R(K).

Let $\delta > 0$, let $\{\Delta_{\delta j}\}_{j=1}^n$ be a cover of K by open discs of radius $\delta$, and let $\{g_{\delta j}\}_{j=1}^n$ be smooth functions such that each $g_j$ (the index $\delta$ is suppressed when convenient) is supported on $\Delta_j$, and $\Sigma g_j = 1$ in a neighborhood of $\partial K$. Extend f continuously to C to have compact support, and let $f_j$ be the (unique) distribution solution of the $\overline{\partial}$-problem

$$\frac{\partial f_j}{\partial \overline{z}} = g_j \frac{\partial f}{\partial \overline{z}}, \qquad f_j(\infty) = 0.$$

The solution $f_j$ has the explicit expression

$$f_j(w) = g_j(w)f(w) + \frac{1}{\pi}\iint \frac{f(z)}{z-w}\frac{\partial g_j}{\partial \overline{z}}\,dxdy$$

$$= \frac{1}{\pi}\iint \frac{f(z)-f(w)}{z-w}\frac{\partial g_j}{\partial \overline{z}}\,dxdy, \qquad w \in C.$$

In particular, $f_j$ is continuous, $f_j$ is analytic off $\Delta_j$, and $f_j$ can be estimated in terms of $\partial g_j/\partial \overline{z}$ and the oscillation of f on $\Delta_j$. Since $\Sigma g_j = 1$ near $\partial K$, we obtain $\partial f/\partial \overline{z} = \partial/\partial \overline{z}(\Sigma f_j)$ near K, and $f - \Sigma f_j$ is analytic near K. To place f in R(K), it suffices then to approximate $\Sigma f_j$. This is done as follows.

Near $\infty$, $f_j$ has a Taylor expansion

$$f_j(z) = \frac{f_j'(\infty)}{z} + \frac{\beta(f_j)}{z^2} + \cdots$$

Suppose we can find continuous functions $h_j$ on C such that $h_j$ is analytic near K, $h_j$ is analytic off $\Delta_j$, $h_j - f_j$ has a triple zero at $\infty$, and $\|h_j\| \le c_0\|f_j\|$. If the $h_j$ satisfy these conditions, and if the $\Delta_j$ and $g_j$ satisfy certain accessory conditions, then there obtains the crucial estimate

$$\Sigma|f_j - h_j| \le c_1\omega_f(\delta),$$

where $\omega_f$ is the modulus of continuity of f, and $c_1$ depends only on $c_0$. Now $F_\delta = f - \Sigma f_j + \Sigma h_j$ is analytic near K, and $\|f - F_\delta\| \le c_1\omega_1(\delta)$. Letting $\delta \to 0$, we obtain $f \in R(K)$.

So our problem boils down to the following: Given a continuous function F on C such that F is analytic on $K^0$, F is analytic off a disc $\Delta$, and $F(\infty) = 0$, we wish to find a function G such that G is analytic near K, G is analytic off $\Delta$, G-F has a triple zero at $\infty$, and $\|G\| \le c_0\|F\|$. In other words, the uniform approximation problem has been reduced to a problem involving the interpolation of two Taylor coefficients of certain functions. It is at this point that analytic capacities enter the picture.

The analytic capacity of a subset E of C is defined by

$$\gamma(E) = \sup\{|f'(\infty)| : f \text{ is analytic off a compact subset of } E,\ |f| \le 1,\ f(\infty) = 0\}.$$

The continuous analytic capacity of E is defined by

$$\alpha(E) = \sup\{|f'(\infty)| : f \text{ as above, } f \text{ continuous on } C\}.$$

The set functions $\gamma$ and $\alpha$ are monotone, and they satisfy

$$\alpha(E) \le \gamma(E) \le \text{diameter}(E).$$

Koebe's theorem can be used to show that $\gamma$ is comparable to diameter for compact connected sets:

(*) $\gamma(E) \le \text{diameter}(E) \le 4\gamma(E)$, E a continuum.

Returning to our interpolation problem, we see that an estimate of the form

$$\gamma(\Delta\setminus K) \leq c\ \alpha(\Delta\setminus K^0)$$

is precisely what is needed to find G as above such that G-F has a double zero at $\infty$. Most difficulties in this business arise from interpolating the remaining coefficient, so that G-F will have a triple zero at $\infty$. For this, we introduce two more set functions:

$$\gamma^*(E) = \sup\{|f''(\infty)| : f \text{ is analytic off a compact subset of } E, \\ |f| \leq 1, f(\infty) = f'(\infty) = 0\}$$

$$\alpha^*(E) = \sup\{|f''(\infty)| : f \text{ as above, } f \text{ is continuous on } \mathbf{C}\}.$$

Again simple estimates yield

$$\alpha^*(E) \leq \gamma^*(E) \leq [\text{diameter}(E)]^2.$$

Modulo the omitted estimates, we arrive on the basis of our remarks so far at the following theorem.

Theorem 1. Suppose there exists $c > 0$ such that

$$\gamma(\Delta\setminus K) \leq c\ \alpha(\Delta\setminus K^0)$$

and

$$\gamma^*(\Delta\setminus K) \leq c\ \alpha^*(\Delta\setminus K^0)$$

for every open disc $\Delta$ centered on $\partial K$. Then $R(K) = A(K)$.

There is one important case in which one obtains the second estimate of Theorem 1 automatically from the first estimate.

Corollary. Suppose there exists $b > 0$ such that

$$\gamma(\Delta_\delta\setminus K) \geq b\delta$$

for every open disc $\Delta_\delta$ of radius $\delta$, centered on $\partial K$. Then $R(K) = A(K)$.

Proof: By considering the squares of the admissible functions for $\gamma$, one sees that $\gamma^*(E) \geq \gamma(E)^2$ for any set E.

Consequently

$$\gamma^*(\Delta_\delta \setminus K) \geq b^2\delta^2.$$

Since always $\alpha(\Delta_\delta \setminus K^0) \leq \alpha(\Delta_\delta) = \delta$ and $\alpha^*(\Delta_\delta \setminus K^0) \leq \alpha^*(\Delta_\delta) = \delta^2$, the estimates of Theorem 1 are satisfied.

Combining this corollary with the estimate (*), we obtain a geometric criterion for approximation, which applies to all finitely connected compact sets.

Corollary. If the diameters of the components of $C\setminus K$ are bounded away from zero, then $R(K) = A(K)$.

The preceding corollary is due to Mergelyan [32]. The proof we have sketched is, in rough outline, identical to Mergelyan's original proof of his well-known theorem.

Before passing to Vitushkin's theorems, we make some observations which will play a role in the approximation theory on product sets in $C^n$. By Whitney's extension construction, the extension of $f \in A(K)$ to $C$ can be chosen to depend linearly and continuously on $f$. The $f_{\delta j}$ depend linearly on (the extended) $f$. Furthermore, each $h_j = h_{\delta j}$ can be chosen to be a linear combination of the same two fixed functions (depending on $\delta$ and $j$, but not on $f$), and the $h_{\delta j}$ can be chosen to depend linearly on $f$. Consequently the approximators $F_\delta$ can be chosen to depend linearly on $f$. Setting $\delta = 1/n$, we obtain a sequence $\{T_n\}_{n=1}^{\infty}$ of finite dimensional linear operators on $C(K)$ such that $\sup\|T_n\| < \infty$, and $T_n f$ converges uniformly to $f$ for all $f \in R(K)$. This yields a theorem of L. Eifler [19], asserting that $R(K)$, and also $A(K)$, have the Banach approximation property.

The same scheme for approximation, together with considerably more effort, yield the following theorem.

Vitushkin's Theorem. The following are equivalent:

(i) $R(K) = A(K)$.

(ii) $\alpha(D\setminus K^0) = \gamma(D\setminus K)$ for every bounded open set $D$.

(iii) For all but at most countably many $z \in \partial K$, there exist $r \geq 1$ and $c > 0$ such that

$$\alpha(\Delta(z;\delta)\setminus K^0) \leq c\ \gamma(\Delta(z;r\delta)\setminus K)$$

for all sufficiently small $\delta > 0$.

Vitushkin's theorem strengthens Theorem 1 in three ways. First, the only estimates required are those for $\alpha$--the estimates for $\alpha^*$ are deduced from those for $\alpha$. Second, the estimates required by Vitushkin's theorem are pointwise estimates, not uniform estimates. Third, there is the introduction of the $r \geq 1$, which enters crucially into the proof of the theorem.

To prove his theorem, Vitushkin first obtains an individual approximation theorem, which is related in some sense to Morera's theorem.

Vitushkin's Individual Theorem. Let f be a continuous compactly supported function on $\mathbb{C}$, and let $a(\delta) > 0$ satisfy $a(\delta) \to 0$ as $\delta \to 0$. If either

(1) there exists $r \geq 1$ such that

$$\left| \iint f(z) \frac{\partial g}{\partial \bar{z}} dxdy \right| \leq \delta a(\delta) \left\| \frac{\partial g}{\partial \bar{z}} \right\|_\infty \gamma(\Delta(z_0;r\delta)\setminus K)$$

whenever g is a $C^2$ function supported by a disc $\Delta(z_0;\delta)$; or

(2) there exists $r \geq 1$ such that

$$\left| \int_{\partial S(z_0;\delta)} f(z)dz \right| \leq a(\delta)\gamma(S(z_0;r\delta)\setminus K)$$

whenever S is a square of center $z_0$ and side $\delta$;

then $f \in R(K)$. Conversely, if $f \in R(K)$, then (1) and (2) are true, with $r=1$ and $a(\delta)$ a fixed multiple of the modulus of continuity $\omega_f(\delta)$ of f.

Actually the conditions (1) and (2) can be weakened somewhat. The weakest condition which is known to serve here is obtained by applying more abstract methods (Davie's theorem).

We return to this point in Section 9.

Curiously, it is not known whether the squares in condition (2) can be replaced by other geometric figures, such as circles.

While satisfying in some respects, Vitushkin's Individual Theorem does not serve to answer the following question raised by J. -E. Bjork: If $f \in C(K)$ satisfies $f^2 \in R(K)$, then is $f \in R(K)$?

One more remark, which may shed some light on the estimate of condition (1), is that the integral $\iint f(z)(\partial g/\partial \bar{z})dxdy$ is the derivative at $\infty$ of the function h which solves the $\bar{\partial}$-problem $\partial h/\partial \bar{z} = g(\partial f/\partial \bar{z})$.

## 2. Negligible Sets

In this section we touch on Melnikov's estimates and other estimates for analytic capacity, and we consider briefly several problems related to Vitushkin's Theorem.

Let E be a subset of $\mathbb{C}$, and consider the following condition:

Estimate (iv). For each $z \in (\partial K)\setminus E$, there exist $r \geq 1$ and $c > 0$ such that

$$\alpha(\Delta(z;\delta)\setminus K^0) \leq c\ \gamma(\Delta(z;r\delta)\setminus K)$$

for all $\delta$ sufficiently small. The first problem is the following.

Problem. Give conditions on E which guarantee that the estimate (iv) implies that $R(K) = A(K)$.

According to the statement of Vitushkin's theorem given in Section 1, the estimate (iv) implies that $R(K) = A(K)$ whenever E is at most countable. On the other hand, if $\alpha(E) > 0$, then one can construct K such that estimate (iv) is valid, while $R(K) \neq A(K)$. (See the stitched disc on p. 221 of [20].) The following problem remains open, and the answer is probably negative.

Problem. If $\alpha(E) = 0$, then does the estimate (iv) imply that $R(K) = A(K)$?

One can ask analogous questions about negligible sets for Vitushkin's Individual Theorem. Here less is known, and the problems appear more difficult.

Another related batch of problems involves estimates for analytic capacities. In view of Vitushkin's theorem, it is clear that such estimates play an important role in approximation theory. Estimates for analytic capacities in terms of geometric quantities yield criteria which guarantee approximability. We have already seen one such estimate, relating the analytic capacity and the diameter of a continuum. Another useful estimate relates capacity to area:

$$\text{Area}(E) \leq \pi\alpha(E)^2, \qquad E \text{ a Borel set},$$

with equality holding whenever E is a disc. There are similar estimates relating $\alpha$ to other Hausdorff measures. This leads into exceptional set theory, and recommended reading includes the lecture notes of L. Carleson [8] and J. Garnett [26].

One class of difficult and useful estimates was obtained by Melnikov (analytic curves) and sharpened by Vitushkin (piecewise $C^2$ curves). As a corollary to certain estimates of Cauchy integrals, they obtain the following theorem.

Theorem 2. Let D be a domain in $\mathbb{C}$ whose boundary consists of a finite number of piecewise - $C^2$ closed curves. Then there is a constant $c(=c(D))$ such that

$$\alpha(J) \leq c[\alpha(J \cap D) + \alpha(J\backslash\overline{D})]$$

for any compact set J.

On the basis of these estimates, they obtain the following result.

Theorem 3. Suppose E is a union of a sequence of $C^2$ curves. Then every continuous function f on $C^*$ can be approximated uniformly be a sequence $\{f_n\}$ of continuous functions on $C^*$ such that each $f_n$ is analytic wherever f is, and $f_n$ is analytic on a neighborhood of E.

The proof runs as follows. It suffices to consider the case in which E is a compact interval lying on a $C^2$ curve; an iteration procedure will yield the general case. Now we apply the original scheme for approximation, choosing the discs $\Delta_j$ appearing there to be centered on E and to cover E. Theorem 2 yields an estimate of the form

$$(**) \quad \alpha(J) \leq c\alpha(J \setminus E), \qquad \text{any set } J,$$

and this estimate allows us to interpolate the first Taylor coefficients of the functions appearing in the approximation scheme. A careful look at the estimates reveals that when the $\Delta_j$ are more or less linearly distributed, the interpolation of the first coefficient is already sufficient to guarantee uniform approximation.

Corollary. If E lies on a countable union of $C^2$ curves, and if estimate (iv) holds at all $z \in (\partial K)\setminus E$, then $R(K) = A(K)$.

Actually, the estimate (**) is already sufficient to guarantee that E has the property of the preceding theorem and corollary. This leads to the following question.

Problem. Is (**) valid whenever E is a compact set satisfying $\alpha(E) = 0$?

There is a stronger and more implausible conjecture, that $\alpha$ is weakly subadditive in the sense that there is a universal constant c such that

$$\alpha(E \cup F) \leq c[\alpha(E) + \alpha(F)]$$

for all reasonable subsets E and F of C. The corresponding problem for the more tractable set function $\gamma$ is already very difficult, and there are available only partial results in that case (see [16]).

## 3. T-invariant Algebras

In this section we introduce T-invariant algebras, and we indicate how the constructive methods of approximation apply to these algebras.

In order that the scheme for approximation developed in Section 1 be applicable to a family A of continuous functions on K, it is necessary that A be an algebra, that A contain all functions analytic in a neighborhood of K, and that A be invariant under the $T_g$-operators

$$(T_g f)(w) = g(w)f(w) + \frac{1}{\pi} \iint \frac{f(z)}{z - w} \frac{\partial g}{\partial \bar{z}} \, dxdy.$$

This leads to the following formal definition.

Definition. A subalgebra A of C(K) is T-invariant if whenever g is a smooth function with compact support, and f is a bounded Borel function such that $f|_K \in A$, then $(T_g f)|_K \in A$.

It is easy to see that if A is a T-invariant closed subalgebra of C(K), then $A \supset R(K)$. Moreover, if $f \in A$ extends to be analytic in a neighborhood of $z_0$, then $[f - f(z_0)]/(z - z_0) \in A$.

Examples of T-invariant algebras abound. The prototypes are the subalgebras R(K) and A(K) of C(K), and the subalgebra A(U) of $C(\bar{U})$. To introduce another class of such algebras, we generalize the definition of R(K) as follows.

For each bounded Borel set E, define R(E) to be the uniform closure in $C(\bar{E})$ of functions of the form

$$f(w) = \iint \frac{h(z)}{z - w} \, dxdy,$$

where h is a bounded compactly supported Borel function on $\mathbb{C}$ which vanishes a.e. (dxdy) on E. Then R(E) is a T-invariant subalgebra of $C(\bar{E})$, which coincides with the usual algebra R(E) whenever E is compact. The measures orthogonal to R(E) are precisely those measures on $\overline{E}$ whose Cauchy transforms vanish a.e. (dxdy) on $C \setminus E$.

More generally, let $h(\delta)$ be an increasing function of $\delta \geq 0$, such that

$$0 \leq h(\delta) \leq c\delta^2,$$

and

$$\int_0^1 \frac{h(\delta)}{\delta^2}\, d\delta < \infty.$$

A measure $\nu$ on C is of growth h if

$$|\nu|(\Delta_\delta) \leq h(\delta)$$

for all $\delta > 0$ and all discs $\Delta_\delta$ of radius $\delta$. The integral condition on h implies that the Cauchy transform

$$\hat{\nu}(w) = \int \frac{d\nu(z)}{z-w}$$

of any compactly supported measure $\nu$ of growth h is continuous. Now let E be a bounded Borel set, and let A be the uniform closure in $C(\overline{E})$ of the Cauchy transforms $\hat{\nu}$ of compactly supported measures $\nu$ such that $|\nu|(E) = 0$ and $\varepsilon\nu$ has growth h for some $\varepsilon > 0$. Then A is a T-invariant uniform algebra on $\overline{E}$. These algebras were introduced in [26].

In order to adapt the approximation scheme to more general algebras, one must associate capacity functions with each algebra. For each subalgebra A of C(K), set

$$\alpha_A(E) = \sup\{|f'(\infty)| \; f \text{ is continuous on } C^*,\ f \text{ is analytic off a compact subset of } E,\ |f| \leq 1, f(\infty) = 0,\ f_K \in A\}.$$

Then $\alpha_A$ is a monotone set function which satisfies

$$\alpha_A(E) \leq \alpha(E).$$

For the most familiar choices of A we have

$$\alpha_{A(K)}(E) = \alpha(E \setminus K^0), \qquad E \text{ a Borel set}$$

$$\alpha_{A(U)}(E) = \alpha(E\setminus U), \qquad E \text{ a Borel set}$$

$$\alpha_{R(K)}(D) = \gamma(D\setminus K), \qquad D \text{ an open set.}$$

The set function $\alpha_A^*$ is defined similarly, by maximizing $|f''(\infty)|$ over the appropriate class of functions. The proof of Theorem 1 yields the following result.

Theorem 4. Let A and B be closed subalgebras of C(K) such that $R(K) \subset B \subset A$, and A is T-invariant. Let V be an open subset of $K^0$ such that all the functions in A are analytic on V. Suppose there is a constant $c > 0$ such that

$$\alpha_A(\Delta) \leq c\alpha_B(\Delta)$$

$$\alpha_A^*(\Delta) \leq c\alpha_B^*(\Delta)$$

whenever $\Delta$ is an open disc centered on $K\setminus V$. Then $B = A$.

Corollary. Let B be a closed subalgebra of C(K), and let V be an open subset of $K^0$ such that all the functions in B are analytic on V. Suppose there exists $c > 0$ such that

$$\alpha_B(\Delta_\delta) \leq c\delta$$

for every disc $\Delta_\delta$ of radius $\delta$ which is centered on $K\setminus V$. Then B consists of precisely the functions in C(K) which are analytic on V.

Analogues of Vitushkin's theorems can also be stated for T-invariant algebras. In Vitushkin's Individual Theorem, for instance, we merely replace R(K) by a T-invariant uniform algebra A on K, we replace $\gamma(\Delta(z_0;r\delta)\setminus K)$ by $\alpha_A(\Delta(z_0;r\delta))$ in condition (1), and we replace $\gamma(S(z_0;r\delta)\setminus K)$ by $\alpha_A(S(z_0;r\delta))$ in condition (2).

The analogue of Melnikov's theorem is valid, with $\alpha$ replaced by $\alpha_A$, providing A is a T-invariant uniform algebra on K. This is proved by a fairly straightforward adaptation of Melnikov's argument.

Corollary. If A is a T-invariant closed subalgebra of C(K), and if E lies on a countable union of analytic curves in $\mathbb{C}$, then every function in A can be approximated uniformly on K by functions in A which extend to be analytic in a neighborhood of E.

The idea of introducing a myriad of capacity functions, depending on the approximation problem, is due to Davie [16]. This idea also occurred to A. O'Farrell [33], who has developed a systematic approach to approximation theory along the lines presented in this section.

## 4. Peak Points

The problem of uniform approximation on nowhere dense sets can be handled very simply and elegantly using Cauchy transforms of orthogonal measures and the notions of peak point and representing measure. The main ingredients are a theorem of Bishop and a peak point criterion of Curtis. In this section we discuss these and related results.

Let A be a uniform algebra on a compact space X, and let $x \in X$. A measure $\mu$ on X is a complex representing measure for x if $f(x) = \int f d\mu$ for all $f \in A$. If in addition $\mu \geq 0$, then $\mu$ is called simply a representing measure for x. The complex representing measures are the continuous extensions of "evaluation at x" from A to C(X), while the representing measures are the positive (=norm-preserving) extensions of "evaluation at x" from A to C(X).

A point $x \in X$ is a peak point for A if there exists $f \in A$ such that $f(x) = 1$, while $|f| < 1$ on $X \backslash \{x\}$. Peak points are characterized as follows.

Theorem 5. Let A be a uniform algebra on a compact metric space X. The following are equivalent for $x \in X$:

(1) x is a peak point for A.

(2) There is a sequence $\{f_n\}$ in A such that $\sup\|f_n\| < \infty$, $f_n(x) = 1$, and $f_n(y) \to 0$ for $y \in X\setminus\{x\}$.

(3) The point mass at x is the only representing measure for x.

(4) Every complex representing measure $\nu$ for x satisfies $\nu(\{x\}) = 1$.

Peak points are related to rational approximation by the following theorem.

Bishop's Theorem. If A is a T-invariant uniform algebra on K, and if almost all (dxdy) points of K are peak points for A, then $A = C(K)$.

Sketch of proof: Let $\nu \in A^{\perp}$. One shows that if $z_0 \in K$ is such that the integral

$$\hat{\nu}(z_0) = \int \frac{d\nu(z)}{z-z_0}$$

is absolutely convergent and nonzero, then $\nu/[\hat{\nu}(z_0)(z-z_0)]$ is a complex representing measure for $z_0$ with no mass at $z_0$, so that $z_0$ is not a peak point. This fact, coupled with the hypothesis, shows that $\hat{\nu} = 0$ a.e. (dxdy), and this is sufficient to guarantee that $\nu = 0$.

In view of Bishop's theorem, it is important to develop criteria for peak points. As a consequence of his estimates, Melnikov obtained a necessary and sufficient criterion, which is analogous to Wiener's criterion for regular points for the Dirichlet problem.

Melnikov's Criterion. Let A be a T-invariant uniform algebra on K. Then $z_0 \in K$ is a peak point for A if and only if

$$\Sigma\, \alpha_A(E_n) = +\infty,$$

where $E_n$ is the annulus $\{1/2^{n+1} < |z-z_0| < 1/2^n\}$.

The proof of Melnikov's criterion is quite difficult. However, there is available a useful sufficient criterion, which is quite easy to establish.

<u>Curtis's Criterion.</u> Let A be a T-invariant uniform algebra on K, and let $z_0 \in K$. If

$$\limsup_{\delta \to 0} \frac{\alpha_A(\Delta(z_0;\delta))}{\delta} > 0,$$

then $z_0$ is a peak point for A.

<u>Proof</u>: The definition of $\alpha_A$ yields $\delta_n \downarrow 0$ and continuous functions $h_n$ on $C^*$ such that $h_n|_K \in A$, $h_n$ is analytic off $\Delta(z_0;\delta_n)$, $|h_n| \le 1$, $h_n(\infty) = 0$, and $|h_n'(\infty)| \ge b > 0$. The functions $f_n = 1 - (z-z_0)h_n/h_n'(\infty)$ then have property (2) of Theorem 5.

Combining Bishop's Theorem and Curtis's Criterion, we obtain an elementary proof of the implication "(iii) $\Rightarrow$ (i)" of the following version of Vitushkin's Theorem. Note that we have only used the implication "(2) $\Rightarrow$ (4)" of Theorem 5, which is quite easy to prove.

<u>Theorem 6.</u> The following are equivalent, for a T-invariant uniform algebra on K:

(i) $A = C(K)$.

(ii) $\alpha_A(\Delta_\delta) = \delta$ for all open discs $\Delta_\delta$ of radius $\delta$.

(iii) $\limsup_{\delta \to 0} \dfrac{\alpha_A(\Delta(z_0;\delta))}{\delta} > 0$, for a.a. (dxdy) $z_0 \in K$.

The details of the proof of Theorem 6 can be found in Chapter VIII of [20]. The condition (iii) can actually be weakened considerably. According to Vitushkin's instability theorem [46], one of the following two alternatives must hold, except for $z_0$ belonging to a set of zero area:

$$\lim_{\delta \to 0} \frac{\alpha_A(\Delta(z_0;\delta))}{\delta} = 1$$

or

$$\lim_{\delta\to 0} \frac{\alpha_A(\Delta(z_0;\delta))}{\delta^2} = 0.$$

Consequently the condition

$$\text{(iv)} \quad \limsup_{\delta\to 0} \frac{\alpha_A(\Delta(z_0;\delta))}{\delta^2} > 0, \text{ for a.a. } (dxdy)\ z_0 \in K,$$

is equivalent to (i), (ii), and (iii) of Theorem 6.

## 5. Gleason Parts

A T-invariant algebra on K need not coincide with C(K), even though K is nowhere dense. Examples of this are the algebras R(K), when K is a Swiss cheese set [20, p. 26]. It turns out, though, that even for such sets one can point to a semblance of an analytic structure in the underlying set K in order to account for the defect of A in C(K). These generalized analytic structures are called Gleason parts. This section is devoted to a discussion of Gleason parts and decomposition theorems for orthogonal measures.

Let A be a uniform algebra on a compact space X. Every nonzero complex-valued homomorphism $\varphi$ of A satisfies $\|\varphi\| = 1 = \varphi(1)$. Consequently the set $M_A$ of such homomorphisms can be regarded as a subset of the unit sphere of the dual space $A^*$ of A. Endowed with the weak-star topology it inherits from $A^*$, $M_A$ becomes a compact space. It is customary to identify each $x \in X$ with the homomorphism "evaluation at x," so that X becomes identified homeomorphically with a closed subset of $M_A$. Writing $f(\varphi)$ for $\varphi(f)$, $\varphi \in M_A$, we find that the functions in A extend continuously from X to $M_A$, and that A can be regarded as a uniform algebra on $M_A$.

In general, $M_A$ is strictly larger than X. However, the following is true.

<u>Arens's Theorem</u>. If A is a T-invariant uniform algebra on K,

then $M_A = K$.

Two homomorphisms $\varphi, \psi \in M_A$ are in the same Gleason part of $M_A$ if there exists $c > 0$ such that Harnack's inequality is valid:

$$c < u(\varphi)/u(\psi) < 1/c, \quad u \in \mathrm{Re}(A), \quad u > 0.$$

Being in the same Gleason part is evidently an equivalence relation, which breaks $M_A$ into disjoint equivalence classes, the Gleason parts of $M_A$.

<u>Theorem 7.</u> The following are equivalent, for $\varphi, \psi \in M_A$:

(1) $\varphi$ and $\psi$ belong to the same Gleason part of $M_A$.

(2) $\|\varphi - \psi\|_{A^*} < 2$.

(3) There exists $c, 0 < c < 1$, such that $|\varphi(f)| < c$ for all $f \in A$ satisfying $\|f\| \leq 1$ and $\psi(f) = 0$.

We can again define representing measures and complex representing measures (on X) for every $\varphi \in M_A$, and the Gleason parts can be characterized in terms of representing measures.

<u>Theorem 8.</u> If $\varphi$ and $\psi$ belong to different Gleason parts of $M_A$, then every representing measure for $\varphi$ is singular to every representing measure for $\psi$. On the other hand, if $\varphi$ and $\psi$ belong to the same Gleason part of $M_A$, and if $\nu$ is a representing measure for $\varphi$, then there is a representing measure $\eta$ for $\psi$ such that $\nu \ll \eta$.

The peak points of A are clearly one-point parts. In general there may be one-point parts which are not peak points, but this does not occur for T-invariant algebras.

<u>Wilken's Theorem.</u> If A is a T-invariant uniform algebra on K, and $z_0 \in K$ is not a peak point for A, then the Gleason part of $z_0$ has positive area.

In particular, the Gleason parts of a T-invariant algebra are the singletons consisting of the peak points, together with an at most countable family $Q_1, Q_2, \ldots$ of subsets of K, each of positive area. Corresponding to this subdivision of K, there is a decomposition of measures orthogonal to A, which is related to the abstract F. and M. Riesz theorem.

Theorem 9. Let $Q_1, Q_2, \ldots$ be the nontrivial Gleason parts of a T-invariant uniform algebra A on K. Then there are Borel subsets $E_1, E_2, \ldots$ of K such that

(1) the $E_j$ are pairwise disjoint;

(2) $Q_j \subset E_j \subset \overline{Q_j}$;

(3) if $\mu \in A^{\perp}$, then $\mu_{E_j} \in A^{\perp}$, and $\mu = \Sigma \mu_{E_j}$;

(4) if $\mu \in A^{\perp}$ is carried by $E_j$, then $\mu$ is absolutely continuous with respect to some representing measure for any fixed point of $Q_j$.

This theorem allows us to express a T-invariant algebra in terms of T-invariant algebras which are primitive in the sense that they have only one nontrivial Gleason part. Indeed, let the $E_j$ be as above, and define $A_j$ to be the set of $f \in C(K)$ such that $\int f d\mu = 0$ for all measures $\mu \in A^{\perp}$ which are carried by $E_j$. It turns out that $A_j$ is a T-invariant uniform algebra on K with precisely one nontrivial Gleason part, namely, $Q_j$. Moreover, $A_j^{\perp}$ consists of precisely the measures in $A^{\perp}$ which are carried by $E_j$, so that

$$A = \cap A_j$$

$$A^{\perp} = \Sigma \oplus A_j^{\perp}.$$

The following theorem is due to B. Øksendal [35].

Theorem 10. Let A be a T-invariant uniform algebra on K, and let S be the set of $z \in K$ which satisfy

$$\lim_{\delta \to 0} \frac{\alpha_A(\Delta(z;\delta))}{\delta} = 1.$$

Then $|\mu|(S) = 0$ for all measures $\mu \in A^{\perp}$.

An elementary estimate shows that the set S above includes all points $z \in K$ at which $\cup Q_j$ has zero area density. Hence S includes almost all (dxdy) peak points of A. Replacing the sets $E_j$ by $E_j \setminus S$, we find that the sets $E_j$ of Theorem 9 can be chosen so that $E_j \setminus Q_j$ has zero area. There remains open the question of whether the $E_j \setminus Q_j$ can be chosen even smaller, in some metric sense.

## 6. Some Counterexamples

In this section, we give two examples which illustrate the limitations of the notion of peak point in dealing with approximation on sets with interior. The first example shows that Gleason parts need not be connected. The second example shows that the natural analogue of Bishop's theorem fails for sets with interior.

The examples are based on a construction, due independently to A. M. Davie and J. Garnett. Garnett's formulation of the central idea is as follows.

Lemma. Let I be a straight line segment, and let $\varepsilon > 0$. Then there exists a rectangle R containing I and bisected by I, and there exist disjoint discs $\Delta_1, \ldots, \Delta_n$ contained in R and centered on I, such that

(1) the sum of the radii of the $\Delta_j$ is less than $\varepsilon$, and

(2) for each $z_0 \in R \setminus (\cup \Delta_j)$, there is a rational function f such that $|f| \leq 1$ off $\cup \Delta_j$, $f(\infty) = 0$, and $|f(z_0)| > 1/100$.

Proof: We can assume that I is the interval $(0, 1 + 1/n)$ on the real axis. Take

$$R = \{z : 0 < \text{Re}z < 1 + 1/n, |\text{Im}z| < 1/n\},$$

and take $\Delta_j$ to have radius $1/(n \log n)$ and center $j/n$, $1 \leq j \leq n$. Then (1) is true, providing n is large. If

$$f_j(z) = \frac{1}{n(\log n)(z - j/n)},$$

then crude estimates yield

$$\Sigma|f_j(z)| < 10, \qquad\qquad z \in C^*\backslash(\cup\, \Delta_j).$$

Fix $z_0 \in R\backslash(\cup\, \Delta_j)$, let the $a_j$ be complex numbers of unit modulus such that $a_j f_j(z_0) > 0$, and set $f = \Sigma\, a_j f_j/10$. Crude estimates, together with the inequality $1 + 1/2 + \ldots + 1/m > \log m$, yield (2).

Example (Davie; Garnett). There is a "string of beads" set K such that R(K) has a disconnected Gleason part.

Proof: Recall [20, p. 146] that a "string of beads" set is a set of the form $K = \overline{\Delta}\backslash(\cup\, \Delta_j)$, where $\Delta$ is the open unit disc, and $\{\Delta_j\}$ is a sequence of disjoint open discs centered on the real axis and contained in $\Delta$, so that $T = [-1,1] \backslash (\cup\, \Delta_j)$ has no linear interior. Then $K^0$ consists of two components, a top half $U_+$ and a bottom half $U_-$, whose boundaries meet along T. Both $U_+$ and $U_-$ are simply connected open sets with rectifiable boundaries, so that harmonic measure for $U_\pm$ is comparable to arc length along $\partial U_\pm$. In particular, if T has positive length, then harmonic measure for $U_+$ is not singular to harmonic measure for $U_-$, so that by Theorem 8, $U_+$ and $U_-$ belong to the same Gleason part of R(K). To obtain a disconnected part, it then suffices to choose the $\Delta_j$ so that T has positive length, while every point of T is a peak point for R(K).

This is accomplished by performing the construction of the lemma on the interval $[-1,1]$, then repeating the construction on the interval between each two consecutive holes, etc., etc. If the radii are chosen sufficiently small at each stage, the sum of the diameters of the excised discs is less than 2, and T has positive length.

Let $z_0 \in T$, and let $\varepsilon > 0$. Then there is a batch of discs excised at some step of the procedure which lies in an $\varepsilon$-neighborhood of $z_0$, and which has $z_0$ in their "rectangle of influence." There is then $f_\varepsilon \in R(K)$ such that $f_\varepsilon$ is analytic off $\Delta(z_0;\varepsilon)$, $|f_\varepsilon| \leq 1$, $f_\varepsilon(\infty) = 0$, and $|f_\varepsilon(z_0)| > 1/100$. As $\varepsilon \to 0$, the $f_\varepsilon$ converge uniformly to 0 on compact subsets of $C^* \backslash \{z_0\}$, by Schwarz's lemma. By Theorem 5, $z_0$ is a peak point for $R(K)$. That does it.

Example (Davie). There is a compact set K such that $R(K) \neq A(K)$, while every point of $\partial K$ is a peak point for $R(K)$.

Proof: The set K will be obtained from the closed unit square S by excising a sequence $\{\Delta_j\}$ of open discs, such that the sum of the radii of the $\Delta_j$ is finite, while the set J $= (\partial K)\backslash(\cup \partial\Delta_j)$ satisfies $\alpha(J) > 0$. Then there is a continuous function f on $C^*$ such that f is analytic off J, $f(\infty) = 0$ and $f'(\infty) \neq 0$. If $\nu$ is the measure dz on $\partial S \cup (\cup \partial\Delta_j)$, oriented properly, then $\nu \perp R(K)$, while $\int f d\nu = \int_{\partial S} f dz = 2\pi i f'(\infty) \neq 0$. Consequently $f \in A(K)$, while $f \notin R(K)$.

The construction proceeds as follows. First excise holes centered on the horizontal bisector of S, as in the Lemma. Next consider two thin rectangles parallel to the x-axis, running the length of the unit square, one located above the excised discs and the other below, but both lying inside the "rectangle of influence" of the excised discs. Divide each of these rectangles into squares of equal size, then perform the same construction on each of the new squares, etc., etc. If the radii of the discs are chosen sufficiently small at each stage, their sum is finite Moreover, if we proceed symmetrically at each stage, $\partial K$ is the union of the $\partial\Delta_j$, together with $\partial S$ and a set of the form $[0,1] \times E$, where E is a perfect subset of $[0,1]$. The argument used in the preceding example shows that each point of $[0,1] \times E$ is a peak point for $R(K)$, so that every point of $\partial K$ is a peak point for $R(K)$. To complete the proof, then, it suffices to establish the following lemma, due to Carleson.

Lemma. If E is any perfect subset of the unit interval [0,1], then $\alpha([0,1] \times E) \geq 1/4$.

Proof: Let g be the conformal map of $C^* \setminus [0,1]$ onto the open unit disc $\Delta$ which satisfies $g(\infty) = 0$. Then $|g'(\infty)| = 1/4$. Let $\tau$ be a probability measure on E with no point masses, and set $f = g * \mu$, i.e., $f(x,y) = \int_E g(x,y-t)d\mu(t)$. Evidently f is analytic off $[0,1] \times E$, $|f| \leq 1$, $f(\infty) = 0$ and $|f'(\infty)| = 1/4$. If $(x,y) \in [0,1] \times E$, and J is a small interval about y, then $\int_E = \int_J + \int_{J \setminus E}$ expresses f as $f_1 + f_2$, where $|f_1| \leq \mu(J)$, and $f_2$ is analytic, hence continuous, near (x,y). It follows that f is continuous at (x,y), so f is continuous on $C^*$, and f is admissible for $\alpha$.

A major unsolved question along these lines is the following problem of Wilken: If Q is a Gleason part for R(K), are the points of $\overline{Q} \setminus Q$ necessarily peak points for R(K)?

## 7. Reducing Bands and Davie's Theorem

Now we turn to the most recent chapter of rational approximation theory. The key to the most recent advances is Davie's theorem, which can be stated in the following form.

Davie's Theorem (First Version). Let A be a T-invariant uniform algebra on K, and let $\lambda_Q$ be the area measure on the set Q of nonpeak points of A. If $f \in H^\infty(\lambda_Q)$, then there is a sequence $\{f_n\}$ in A such that $\|f_n\| \leq \|f\|$, and $f_n$ converges a.e. $(d\lambda_Q)$ to f.

There is another version of Davie's theorem, which is formulated in terms of the second dual $A^{**}$ of A. In order to set the stage for this version, we take a detour and describe an attractive approach which has been developed by H. König, G. Seever and B. Cole.

A band of measures $\mathcal{B}$ on X is a closed subspace of M(X) such that $\nu \in \mathcal{B}$ and $\eta \ll \nu$ imply $\eta \in \mathcal{B}$. Any band has a complementary band $\mathcal{B}'$, consisting of the measures on X singular to each measure in $\mathcal{B}$. The Lebesgue decomposition theorem becomes

$$M(X) = \mathcal{B} \oplus \mathcal{B}'.$$

Any band $\mathcal{B}$ can be represented as an inductive limit (union) of the spaces $L^1(\nu)$, for $\nu \in \mathcal{B}$:

$$\mathcal{B} = \overrightarrow{\lim} \{L^1(\nu) : \nu \in \mathcal{B}\}.$$

Consequently the dual $\mathcal{B}^*$ of $\mathcal{B}$ can be represented as a projective limit:

$$\mathcal{B}^* = L^\infty(\mathcal{B}) = \overleftarrow{\lim} \{L^\infty(\nu) : \nu \in \mathcal{B}\}.$$

In other words, an object $F \in L^\infty(\mathcal{B})$ determines, for each measure $\nu \in \mathcal{B}$, a function $F_\nu \in L^\infty(\nu)$, and these functions $F_\nu$ satisfy a compatibility condition, that $F_\nu = F_\eta$ a.e. $(d\nu)$, whenever $\nu \ll \eta$. Moreover,

$$\|F\| = \sup\{\|F_\nu\|_{L^\infty(\nu)} : \nu \in \mathcal{B}\}, \qquad F \in L^\infty(\mathcal{B}).$$

With the natural multiplication induced from the $L^\infty(\nu)$, $L^\infty(\mathcal{B})$ becomes a commutative $B^*$-algebra.

In the case $\mathcal{B} = M(X)$, this yields the representation $C(X)^{**} \cong L^\infty(M(X))$ obtained by Arens [2].

To each band $\mathcal{B}$ there corresponds an idempotent $P \in C(X)^{**}$, defined by setting $P_\nu = 1$ if $\nu \in \mathcal{B}$, and $P_\eta = 0$ if $\eta \in \mathcal{B}'$. The correspondence $\mathcal{B} \to P$ is a one-to-one correspondence between bands in $M(X)$ and idempotents in $C(X)^{**}$. It is customary to identify $L^\infty(\mathcal{B})$ with the closed subspace $PC(X)^{**}$ of $C(X)^{**}$.

Let A be a uniform algebra on X. Recall that $H^\infty(\nu)$ is the weak-star closure of A in $L^\infty(\nu)$, $\nu \in M(X)$. Now $A^{**}$ is isometrically isomorphic to the weak-star closed subspace $(A^\perp)^\perp$ of $C(X)^{**}$. It is easy to check that $F \in C(X)^{**}$ belongs to $A^{**}$ if and only if $F_\nu \in H^\infty(\nu)$ for all $\nu \in M(X)$. In other words,

$$A^{**} = \overleftarrow{\lim} \{H^\infty(\nu) : \nu \in M(X)\}.$$

A band $\mathcal{B}$ is <u>reducing</u> if whenever $\nu \in A^\perp$, then the projection of $\nu$ onto $\mathcal{B}$ also belongs to $A^\perp$. Hence $\mathcal{B}$ is reducing if and only if

$$A^\perp = (A^\perp \cap \mathcal{B}) \oplus (A^\perp \cap \mathcal{B}').$$

This occurs if and only if the idempotent P of $\mathcal{B}$ belongs to $A^{**}$, and in this case we obtain

$$A^{**} \cong H^\infty(\mathcal{B}) \oplus H^\infty(\mathcal{B}'),$$

where

$$H^\infty(\mathcal{B}) = PA^{**} = \overleftarrow{\lim} \{H^\infty(\nu) : \nu \in \mathcal{B}\}.$$

One example of a reducing band is the <u>singular band</u> $\mathcal{S}$, consisting of all $\tau \in M(X)$ such that $\tau$ is singular to all measures in $A^\perp$. The complementary band $\mathcal{S}'$ of $\mathcal{S}$ is then the band generated by the measures in $A^\perp$. Evidently $H^\infty(\mathcal{S}) = L^\infty(\mathcal{S})$, so that

$$A^{**} \cong L^\infty(\mathcal{S}) \oplus H^\infty(\mathcal{S}').$$

Minimal reducing bands are defined in the obvious way, and these are in one-to-one correspondence with the minimal idempotents in $A^{**}$. The following theorem provides examples of reducing bands which are minimal.

<u>Abstract F. and M. Riesz Theorem.</u> Let A be a uniform algebra on X, and let $\varphi \in M_A$. Let $\mathcal{B}_\varphi$ be the band generated by the representing measures on X for $\varphi$. Then $\mathcal{B}_\varphi$ is a minimal reducing band.

By Theorem 8, $\mathcal{B}_\varphi = \mathcal{B}_\psi$ whenever $\varphi, \psi \in M_A$ belong to the same Gleason part, while $\mathcal{B}_\varphi$ and $\mathcal{B}_\psi$ are singular whenever $\varphi$ and $\psi$ belong to distinct Gleason parts. More generally, any two distinct minimal reducing bands are singular. Cole has observed that there may be minimal reducing bands which are not of the form $\mathcal{B}_\varphi$.

As a special case of the bands $\mathcal{B}_\varphi$, there are the bands $L^1(\delta_q)$ consisting of all multiples of the point mass at a (generalized) peak point q. These are the <u>trivial minimal reducing bands</u>, and they are subbands of $\mathcal{S}$. The nontrivial minimal reducing bands are singular to $\mathcal{S}$.

Now let $\mathcal{J}$ be the band of measures singular to every minimal reducing band and to $\mathcal{S}$. Then

$$M(X) = \mathcal{S} \oplus \mathcal{J} \oplus \Sigma \oplus \{\mathcal{B} : \mathcal{B} \text{ is a nontrivial minimal reducing band}\}.$$

The band $\mathcal{J}$ is a reducing band, and Cole has proved that it has the following property.

<u>Theorem 11</u>. If $\nu \in A^\perp \cap \mathcal{J}$ and $\varepsilon > 0$, then there are reducing bands $\mathcal{J}_1,\ldots,\mathcal{J}_n$ such that $\mathcal{J} = \mathcal{J}_1 \oplus \cdots \oplus \mathcal{J}_n$, and the projection of $\nu$ into each $\mathcal{J}_j$ has norm less than $\varepsilon$.

As an example of this theory, let $\Delta$ be the open unit disc, let $X = \partial\Delta$, and let A be the disc algebra on $\partial\Delta$, consisting of functions in $C(\partial\Delta)$ which extend analytically to $\Delta$. The band of measures on $\partial\Delta$ generated by $A^\perp$ coincides with $L^1(d\theta)$, so that the singular band $\mathcal{S}$ consists of all measures on X singular to $d\theta$, and $A^{**} = L^\infty(\mathcal{S}) \oplus H^\infty(d\theta)$. From Fatou's theorem we obtain

$$A^{**} \cong L^\infty(\mathcal{S}) \oplus H^\infty(\Delta).$$

The following more general example is due to B. Cole and M. Range [10], and it is related to work of G. M. Henkin [29] and N. Kerzman [30]. Let D be a strictly pseudoconvex domain in $C^n$ with smooth boundary, let $X = \partial D$, and let A be the algebra of functions in $C(\partial D)$ which extend to be continuous on $\overline{D}$ and analytic on D. Fix $z \in D$, and let $\mathcal{B} = \mathcal{B}_z$ be the band of measures absolutely continuous with respect to representing measures on $\partial D$ for z. Then $A^{**} = L^\infty(\mathcal{S}) \oplus H^\infty(\mathcal{B})$. Moreover, the natural map $H^\infty(\mathcal{B}) \rightarrow H^\infty(D)$ is an isometric isomorphism, so that again

$$A^{**} \cong L^{\infty}(S) \oplus H^{\infty}(D).$$

Now Cole knew that there is a similar decomposition for $A(U)^{**}$ whenever $A(U)$ is pointwise boundedly dense in $H^{\infty}(U)$, and the question was whether there always is such a decomposition. This question was answered surprisingly in the affirmative by Davie.

Let $A$ be a T-invariant algebra on $K$, let $Q$ be the set of nonpeak points of $A$, and let $\mathcal{B}_Q$ be the band generated by the representing measures on $K$ for points of $Q$. It can be shown that $\mathcal{B}_Q'$ coincides with the singular band $S$, so that

$$A^{**} \cong L^{\infty}(S) \oplus H^{\infty}(\mathcal{B}_Q).$$

Moreover, the area measure $\lambda_Q$ on $Q$ belongs to $\mathcal{B}_Q$.

<u>Davie's Theorem (Second Version)</u>. The projection $H^{\infty}(\mathcal{B}_Q) \to H^{\infty}(\lambda_Q)$ is an isometric isomorphism and (consequently) a weak-star homeomorphism. Hence

$$A^{**} \cong L^{\infty}(S) \oplus H^{\infty}(\lambda_Q).$$

The two versions of Davie's theorem are essentially equivalent. Davie's original proof proceeds by establishing (implicitly) the second version, and then deducing the first version from the second and a separation argument. The proof depends on Vitushkin's approximation scheme.

In order to make the analogy with the earlier examples more explicit, we mention that the functions in $H^{\infty}(\lambda_Q)$ can be realized as point functions on $Q$. To see this, it suffices to show that if $\{f_n\}$ is a bounded sequence in $A$ which converges a.e. $(d\lambda_Q)$, then in fact $f_n(q)$ converges for all $q \in Q$. This follows easily from the Browder metric density theorem. It also follows from the weak-star homeomorphism of $H^{\infty}(\mathcal{B}_Q)$ and $H^{\infty}(\lambda_Q)$, and the fact that the point masses at points of $Q$ belong to $\mathcal{B}_Q$.

## 8. Distance Estimates

Let A be a uniform algebra on X, and let $\sigma$ be a positive measure on X. In this section we discuss the validity of distance estimates of the form

$$(*) \quad d(h,A) = d(h,H^\infty(\sigma)), \qquad \text{all } h \in C(X).$$

Here "d" denotes "distance," estimated in the norms of C(X) and $L^\infty(\sigma)$ respectively. There is a simple dual criterion which is equivalent to (*), namely,

$$(**) \quad \text{ball}(A^\perp \cap L^1(\sigma)) \text{ is weak-star dense in ball } A^\perp.$$

Now A,C(X) and $A^{**}$ can all be regarded as closed subspaces of $C(X)^{**}$. Moreover, the double dual of C(X)/A is isometrically isomorphic to $C(X)^{**}/A^{**}$. That the embedding $C(X)/A \to C(X)^{**}/A^{**}$ is an isometry can be expressed in terms of the distance function d:

$$d(h,A) = d(h,A^{**}), \qquad \text{all } h \in C(X).$$

In the preceding section we discussed algebras for which there is a positive measure $\sigma$ satisfying

$$A^{**} \cong L^\infty(S) \oplus H^\infty(\sigma).$$

In these cases, it would seem reasonable to expect (*) to be valid. Unfortunately, it is not known whether (*) is true whenever $A^{**}$ can be decomposed in the above manner. However, (*) is valid in each of the cases discussed in the preceding section. In the case of the disc algebra, this amounts to Sarason's theorem [39]:

$$d(h,A(\Delta)) = d(h,H^\infty(d\theta)), \quad \text{all } h \in C(\partial\Delta).$$

In this case, $A^\perp = H^1(d\theta) \subset L^1(d\theta)$, so that the dual version (**) is trivially valid. The following theorem is proved in [25].

<u>Theorem 12</u>. Let A be a T-invariant uniform algebra on K, and

let $\lambda_Q$ be the area measure on the set Q of nonpeak points of A. Then

$$d(h,A) = d(h,H^\infty(\lambda_Q)), \qquad \text{all } h \in C(K).$$

Proof: The shortest proof, due to Cole, runs along the following lines. Let $f \in H^\infty(\lambda_Q)$, and let $F \in H^\infty(\mathcal{B}_Q)$ be the element corresponding to f under the isometry $H^\infty(\mathcal{B}_Q) \cong H^\infty(\lambda_Q)$. The problem boils down to showing that if $|f| \leq c$ a.e. $(d\lambda_Q)$ on some open disc $\Delta_0$ centered at $q \in \overline{Q}$, and if $\tau \in A^\perp$, then $|F| \leq c$ a.e. $(d\tau)$ on some neighborhood of q. For this, we apply the Bishop splitting lemma, to write $\tau = \tau_0 + \tau_1$, where $\tau_0$ and $\tau_1$ belong to $A^\perp$, $\tau_0$ is carried by $\Delta_0$, and $\tau_1$ has no mass near q. Davie's theorem, applied to the closure of A in $C(\overline{\Delta}_0 \cap K)$, shows that the map $H^\infty(\lambda_{Q\cap\Delta_0} + |\tau_0|) \to H^\infty(\lambda_{Q\cap\Delta_0})$ is an isometry. That yields immediately the required assertion.

An immediate consequence of the preceding theorem is the following approximation theorem.

Theorem 13. Let A be a T-invariant uniform algebra on K, and let $\lambda_Q$ be the area measure on the set Q of nonpeak points of A. Then

$$C(K) \cap H^\infty(\lambda_Q) = A.$$

In particular, if $f \in C(K)$, and if there is a bounded sequence in R(K) which converges a.e. (dxdy) to f on K, then $f \in R(K)$. This concrete approximation theorem does not seem as yet to be accessible from the purely constructive methods.

In the case of the algebra A(U), almost all (dxdy) points of $\partial U$ are peak points for A(U), so that the measure $\lambda_Q$ coincides with the area measure $\lambda_U$ on U. The distance estimate becomes

$$d(h,A(U)) = d(h,H^\infty(\lambda_U)), \qquad \text{all } h \in C(\overline{U}).$$

Note that $H^\infty(\lambda_U)$ is a weak-star closed subalgebra of $H^\infty(U)$, consisting of all $f \in H^{\infty}(U)$ which can be approximated pointwise on U by a bounded sequence in A(U). The following theorem gives a condition for the coincidence of $H^{\infty}(\lambda_U)$ and $H^{\infty}(U)$.

Theorem 14. A(U) is pointwise boundedly dense in $H^{\infty}(U)$ if and only if

$$d(h,A(U)) = d(h,H^{\infty}(U)), \qquad \text{all } h \in C(\overline{U}).$$

Proof: The forward implication is a special case of Theorem 12, and the reverse implication is proved in [17].

The estimates above lead easily to estimates of the form

$$d(h,A(U)) = d(h,H^{\infty}(\mu)), \qquad \text{all } h \in C(\partial U),$$

where $\mu$ is the harmonic measure on $\partial U$. This latter estimate includes Sarason's theorem. Moreover, as in the case of the disc algebra, the linear space $H^{\infty}(\mu) + C(\partial U)$ turns out to be a closed subalgebra of $L^{\infty}(\mu)$. These and related questions are studied in detail in [17]. Here we present a sample from [17], which generalizes the preceding distance estimate. Analogous results have been obtained for pseudoconvex domains in $C^n$ by M. Range [37].

For each subset E of $\partial U$, let $L^{\infty}_E(\mu)$ be the subspace of $f \in L^{\infty}(\mu)$ which are (essentially) continuous at each point of E, and set $H^{\infty}_E(\mu) = H^{\infty}(\mu) \cap L^{\infty}_E(\mu)$.

Theorem 15. Let $\mu$ be the harmonic measure on $\partial U$, and let $H^{\infty}(\mu)$ be the weak-star closure of A(U) in $L^{\infty}(\mu)$. If E is a subset of $\partial U$, then

$$d(h,H^{\infty}(\mu)) = d(h,H^{\infty}_E(\mu)), \qquad \text{all } h \in L^{\infty}_E(\mu).$$

This type of quantitative approximation theorem leads immediately to uniform approximation theorems.

Corollary. Every $f \in H^{\infty}_E(\mu)$ can be approximated uniformly by functions in $H^{\infty}(\mu)$ which are continuous on a neighborhood of E.

Proof: Let $\varepsilon > 0$. Since $f \in L_E^\infty(\mu)$, there are an open subset $V$ of $\partial U$ and an $h \in L_V^\infty(\mu)$ such that $\|h-f\| < \varepsilon/2$. By Theorem 15, there is $g \in H_V^\infty(\mu)$ such that $\|g-h\| < \varepsilon/2$. This $g$ approximates $f$ within $\varepsilon$.

From Theorem 12 we deduce that $\text{ball}(A^\perp \cap L^1(\lambda_Q))$ is weak-star dense in ball $A^\perp$. The following approximation theorem of a somewhat different nature can also be obtained.

Theorem 16. Let $A$, $Q$ and $\lambda_Q$ be as above, and let $q \in Q$. Then the representing measures for $q$ in $L^1(\lambda_Q)$ are weak-star dense in the set of all representing measures for $q$.

For a proof of this theorem, in a more general setting, see [25]. Note that the existence of at least one representing measure in $L^1(\lambda_Q)$ follows from the fact that the map $H^\infty(\beta_Q) \to H^\infty(\lambda_Q)$ is a weak-star homeomorphism.

## 9. Pointwise Bounded Approximation

In this section we discuss theorems on pointwise bounded approximation which are analogous to Vitushkin's theorems on uniform approximation.

Since $H^\infty(\lambda_Q) \cap C(K) = A$, any theorem on pointwise bounded approximability, when coupled with a continuity hypothesis, yields a theorem on uniform approximability. Individual theorems for pointwise bounded approximation can then be regarded as the most inclusive theorems connecting rational approximation and analytic capacities. The following such theorem is obtained in [25], and the proof depends crucially on Davie's theorem.

Theorem 17. Let $A$ be a $T$-invariant uniform algebra on $K$, and let $\lambda_Q$ be the area measure on the set $Q$ of nonpeak points of $A$. The following are equivalent, for a bounded Borel function $f$ on $C$:

(i) $f \in H^\infty(\lambda_Q)$.

(ii) $\left|\iint f(z)\,\frac{\partial g}{\partial \bar z}\,dxdy\right| \le 2\pi\delta\, \mathrm{osc}(f,\Delta_\delta)\left\|\frac{\partial g}{\partial \bar z}\right\|_\infty \alpha_A(\Delta_\delta)$

for every smooth complex-valued function g supported on a disc $\Delta_\delta$ of radius $\delta$.

(iii) For all but at most countably many $z_0 \in \bar{Q}$, there exist $r \ge 1$ and $c > 0$ satisfying

$$\left|\iint f(z)\,\frac{\partial g}{\partial \bar z}\,dxdy\right| \le c\delta\left\|\frac{\partial g}{\partial \bar z}\right\|_\infty \alpha_A(\Delta(z_0;r\delta))$$

for all $\delta > 0$ sufficiently small, and all smooth complex-valued functions g supported by $\Delta(z_0;\delta)$.

Here $\mathrm{osc}(f,T)$ is the oscillation of f on a set T:

$\mathrm{osc}(f,T) = \sup\{|f(z)-f(w)| : z,w \in T\}$.

If f is continuous on $\mathbb{C}$, and $f \in H^\infty(\lambda_Q)$, then the estimate (ii) yields

$$\left|\iint f\,\frac{\partial g}{\partial \bar z}\,dxdy\right| \le 2\pi\delta\omega_f(\delta)\left\|\frac{\partial g}{\partial \bar z}\right\|_\infty \alpha_A(\Delta_\delta).$$

Vitushkin's Individual Theorem then places $f|_K$ in A. This yields a second proof of the identity

$$H^\infty(\lambda_Q) \cap C(K) = A.$$

As another corollary to Theorem 17, one can sharpen Vitushkin's Individual Theorem, replacing a "little oh" by a "big oh."

<u>Corollary</u>. Let A be a T-invariant uniform algebra on K, and suppose f is continous on $\mathbb{C}$. If either

(1) for all but at most countably many $z_0 \in K$ there exist $r \ge 1$ and $c > 0$ such that

$$\left|\iint f(z)\,\frac{\partial g}{\partial \bar z}\,dxdy\right| \le c\delta\left\|\frac{\partial g}{\partial \bar z}\right\|_\infty \alpha_A(\Delta(z_0;r\delta))$$

for all sufficiently small $\delta > 0$ and all smooth complex-valued functions g supported on $\Delta(z_0;\delta)$; or

(2) there exist $r \geq 1$ and $c > 0$ such that

$$\left| \int_{\partial S(z_0;\delta)} f(z)dz \right| \leq c\alpha_A(S(z_0;r\delta))$$

for all sufficiently small $\delta > 0$, and for all squares $S(z_0;\delta)$ of side $\delta$ and center $z_0 \in K$;

then $f \in A$.

The proof of Theorem 17 follows the lines of the proof of Vitushkin's Individual Theorem, although several additional estimates are required, and Davie's theorem provides the margin of victory. One of the features of the proof is that it produces a sequence $\{T_n\}$ of finite dimensional linear operators from $H^\infty(\lambda_Q)$ to A such that $\sup\|T_n\| < \infty$, and $T_n f$ converges a.e. $(d\lambda_Q)$ to f for all $f \in H^\infty(\lambda_Q)$. This property is bound to have further consequences for the algebra $H^\infty(\lambda_Q)$.

The following analogue for pointwise bounded approximation of Vitushkin's Theorem is a simple consequence of Theorem 17.

Theorem 18. The following are equivalent:

(i) A(U) is pointwise boundedly dense in $H^\infty(U)$.

(ii) $\gamma(D\setminus U) = \alpha(D\setminus U)$ for every open set D.

(iii) For all but at most countably many $z_0 \in \partial U$, there exist $r \geq 1$ and $c > 0$ such that

$$\gamma(\Delta(z_0;\delta)\setminus U) \leq c\alpha(\Delta(z_0;r\delta)\setminus U)$$

for all sufficiently small $\delta > 0$.

This theorem was originally proved in [23], and the results there show that the estimates of (iii) need hold only at points of $(\partial U)\setminus E$, where E is a Borel set of zero length lying on a countable union of $C^2$ curves. These sets E form a class of sets which are "negligible" for pointwise bounded approximation The analogue of the conjecture discussed in Section 2 is that every compact set E satisfying $\gamma(E) = 0$ is negligible for pointwise bounded approximation.

There is no known characterization of those compact sets $K$ such that $R(K)$ is pointwise boundedly dense in $H^\infty(K^0)$. The difficulty is that the nonpeak points of $R(K)$ may meet $\partial K$ in a set of positive area. However, it can be shown that if $R(K)$ is pointwise boundedly dense in $H^\infty(K^0)$, and if $R(\partial K) = C(\partial K)$, then $R(K) = A(K)$. This fact can sometimes be used to reduce the $R(K)$ case to the $A(K)$ case. One obtains, for instance, the following result.

Theorem 19. The following are equivalent:

(i) $R(K)$ is pointwise boundedly dense in $H^\infty(K^0)$, and $R(\partial K) = C(\partial K)$.

(ii) $\gamma(D\backslash K) = \gamma(D\backslash K^0)$ for every open set $D$.

(iii) For all but at most countably many $z_0 \in K$, there exist $c > 0$ and $r \geq 1$ such that

$$\gamma(\Delta(z_0;\delta)\backslash K^0) \leq c\gamma(\Delta(z_0;r\delta)\backslash K)$$

for all sufficiently small $\delta > 0$.

This theorem, together with the estimate relating the analytic capacity of a continuum to its diameter, yields the Farrell-Rubel-Shields theorem. In fact, one obtains the following.

Corollary. If the diameters of the components of $\mathbb{C}\backslash K$ are bounded away from zero, then $R(K)$ is pointwise boundedly dense in $H^\infty(K^0)$.

## 10. Dirichlet Algebras

In this section, we consider the problem of approximability by the real parts of analytic functions. This leads in the abstract context to the notion of Dirichlet algebra. We begin by citing some abstract results, and then we turn to approximation in the plane.

A uniform algebra A on X is a Dirichlet algebra on X if the space ReA of real parts of functions in A is uniformly dense in $C_R(X)$. This occurs if and only if there is no nonzero real measure on X which is orthogonal to A. The following theorem is elementary.

Theorem 20. If A is a Dirichlet algebra on a compact metric space X, then every point of X is a peak point for A.

A subset P of $M_A$ is an analytic disc if there is a continuous one-to-one map $\Phi$ of the open unit disc $\Delta$ onto P such that $f \circ \Phi$ is analytic on $\Delta$ for all $f \in A$. The following two theorems are deeper.

Wermer Embedding Theorem. Let A be a Dirichlet algebra on X. Then every Gleason part of A is either an analytic disc or a one-point part.

Theorem 21. Let A be a Dirichlet algebra on X, and let $\{P_j\}_{j\geq 1}$ be an at most countable family of Gleason parts of A which are analytic discs, with coordinate functions $\Phi_j:\Delta \to P_j$. If $\{h_j\}$ is a bounded sequence in $H^\infty(\Delta)$, then there is a sequence $\{f_k\}_{k=1}^\infty$ in A such that $\|f_k\| \leq \sup\|h_j\|_\Delta$, and $\{f_k\}$ converges pointwise to $h_j \circ \Phi_j^{-1}$ on $P_j$, $j \geq 1$.

Theorem 21 is an abstract version of the Farrell-Rubel-Shields Theorem. For an elegant treatment of these theorems, together with an abstract proof of Mergelyan's theorem, there are the lecture notes of Wermer [47].

Now let A be a T-invariant uniform algebra on K, and let E be a closed subset of K such that E is a boundary for A (that is, $\|f\|_E = \|f\|_K$ for all $f \in A$, so that A can be regarded as a uniform algebra on E), and such that A is a Dirichlet algebra on E. Evidently each peak point for A lies in E, so that by Theorem 20, E coincides with the set of peak points for A. From the Wermer Embedding Theorem, we conclude that the nontrivial Gleason parts of A are the components of $U = K\setminus E$, and each of these is simply connected. By Theorem 21, A is pointwise boundedly dense in $H^\infty(U)$. Since $C(K) \cap H^\infty(\lambda_U) = A$, we conclude that A consists of the functions in A(U), extended in all possible continuous ways to K. Consequently we may as well focus our attention on the algebra A(U), and the main result in this case is as follows.

Theorem 22. Suppose each component of U is simply connected. Then the following are equivalent:

(i) A(U) is a Dirichlet algebra on $\partial U$.

(ii) A(U) is pointwise boundedly dense in $H^\infty(U)$.

(iii) $\alpha(\Delta(z;\delta)\setminus U) \geq \delta/4$ for each $z \in \partial U$ and each $\delta > 0$.

(iv) For all but at most countably many $z \in \partial U$,

$$\liminf_{\delta \to 0} \frac{\alpha(\Delta(z;\delta)\setminus U)}{\delta} > 0.$$

(v) If $\psi_j$ is the conformal map from the unit disc onto the $j^{th}$ component of U, and $\tilde{\psi}_j$ denotes the (almost everywhere) boundary value function of $\psi_j$, then there is a Borel subset E of $\partial\Delta$ of full arc length measure, such that $\tilde{\psi}_j$ is one-to-one on E, and the sets $\tilde{\psi}_j(E)$ are disjoint $(j \geq 1)$.

The equivalence of (ii), (iii) and (iv) follows from Theorem 18, together with the estimate relating analytic capacity and diameter of a continuum. The constant 1/4 appearing here is Koebe's constant, and it is sharp. The fact that (i) implies (ii) follows from Theorem 21, while the reverse implication was proved by Garnett and the author in [23]. The equivalence of (v) with the other conditions is due to Davie [14].

Theorem 22 is also valid for open subsets of the extended complex plane $C^*$. A similar theorem describes when the uniform closure of ReA(U) has finite codimension in $C_R(\partial U)$. The precise results are announced in [22]. An application of Theorem 22 to a problem in quasiconformal mapping is given in [23].

If we translate parts (i) through (iv) to the algebra R(K), we obtain the following.

Theorem 23. Suppose each component of K is simply connected. Then the following are equivalent:

(i) R(K) is a Dirichlet algebra on $\partial K$.

(ii) R(K) is pointwise boundedly dense in $H^\infty(K^0)$, and $R(\partial K) = C(\partial K)$.

(iii) $\gamma(\Delta_\delta \setminus K) \geq \delta/4$ for all discs $\Delta_\delta$ of radius $\delta$ centered on $\partial K$.

(iv) For all but at most countably many $z \in \partial K$,

$$\liminf_{\delta \to 0} \frac{\gamma(\Delta(z;\delta)\setminus K)}{\delta} > 0.$$

Moreover, if these equivalent assertions are fulfilled, then $R(K) = A(K)$.

The problem arises of determining just how large can be the exceptional set of criterion (iv), at which the pointwise estimate need not be required. In connection with this problem, Davie has shown (cf. [11]) that if every component of $K^0$ is simply connected, and if the inner boundary of K has zero 1/2-dimensional Hausdorff measure, then R(K) is a Dirichlet algebra on $\partial K$. It is conjectured that "1/2" can be replaced here by "1".

Suppose now that $\mathbb{C}\setminus K$ is connected. The analytic polynomials are then dense in R(K), by Runge's theorem, and the estimate relating the analytic capacity and the diameter of a continuum shows that condition (iii) is fulfilled at each $z \in \partial K$. That (iii) implies (i) then yields the Walsh-Lebesgue Theorem, asserting that harmonic polynomials in x and y are dense in

$C_R(\partial K)$. That (iii) implies (ii) yields the Farrell-Rubel-Shields Theorem, and the conclusion $R(K) = A(K)$ is Mergelyan's Theorem. While Theorem 23 is stated as a concrete theorem in approximation theory, its proof involves a good deal of functional analysis.

There are two more technical consequences of Theorem 23, which were proved by D. Sarason and which will be referred to in the next section.

Corollary. If $\{K_n\}$ is a decreasing sequence of compact subsets of $\mathbb{C}$ such that $R(K_n)$ is a Dirichlet algebra on $\partial K_n$, and if $K = \cap K_n$, then $R(K)$ is a Dirichlet algebra on $\partial K$.

Proof: The validity of the estimate (iii) for the $K_n$ easily yields the estimate (iv) for K.

Corollary. If R(K) is a Dirichlet algebra on $\partial K$, and if L is a compact subset of $\mathbb{C}$ such that $L \subset K$, and every component of $K \backslash L$ reaches out to $\partial K$, then R(L) is a Dirichlet algebra on $\partial L$.

Proof: If $z \in (\partial L) \cap (\partial K)$, then we obtain the estimate (iii) for L from the corresponding estimate for K. If $z \in (\partial L) \backslash \partial K$, then we obtain the estimate (iv) for L from the estimate comparing analytic capacity and diameter.

## 11. Weak-Star Approximation by Polynomials

This section is devoted to the following problem: Given a compactly supported measure $\nu$ on $\mathbb{C}$, what is the weak-star closure of the analytic polynomials in $L^\infty(\nu)$? This problem was solved by Sarason [42], and we will sketch his solution. The more difficult problem of determining when the analytic polynomials are dense in $L^2(\nu)$ remains unsolved.

Actually, Sarason's method of proof applies to a more general problem, namely, to describe the weak-star closure of R(K) in $L^\infty(\nu)$, where R(K) is a Dirichlet algebra, and $\nu$ is a measure on K. The polynomial case can be obtained from the general case by taking K to be a closed disc containing the support of $\nu$.

Fix a K such that R(K) is a Dirichlet algebra on $\partial K$. Let $Q_1, Q_2, \ldots$ be the components of $K^0$, and let $\mu_j$ be the harmonic measure on $\partial Q_j$ for some suppressed point of $Q_j$. Since there are no real measures on $\partial K$ orthogonal to R(K), $\mu_j$ is the unique representing measure on $\partial K$ for that fixed point of $Q_j$. From this, it is not difficult to see that the minimal reducing band of measures on K corresponding to the Gleason part $Q_j$ consists of precisely the measures $\nu$ on $\overline{Q}_j$ such that $\nu|_{\partial Q_j} \ll \mu_j$. The direct sum of these bands coincides with the band generated by the measures in K orthogonal to R(K), and this band will be denoted by $\mathcal{B}_K$.

For $\nu \in \mathcal{B}_K$, define $H^\infty(K^0,\nu)$ to be the linear subspace of $L^\infty(\nu)$ consisting of those functions which are a.e. $(d\nu)$ limits of bounded sequences in R(K). Since ball R(K) is weak-star dense in ball $R(K)^{**}$, it is easy to see that $H^\infty(K^0,\nu)$ is the projection of $R(K)^{**}$ into $L^\infty(\nu)$. If $\eta$ is a measure on K, and if $\eta_a$ is the component of $\eta$ in $\mathcal{B}_K$, then the space of pointwise bounded limits of R(K) in $L^\infty(\eta)$ coincides with $L^\infty(\eta - \eta_a) \oplus H^\infty(K^0,\eta_a)$.

The <u>hull</u> of a measure $\nu \in \mathcal{B}_K$ is defined to be the union of the closed support of $\nu$ and the set of $z \in K^0$ such that $|f(z)| \le \|f\|_{L^\infty(\nu)}$ for all $f \in H^\infty(K^0,\nu)$. The hull of $\nu$ is a closed subset of K, and every component of K\hull $\nu$ reaches out to $\partial K$.

Now we are ready to describe Sarason's solution. We are assuming that R(K) is a Dirichlet algebra on $\partial K$, and that $\nu$ is a measure on K. For each ordinal number $\omega$, we define a compact set $K_\omega$ such that $R(K_\omega)$ is a Dirichlet algebra on $\partial K_\omega$, and a measure $\nu_\omega$ on $K_\omega$, as follows:

(1) $\nu_\omega$ is the component of $\nu$ in $\mathcal{B}_{K_\omega}$.

(2) $K_1$ is the $R(K)$-convex hull of the closed support of $\nu$, that is, $K_1$ consists of all $z \in K$ such that $|f(z)| \leq \|f\|_{L^\infty(\nu)}$ for all $f \in R(K)$.

(3) If $\omega > 1$ has an immediate predecessor $\sigma$, then $K_\omega$ is the hull of $\nu_\sigma$.

(4) If $\omega$ is a limit ordinal, then $K_\omega = \cap\{K_\sigma : \sigma < \omega\}$.

Note that the compact sets $K_\omega$ are decreasing, while the bands $\mathcal{B}_{K_\omega}$ are also decreasing. The fact that $R(K_\omega)$ is a Dirichlet algebra is proved by induction, using the second corollary to Theorem 23 when $\omega$ has an immediate predecessor, and the first corollary to Theorem 23 when $\omega$ is a limit ordinal.

For each ordinal $\omega > 1$, let $B_\omega$ be the subspace of $L^\infty(\nu)$ defined by induction to consist of all $f \in L^\infty(\nu)$ for which there is a sequence in $\cup\{B_\sigma : \sigma < \omega\}$ which is bounded in $L^\infty(\nu)$, and which converges a.e. $(d\nu)$ to $f$. The definition of $B_1$ is the same, except that $\cup\{B_\sigma : \sigma < \omega\}$ is replaced by $R(K)$. The $\{B_\omega\}$ form an increasing family of linear subspace of $L^\infty(\nu)$, and each $B_\omega$ is contained in the weak-star closure of $R(K)$ in $L^\infty(\nu)$. The main theorem is the following.

<u>Theorem 24.</u> Suppose $R(K)$ is a Dirichlet algebra on $\partial K$, and that $\nu$ is a measure on $K$. If $K_\omega, \nu_\omega$ and $B_\omega$ are defined as above, then

$$B_\omega = L^\infty(\nu - \nu_\omega) \oplus H^\infty(K_\omega^0, \nu_\omega)$$

for each ordinal $\omega$. If $\tau$ is the first ordinal such that $K_\tau = K_{\tau+1}$, then $\tau$ is at most countable, and the weak-star closure of $R(K)$ in $L^\infty(\nu)$ coincides with $B_\tau$:

$$H^\infty(\tau) = L^\infty(\nu - \nu_\tau) \oplus H^\infty(K_\tau^0, \nu_\tau).$$

It turns out that this procedure can terminate at any

countable ordinal $\tau$. For details and related items, see [38], [40], [41] and [42].

As Sarason pointed out, there are two cases of especial interest. Recall that a subset S of U is a <u>dominating set for</u> $H^\infty(U)$ if $\|f\|_U = \|f\|_S$ for all $f \in H^\infty(U)$. Also, the weak-star topology of $H^\infty(U)$ is understood to be the weak-star topology inherited from $L^\infty(\lambda_U)$, where $\lambda_U$ is the area measure on U. The two special cases are obtained by applying Theorem 24 to the area measure $\lambda_U$ and to a discrete measure. The respective results are as follows.

<u>Corollary</u>. Let V be a bounded open subset of C, and let W be the largest open subset of C such that V is a dominating set for $H^\infty(W)$. Then the weak-star closure of the analytic polynomials in $H^\infty(V)$ coincides with the restriction of $H^\infty(W)$ to V.

<u>Corollary</u>. Let S be a bounded sequence in C and let W be the largest open subset of C such that $S \cap W$ is a dominating set for $H^\infty(W)$. Then the weak-star closure of the analytic polynomials in $\ell^\infty(S)$ is isomorphic to $\ell^\infty(S\backslash W) \oplus H^\infty(W)$.

The results of this section can be applied to operator theory. They have a bearing on the problem of determining which normal operators on Hilbert space are completely normal, in the sense that the complement of every invariant subspace is also invariant.

## 12. Approximation on Product Sets

In this section, let $K_1$ and $K_2$ be fixed compact subsets of C, and let $A_1$ and $A_2$ be fixed T-invariant uniform algebras on $K_1$ and $K_2$ respectively. We will discuss the tensor product $A = A_1 \otimes A_2$, regarded as a uniform algebra on $K_1 \times K_2$. The main theorems are due to B. Cole (bidisc algebra) and O. Bekken (general case, see [4]). Analogous results, sometimes more complicated to state, are valid for tensor products of an arbitrary finite number of T-invariant uniform algebras.

By definition, A is the uniform closure in $C(K_1 \times K_2)$ of finite linear combinations of functions of the form $f(z)g(w)$, where $f \in A_1$ and $g \in A_2$. Related to the tensor product, there is the slice algebra $A_1 \# A_2$, consisting of functions $f \in C(K_1 \times K_2)$ such that $f(\cdot, w) \in A_1$ for each fixed $w \in K_2$, and $f(z, \cdot) \in A_2$ for each fixed $z \in K_1$. As indicated in Section 1, the constructive techniques show that T-invariant algebras have the Banach approximation property. This is Eifler's theorem, and a corollary is that

$$A_1 \otimes A_2 = A_1 \# A_2.$$

This identity can also be deduced from a vector-valued version of Vitushkin's Theorem (see [21]).

Let $Q_1$ and $Q_2$ denote the set of nonpeak points for $A_1$ and $A_2$ respectively, and set $Q = Q_1 \times Q_2$. Consider the following five bands of measures on $K_1 \times K_2$:

(1) $\mathcal{B}_Q$ is the band generated by representing measures for points of Q.

(2) $\mathcal{P}_1$ is the band generated by measures in $A^\perp$ of the form $\delta_p \times \eta$, where p is a peak point for $A_1$, and $\eta \in A_2^\perp$. This coincides with the band generated by the representing measures for points of $(K_1 \setminus Q_1) \times Q_2$.

(3) $\mathcal{J}_1$ is the band generated by measures $\nu \in A^\perp$ such that $|\nu|(\{p\} \times K_2) = 0$ for all $p \in K_1$; and $\nu$ is carried by a set of the form $E \times K_2$, where E is a subset of $K_1$ which has zero mass for all measures in $A_1^\perp$.

(4 and 5) $\mathcal{P}_2$ and $\mathcal{J}_2$ are defined as $\mathcal{P}_1$ and $\mathcal{J}_1$, except that the roles of the variables are interchanged.

Using peak set theory and results on Gleason parts, it is easy to see that these five bands are disjoint reducing bands. The idea of the following theorem is due to Cole, who studied the bidisc algebra. The extension of Cole's results to T-invariant algebras is due to O. Bekken [4].

Theorem 25. The band generated by the measures in $A^\perp$ coincides with $\mathcal{B}_Q \oplus \mathcal{P}_1 \oplus \mathcal{P}_2 \oplus \mathcal{J}_1 \oplus \mathcal{J}_2$.

The band $\mathcal{B}_Q \oplus \mathcal{P}_1 \oplus \mathcal{P}_2$ is the band generated by the representing measures for the nonpeak points of A. The band $\mathcal{J}_1 \oplus \mathcal{J}_2$ coincides with the "infinitely divisible" band $\mathcal{J}$ of Theorem 11. From that theorem one can deduce the following corollary, which has some abstract significance.

Corollary. Every extreme point of ball $A^\perp$ is absolutely continuous with respect to a representing measure for some nonpeak point of A.

Now let $\lambda_Q = \lambda_{Q_1} \times \lambda_{Q_2}$ be the volume measure on $Q_1 \times Q_2$. The analogue of Davie's theorem is valid in this setting, as are the distance estimates.

Theorem 26. The projection $H^\infty(\mathcal{B}_Q) \to H^\infty(\lambda_Q)$ is an isometric isomorphism and a weak-star homeomorphism. If $f \in H^\infty(\lambda_Q)$, then there is a sequence $\{f_n\}$ in A such that $\|f_n\| \leq \|f\|$, and $\{f_n\}$ converges a.e. $(d\lambda_Q)$ to f.

Theorem 27. Suppose $Q_1$ and $Q_2$ are dense in $K_1$ and $K_2$ respectively. Then

$$d(h,A) = d(h,H^\infty(\lambda_Q)), \qquad \text{all } h \in C(K_1 \times K_2).$$

Equivalently, ball$(A^\perp \cap L^1(\lambda_Q))$ is weak-star dense in ball $A^\perp$.

As before, the algebra $H^\infty(\lambda_Q)$ can be identified with an algebra of point functions in Q, namely, the algebra B(Q) of all functions f on Q which are pointwise limits on Q of bounded sequences in A. Bekken has characterized the functions in B(Q) as follows.

Theorem 28. Let f be a bounded Borel function on Q. Then

$f \in E(Q)$ if and only if $f(\cdot,w) \in H^\infty(\lambda_{Q_1})$ for all $w \in Q_2$, and $f(z,\cdot) \in H^\infty(\lambda_{Q_2})$ for all $z \in Q_1$.

The analogues of the preceding theorems are also valid for the algebra $A(U \times V)$, regarded as a uniform algebra on $\partial U \times \partial V$, with $\lambda_U \times \lambda_V$ replaced by the product of harmonic measures on $\partial U$ and $\partial V$. There are also theorems for $R(K_1 \times K_2)$, regarded as a uniform algebra on $\partial K_1 \times \partial K_2$.

In the case of the bidisc algebra $A(\Delta \times \Delta)$, regarded as a subalgebra of $C(\partial\Delta \times \partial\Delta)$, one can say more about the measures in the slice bands. A combination of the F. and M. Riesz theorem and the technique of disintegration of measures leads to the following result. The generic point of $\partial\Delta \times \partial\Delta$ is denoted by $(e^{i\theta}, e^{i\psi})$.

Theorem 29. Suppose $\eta$ is a measure on $\partial\Delta \times \partial\Delta$ which is orthogonal to the bidisc algebra $A(\Delta \times \Delta)$. Then

$$\eta = \eta_0 + \eta_1 + \eta_2,$$

where $\eta_0, \eta_1, \eta_2$ are mutually singular measures in $A(\Delta \times \Delta)^\perp$; $\eta_0 \in \mathcal{B}_{\Delta\times\Delta}$ [that is, $\eta_0$ is absolutely continuous with respect to some representing measure for $(0,0)$]; $\eta_1 = h(\theta,\psi)d\tau(\theta)d\psi$, where $\tau$ is a measure on $\partial\Delta$ carried by a set of zero arc length, $h \in L^1(\tau \times d\psi)$, and $h(\theta,\cdot)$ belongs to the Hardy space $H_0^1(d\psi)$ for each fixed $\theta$; and $\eta_2$ has a description similar to $\eta_1$, except that the roles of $\theta$ and $\psi$ are interchanged.

One consequence of this theorem is that the only real measures on $\partial\Delta \times \partial\Delta$ orthogonal to $A(\Delta \times \Delta)$ lie in the band $\mathcal{B}_{\Delta\times\Delta}$. As another consequence, one obtains a decomposition theorem for real measures $\nu$ on $\partial\Delta \times \partial\Delta$ which satisfy

$$\int z^n w^m d\nu = 0, \qquad n,m \geq 1.$$

In fact, applying the preceding decomposition to $\eta = zw\nu$, one finds that any such $\nu$ can be written as

$$\nu = \nu_0 + (\tau_1 \times d\psi) + (d\theta \times \tau_2),$$

where $\nu_0 \in \mathcal{B}_{\Delta\times\Delta}$, and $\tau_1$ and $\tau_2$ are real measures carried by sets of zero arc length. A change of variables $(z,w) \rightarrow (z,\overline{w})$ leads to the Ahern decomposition [1] of RP measures on $\partial\Delta \times \partial\Delta$.

These ideas can be applied to operator theory, and one application to operator model theory is given by D. Clark [9]. In another application, E. Briem, Davie and Øksendal [6] have obtained a functional calculus for a pair of commuting completely nonunitary contractions S, T on a Hilbert space. The point is that the joint spectral measure of a joint unitary dilation of S and T, while not necessarily absolutely continuous with respect to area measure $d\theta \times d\psi$, does lie in the band $\mathcal{B}_{\Delta\times\Delta}$. The isomorphism $H^\infty(\mathcal{B}_{\Delta\times\Delta}) \cong H^\infty(\Delta \times \Delta)$ then allows one to define $f(S,T)$ for any $f \in H^\infty(\Delta \times \Delta)$, and the correspondence $f \rightarrow f(S,T)$ has desirable properties.

Acknowledgments

I would like to acknowledge several conversations with B. Cole and several letters from B. Cole. It is his point of view which permeates the latter half of this report.

[1]Partially supported by N.S.F. Grant # GP 33693X.

## References

As a rule, the primary sources which are referenced in [20] are not included in the following list.

[1] Ahern, P., Inner functions on the polydisc and measures on the torus. Michigan Math. J., 20 (1973), 33-37.

[2] Arens, R., Operations induced in function classes. Monatsh. Math. 55 (1951), 1-19.

[3] Bekken, O., Products of algebras on plane sets. U.C.L.A. Thesis, 1972.

[4] Bekken, O., Rational approximation on product sets.

Preprint.

[5] Brennan, J., Invariant subspaces and weighted polynomial approximation. Arkiv för Matematik. To appear.

[6] Briem, E., A. Davie and B. Øksendal, A functional calculus for pairs of commuting contractions. Preprint.

[7] Browder, A., Introduction to Function Algebras. Benjamin, 1969.

[8] Carleson, L., Selected Problems on Exceptional Sets. Van Nostrand, 1967.

[9] Clark, D., Some star-invariant subspaces in two variables. Duke Math. J. 39 (1972), 539-550.

[10] Cole, B. and M. Range, A-measures on complex manifolds and some applications. J. Functional Analysis 11 (1972), 393-400.

[11] Davie, A., Real annihilating measures for R(K). J. Functional Analysis 6 (1970), 357-386.

[12] Davie, A., Bounded approximation and Dirichlet sets. J. Functional Analysis 6 (1970), 460-467.

[13] Davie, A., An example on rational approximation. Bull. London Math. Soc. 2 (1970), 83-86.

[14] Davie, A., Dirichlet algebras of analytic functions. J. Functional Analysis, 6 (1970), 348-356.

[15] Davie, A., Bounded limits of analytic functions. Proc. A.M.S. 32 (1972), 127-133.

[16] Davie, A., Analytic capacity and approximation problems. Trans. A.M.S. To appear.

[17] Davie, A., T. Gamelin and J. Garnett, Uniform approximation of bounded analytic functions. Trans. A.M.S. To appear.

[18] Davie, A. and D. Wilkin, An extension of Melnikov's theorem. J. Functional Analysis 11 (1972), 179-183.

[19] Eifler, L., The slice product of function algebras. Proc. A.M.S. 23 (1969), 559-564.

[20] Gamelin, T., Uniform Algebras. Prentice-Hall, 1969.

[21] Gamelin, T. and J. Garnett, Constructive techniques in rational approximation. Trans. A.M.S. 143 (1969), 187-200.

[22] Gamelin, T. and J. Garnett, Pointwise bounded approximation and hypodirichlet algebras. Bull. A.M.S. 77 (1971), 137-141.

[23] Gamelin, T. and J. Garnett, Pointwise bounded approximation and Dirichlet algebras. J. Functional Analysis 8 (1971), 360-404.

[24] Gamelin, T. and J. Garnett, Uniform approximation to bounded analytic functions. Rev. Unión Mat. Arg. 25 (1970), 87-94.

[25] Gamelin, T. and J. Garnett, Bounded approximation by rational functions. Pacific J. Math. To appear.

[26] Garnett, J., Analytic capacity and measure. Lecture Notes in Math., Vol. 297, Springer, 1972.

[27] Hedberg, L., Approximation in the mean by analytic functions. Trans. A.M.S. 163(1972), 157-171.

[28] Hedberg, L., Approximation in the mean by solutions of elliptic equations, Uppsala Univ. Math. Dept. Report No. 38(1972)

[29] Henkin, G. M., Integral representations of holomorphic functions in strongly pseudoconvex domains and certain applications. Mat. Sbornik 78 (120), (1969), 611-632; Math. U.S.S.R., Sb. April 1969, 7 (4), 597-616.

[30] Kerzman, N., Hölder and $L^p$ estimates for solutions of $\bar{\partial}u = f$ in strongly pseudoconvex domains. Comm. Pure and Appl. Math. 24 (1971), 301-379.

[31] Konig, H. and G. Seever, The abstract F. and M. Riesz theorem. Duke Math. J. 36 (1969), 791-797.

[32] Mergelyan, S. N., Uniform approximation to functions of a complex variable. Uspehi Mat. Nauk 7 (1952), no. 2 (48), 31-122, A.M.S. Transl., Ser. 1, Vol. 101.

[33] O'Farrell, A., Thesis. Brown University, 1973.

[34] Øksendal, B., R(X) as a Dirichlet algebra and representation of orthogonal measures by differentials. Math. Scand. 29 (1971), 87-103.

[35] Øksendal, B., Null sets of measures orthogonal to R(X). Amer. J. Math. 94 (1972), 331-342.

[36] Polking, J., Approximation in $L^p$ by solutions of elliptic partial differential equations. Preprint.

[37] Range, M., Approximation to bounded holomorphic functions on strictly pseudoconvex domains. Pacific J. Math. 41 (1972), 203-213.

[38] Sarason, D., Weak-star generators of $H^\infty$. Pacific J. Math. 17 (1966), 519-528.

[39] Sarason, D., Generalized interpolation in $H^\infty$. Trans. A.M.S., 127 (1967), 179-203.

[40] Sarason, D., On the order of a simply connected domain. Michigan Math. J. 15 (1968), 129-133.

[41] Sarason, D., A remark on the weak-star topology of $\ell^\infty$. Studia Math. 30 (1968), 355-359.

[42] Sarason, D., Weak-star density of polynomials. J. Reine Angew. Math. 252 (1972), 1-15.

[43] Seever, G., Algebras of continuous functions on hyper-stonian spaces. Preprint.

[44] Stout, L., The Theory of Uniform Algebras. Bogden and Quigley, 1971.

[45] Vitushkin, A. G., Estimates of the Cauchy integral. Mat. Sbornik 71 (113), (1966), 515-535, A.M.S. Transl. Ser. 2, 80 (1969), 257-278.

[46] Vitushkin, A. G., Analytic capacity of sets and problems in approximation theory. Uspehi Mat. Nauk 22, No. 6 (138), (1967), 141-199. Russian Math. Surveys 22 (1967), 139-200.

[47] Wermer, J., Seminar über Funktionen-Algebren. Lecture Notes in Math., Vol. 1, Springer Verlag, 1964.

[48] Zalcman, L., Analytic Capacity and Rational Approximation. Lecture Notes in Math., Vol. 50, Springer Verlag, 1968.

[49] Davie, A., Bounded approximation by analytic functions. J. Approx. Theory 6 (1972), 316-319.

Department of Mathematics
University of California, Los Angeles
Los Angeles, California 90024

# TOPICS IN MULTIVARIATE APPROXIMATION THEORY

Joseph W. Jerome[1]

## 0. Introduction

In the first of the six sections of this paper are discussed generalized Peano kernel theorems, including the moment matching and change of scale results and their implications. Methods of spline approximation by piecewise polynomials of fixed degree in each variable, employing local quasi-interpolation methods and multidimensional thin support splines, follow in section two. In section three are considered the approximation theory of the finite element method and associated strictly local approximation schemes. The nodal finite element method is sketched. This is followed in section four by the topics of approximation in $R^N$ via rectangular grids and extrapolation. Included here are equivalence theorems via Fourier analysis for optimal convergence as well as spline and generalized cardinal interpolation and their connections. In addition, extension theorems related to the Calderon and Whitney results are discussed. In section five, estimates involving the exact asymptotic order of the widths and entropies of the unit balls of Sobolev spaces imbedded in spaces of lower (possibly negative) order are presented. Finally, in section six, a brief discussion is given of estimates for the convergence of Galerkin methods for the approximate solution of boundary value and eigenvalue problems, as well as certain elliptic generalizations of univariate spline functions.

Additional topics are discussed, all with appropriate

references. The format of sectional bibliographies is in keeping with that of the excellent monograph of Richard Varga [0.18]. We mention here a number of interesting papers which did not naturally fit into the sectional scheme or were not available to the writer, including the interpolation results of Barnhill and Wixom [0.1], Barnhill, Birkhoff and Gordon [0.2], Ehlich and Haussmann [0.4], Ferguson [0.5], Guenther and Roetman [0.7], Haussmann [0.8,0.9], Salzer [0.13, 0.14], Thacher [0.16] and Thacher and Milne [0.17]. We cite also the multidimensional smoothing contributions of Munteanu [0.10] and Nielson [0.11], the least squares methods of Buchanen and Thomas [0.3] and Greville [0.6], the Tchebycheff approximation results of Rice [0.12] and the related work of Weinstein [0.19], as well as the applications of Sakai [0.15].

The writer wishes to acknowledge gratefully a number of helpful conversations prior to the preparation of this manuscript, viz., with Garrett Birkhoff, Carl de Boor, James Bramble, Michael Golomb and Gilbert Strang. Also, Larry Schumaker generously made available to the writer his bibliography of over 1,200 items which proved extremely helpful in the literature search. There follows a listing and explanation of symbols.

$\alpha = (\alpha_1,\ldots,\alpha_N)$ and $\gamma = (\gamma_1,\ldots,\gamma_N)$ are standard multi-integers with nonnegative entries.

$j = (j_1,\ldots,j_N)$ is always a multi-integer.

$D^\alpha = \left(\frac{\partial}{\partial x_1}\right)^{\alpha_1} \cdots \left(\frac{\partial}{\partial x_N}\right)^{\alpha_N}$ is taken in the sense of distributions; $x^\gamma = x_1^{\gamma_1} \cdots x_N^{\gamma_N}$, $x \in R^N$, and $|\alpha| = \sum_i \alpha_i$.

$P_k$ denotes the (real) polynomials of degree $\leq k$. For $m \geq 0$, $1 \leq p \leq \infty$ and $\Omega$ a domain in $R^N$, the (usually real) Sobolev spaces of integral order are defined:

(0.1i) $\quad W^{m,p}(\Omega) = \{f : D^\alpha f \in L^p(\Omega),\ 0 \le |\alpha| \le m\}$,

(0.1ii) $$|f|_{\ell,p} = \begin{cases} \{\int_\Omega \sum_{|\alpha|=\ell} |D^\alpha f|^p\}^{1/p}, & 1 \le p < \infty, \\ \max_{|\alpha|=\ell} (\operatorname{ess\,sup}_{x \in \Omega} |D^\alpha f(x)|), & p = \infty, \end{cases}$$

(0.1iii) $$\|f\|_{m,p} = \begin{cases} \left\{\sum_{\ell=0}^{m} |f|^p_{\ell,p}\right\}^{1/p}, & p < \infty, \\ \max_{0 \le \ell \le m} |f|_{\ell,\infty}, & p = \infty. \end{cases}$$

$W_0^{m,p}(\Omega)$ is the completion in $W^{m,p}(\Omega)$ of the class $C_0^\infty(\Omega)$ of infinitely differentiable functions with compact support in $\Omega$ and $W_c^{m,p}(\Omega)$ consists of the linear subspace of $W^{m,p}(\Omega)$ of compact support functions. For $s \ge 0$ and s integral, we define $W^{s,p}(\Omega)$ by (0.1), whereas for nonintegral s and $1 \le p < \infty$

(0.2i) $\quad W^{s,p}(\Omega) = \{f \in W^{[s],p}(\Omega) : |f|_{s,p} < \infty\}$,

(0.2ii) $$|f|_{s,p} = \left\{\int_\Omega \int_\Omega \sum_{|\alpha|=[s]} \frac{|D^\alpha f(x) - D^\alpha f(y)|^p}{|x-y|^{p(s-[s])+N}}\, dxdy\right\}^{1/p},$$

(0.2iii) $$\|f\|_{s,p} = \{\|f\|^p_{[s],p} - |f|^p_{s,p}\}^{1/p}$$

$W^{s,\infty}(\Omega)$ and $W_0^{s,p}(\Omega)$ are defined in the obvious way. For $s<0$ and $1/p + 1/q = 1$ we define the duality relation

(0.3) $\quad W^{s,p}(\Omega) = (W_C^{-s,q}(\Omega))'$.

For $r \ge 0$ we define

(0.4i) $\quad W^{r,2}(R^N) = \{f : \|f\|_{r,2} < \infty\}$,

(0.4ii) $\quad \|f\|_{r,2} = \{\int_N (1 + |x|^2)^r |\hat{f}(x)|^2 dx\}^{1/2}$

where ^ denotes Fourier transform (cf. section 4).

$W^{s,2}(R^N) = W^{r,2}(R^N)$ as a set with equivalent norms if $s=r$.

$C^k(\Omega)$, $k \geq 0$, is the usual space of k-times continuously differentiable functions on $\Omega$. Moreover,

$$(0.5) \qquad C^{m_1,\ldots,m_N} = \{f : D^\alpha f \in C(\Omega),\ 0 \leq \alpha_i \leq m_i,\ 1 \leq i \leq N\}.$$

Finally, Z is the set of integers and S is the space of functions rapidly decreasing at infinity.

## References


[0.1] Barnhill, R. E. and J. A. Wixom, An error analysis for the bivariate interpolation of analytic functions, SIAM J. Numer. Anal. 6 (1969), 450-457.

[0.2] Barnhill, R.E., G. Birkhoff and W. J. Gordon, Smooth interpolation in triangles, J. Approx. Theory. To appear.

[0.3] Buchanen, J.E. and D. H. Thomas, On least squares fitting of two-dimensional data with a special structure. SIAM J. Numer. Anal. 5 (1968), 252-257.

[0.4] Ehlich, H. and W. Haussmann, Konvergenz mehrdimensionaler Interpolation, Numer. Math. 15 (1970), 165-174.

[0.5] Ferguson, J., Multivariable curve interpolation. J. Asso. Comp. Mach. 11 (1964), 221-228.

[0.6] Greville, T. N. E., Note on fitting of functions of several independent variables. J. Soc. Indust. Appl. Math. 9 (1961), 109-115.

[0.7] Guenther, R. B. and E. L. Roetman, Some observations on interpolation in higher dimensions. Math. Comp. 24 (1970), 517-522.

[0.8] Haussmann, W., Tensorprodukte und mehrdimensionale Interpolation. Math. Z. 113 (1970), 17-23.

[0.9] __________, Tensorproduktmethoden bei mehrdimensionaler Interpolation. Math. Z. 124 (1972), 191-198.

[0.10] Munteanu, M. J., Contributions à la Theorie des Fonctions Splines a Une et a Plusieurs Variables, dissertation, University of Louvain, Belgium, 1970.

[0.11] Nielson, G. M., Surface approximation and data smoothing using generalized spline functions, dissertation, University of Utah, Salt Lake City, Utah, 1970.

[0.12] Rice, J. R., Tchebycheff approximation in several variables. Trans. Amer. Math. Soc. 109 (1963), 444-466.

[0.13] Salzer, H. E., Note on multivariate interpolation for unequally spaced arguments, with an application to double summation. J. Soc. Indust. Appl. Math. 5 (1957), 254-262.

[0.14] ________, Formulas for bivariate hyperosculatory interpolation. Math. Comp. 25 (1971), 119-133.

[0.15] Sakai, M., Multidimensional cardinal spline function and its applications. Mem. Fac. Sc. Kyushu Univ. 24 (1970), 40-46.

[0.16] Thacher, H. C., Derivation of interpolation formulas in several independent variables. Annals N. Y. Acad. Sc. 86 (1960), 758-775.

[0.17] Thacher, H. C. and W. E. Milne, Interpolation in several variables. J. Soc. Indust. Appl. Math. 8 (1960), 33-42.

[0.18] Varga, R. S., Functional Analysis and Approximation Theory in Numerical Analysis, CBMS Regional Conf. Ser. in Appl. Math., SIAM (publ.), 1971.

[0.19] Weinstein, S. E., Product approximations of functions of several variables. SIAM J. Numer. Anal. 3 (1971), 178-189.

## 1. Generalized Peano Kernel Theorems, Remainders and Convergence

Remainders and their representations, or better, their estimation lie at the heart of approximation theory. Sard, in a series of important papers [1.12, 1.13, 1.14, 1.15, 1.16, 1.17] has examined function classes defined in terms of generalized Taylor representations with integral remainders, determined their duals and, somewhat more generally, has investigated factorizations of a prescribed linear operator R into the product QU, where U is a specified operator, frequently involving differentiation, and where kernel R $\supset$ kernel U. If R is in fact the difference between a given linear operator T and an approximation exact on the kernel of U, then the problem of best approximation involves the optimal choice of Q. This approach, which is equivalent to the hypercircle approach of Golomb and Weinberger [1.6], has been intensively investigated and we refer the reader to the blending methods employed by Barnhill and Mansfield [1.1] and by Gordon [1.7] and the reproducing kernel methods of Mansfield [1.8, 1.9] as well as to the approaches of Ritter [1.11] and Stancu [1.18]. The bibliography of [1.18] contains additional references not mentioned here. We may refer to such methods, which utilize explicit representations for remainder analysis, as methods of Peano kernel type. We wish to devote the rest of this section to remainder analysis achievable without representation theory for very broad function classes, viz., the Sobolev classes. This involves what has come to be called the "Bramble-Hilbert Lemma," i.e., a generalized Peano kernel theorem, and we shall give the background for this now.

Let $\Omega$ be a bounded strongly Lipschitz domain in $R^N$ in the sense of Morrey [1.10, p. 72]. This is equivalent to the notion of regular in the sense of Calderon, defined in terms of a certain cone condition. Included, for example, are bounded, open convex domains and certain unions of such domains. Now if h is the diameter of $\Omega$, $p\geq 1$, and $P = Tu$ is the unique polynomial of total degree k-1 defined by,

(1.1) $\int_\Omega D^\alpha P = \int_\Omega D^\alpha u,\ 0\leq|\alpha|\leq k-1,\ u \in W^{k,p}(\Omega),$

then the estimate

(1.2) $|u-Tu|_{\ell,p} \leq Ch^{k-\ell}|u|_{k,p},\ 0\leq\ell\leq k-1,$

holds [1.10, p. 85]. Estimates of the form (1.2) are sometimes called "change of scale" estimates after the method often used to obtain them. They are even obtainable for nonintegral Sobolev norms. For example, in the case where $\Omega$ is a hypercube, and $s>0$, Birman and Solomjak [1.2] have approximated $W^{s,p}(\Omega)$ functions by polynomials $P = Tu$ of degree $\nu = s-1$ for integral s and $\nu = [s]$ for nonintegral s. The derivative matching conditions of (1.1) may be replaced by moment matching conditions

$$\int_\Omega P(x)x^\alpha dx = \int_\Omega u(x)x^\alpha dx,\ 0\leq|\alpha|\leq\nu,$$

as in [1.2]. The estimate

(1.3) $\|u-Tu\|_{0,p} \leq Ch^s|u|_{s,p},\ u \in W^{s,p}(\Omega),$

and even more general results involving the $L^q(\Omega)$ norm on the left hand side of (1.3) are obtained in [1.2, Lemmas 3.1, 3.2]. Estimates such as (1.3) and its generalizations, together with appropriate partitioning results, are used in [1.2] to obtain upper bounds for widths and entropies of weighted Sobolev classes (see section 5).

Bramble and Hilbert [1.3] used estimate (1.2) to obtain the equivalence of the factor norm on $W^{k,p}(\Omega)/P_{k-1}$ with

$h^k|u|_{k,p}$, where u is any representer of a given factor class. This equivalence was then used to obtain the following

<u>Lemma</u> (Bramble and Hilbert). Let F be a continuous linear functional on $W^{k,p}(\Omega)$, $p \geq 1$, such that $F(P) = 0$ for all $P \in \mathcal{P}_{k-1}$ and $\|F\|$ is independent of h. Then the estimate

(1.4) $\quad |F(u)| \leq Ch^k|u|_{k,p}, \; u \in W^{k,p}(\Omega),$

holds for some positive constant C independent of h.

Extensions of this result to spaces of smooth functions, including Lipschitz classes, were also given by Bramble and Hilbert. Perhaps the most significant extension, however, was given in a sequel [1.4] to [1.3]. Here it was shown that, if the linear functional F annihilates a class $\mathcal{P}$ intermediate between $\mathcal{P}_{k-1}$ and $\mathcal{P}_k$, then certain derivatives with indices, corresponding to monomials annihilated by F, do not appear in the semi-norm on the right hand side of (1.4). Of course, the monomials $x_\nu^k$, $\nu = 1,\ldots,N$, are not permitted to belong to $\mathcal{P}$.

Now let $\Pi$ be a continuous linear mapping of $W^{k,p}(\Omega)$ into $W^{m,p}(\Omega)$, $0 \leq m \leq k$, such that $\Pi$ is exact on $\mathcal{P}_{k-1}$. Here $1 \leq p \leq \infty$. Let $\hat{\Omega}$ be a fixed bounded, regular domain from which $\Omega$ is obtained via a nonsingular linear transformation. Let $\rho$ be the supremum of the diameters of the spheres contained in $\Omega$ with a similar meaning for $\hat{\rho}$. Ciarlet and Raviart [1.5], using essentially the fundamental lemma, have obtained the estimate

(1.5) $\quad \|u - \Pi u\|_{m,p} \leq Ch^k/\rho^m |u|_{k,p}$ ,

where $C = C(N,k,p,\hat{\Omega})\,(\hat{h}^m/\hat{\rho}^k)\,\|I - \hat{\Pi}\|$. They have observed also that, when $\hat{\Omega}$ is an equilateral N-simplex, the identity

(1.6) $\quad \hat{h}/\hat{\rho} = [N(N+1)/2]^{1/2}$

holds. Estimates such as (1.5) play a significant role in the approximation theory of the finite element method (see section 3).

## References


[1.1] Barnhill, R. E. and L. E. Mansfield, Spline-blended approximation of linear operators, to appear.

[1.2] Birman, M. S. and M. Z. Solomjak, Piecewise-polynimial approximations of functions of the classes $W_p^\alpha$. Math. USSR-Sbornik 2 (1967), 295-317.

[1.3] Bramble, J. H. and S. R. Hilbert, Estimation of linear functionals on Sobolev spaces with application to Fourier transforms and spline interpolation. SIAM J. Numer. Anal. 7 (1970), 112-124.

[1.4] ________, Bounds for a class of linear functionals, with applications to Hermite interpolation. Numer. Math. 16 (1971), 362-369.

[1.5] Ciarlet, P. G. and P. A. Raviart, General Lagrange and Hermite interpolation in $R^n$ with applications to finite element methods. Archive for Rational Mechanics and Analysis 46 (1972), 177-199.

[1.6] Golomb, M. and H. F. Weinberger, Optimal approximation and error bounds. In On Numerical Approximation, R. E. Langer, ed., Univ. of Wisconsin Press, Madison, Wisc., 1959, pp. 117-190.

[1.7] Gordon, W. J., Blending-function methods of bivariate and multivariate interpolation and approximation. SIAM J. Numer. Anal. 9 (1971), 158-177.

[1.8] Mansfield, L. E., Optimal approximation and error bounds in spaces of bivariate functions, J. Approx. Th. 5 (1972), 77-96.

[1.9] ________, On the optimal approximation of linear functionals in spaces of bivariate functions. SIAM J. Numer. Anal. 8 (1971), 115-126.

[1.10] Morrey, C. B., Multiple Integrals in the Calculus of Variations. Springer-Verlag, New York, 1966.

[1.11] Ritter, K., Two-dimensional spline functions and best approximation of linear functionals. J. Approx. Th. 3 (1970), 352-368.

[1.12] Sard, A., Remainders: functions of several variables. Acta Math. (Sweden) 84 (1951), 319-346.

[1.13] ________, Remainders as integrals of partial derivatives. Proc. Amer. Math. Soc. 3 (1952), 732-741.

[1.14] ________, New function spaces and their adjoints. Ann. New York Acad. Sc. 86 (1960), 700-757.

[1.15] ________, Function spaces. Bull. Amer. Math. Soc. 71 (1965), 397-418.

[1.16] ________, Optimal approximation. J. Funct. Anal. 1 (1967), 222-244; Addendum, 2 (1968), 368-369.

[1.17] ________, Approximation based on nonscalar observations. To appear, J. Approx. Th.

[1.18] Stancu, D. D., The remainder of certain linear approximation formulas in two variables. SIAM J. Numer. Anal. (Ser. B) 1 (1964), 137-163.

## 2. Spline Approximation on Bounded Sets by Piecewise Polynomials of Fixed Degree in Each Variable

Let $\Omega \subset R^N$ be a bounded, open connected set and let a rectangular grid be specified in $R^N$. With each point of the grid is associated at least one N-dimensional normalized B-spline which is the tensor product of one-dimensional B-splines of degree k-1, $k \geq 1$, where confluent knots of multiplicity $\leq k$ are permitted in the one-dimensional splines. Thus, if the grid points $(x_{1,j_1},\ldots,x_{N,j_N})$ are written according to multiplicity, the spline $N_k(j;\cdot)$ has support in the parallelepiped $\prod_{i=1}^{N} [x_{i,j_i}, x_{i,j_i+k}]$. We single out a finite subset $\pi$ of the grid points, whose convex hull is a rectangular polygon by the criterion that the support of $N_k(j,\cdot)$ intersect $\Omega$. The linear span of all such multivariate B-splines will comprise the finite dimensional space $S$ of approximants. De Boor and Fix [2.13] have constructed a linear approximation operator $F_\pi$ from $C^{k-1}(\Omega) \cap W^{k-1,\infty}(\Omega)$ into $S$ satisfying the following properties:

(2.1i) $F_\pi f$ is local in the sense that its value at a point x depends only on the values of f in a uniformly small neighborhood of x;

(2.1ii) $F_\pi(x^\gamma) = x^\gamma$ for $|\gamma| < k$;

(2.1iii) $\|D^\alpha(f-F_\pi f)\|_{0,\infty} \leq K_\alpha |\pi|^{k-1-|\alpha|}\omega_{k-1}(f,|\pi|)$ for $0 \leq |\alpha| \leq k-1$ provided $\Omega$ satisfies a local convexity condition, which holds for convex sets and rectangular polygons. Here

$$\omega_{k-1}(f,|\pi|) = \max_{|\gamma|=k-1} \omega(D^\gamma f;|\pi|),$$

$\omega$ denotes the modulus of continuity and $|\pi|$ is the partition modulus, measured in terms of maximum coordinate mesh lengths.

The mapping $F_\pi$ has the simple form

$$(2.2) \quad F_\pi f(x) = \sum_j \sum_{|\alpha|<k} \omega_{j,\alpha} D^\alpha f(\tau_j) N_k(j;x)$$

where the points $\tau_j$ lie in the support of $N_k(j;\cdot)$ and $\omega_{j,\alpha}$ are appropriate weights. $F_\pi f$ is called a quasi-interpolant by de Boor and Fix. If the points $\tau_j$ are appropriately centered, the estimate (2.1iii) holds with no mesh ratio restriction for $\alpha_\nu \leq [k/2]$, $1 \leq \nu \leq N$; otherwise locally bounded mesh ratio assumptions are required. Estimates are also obtained in [2.13] for functions in $W^{k,p}(\Omega)$; in this case a modified quasi-interpolant, involving derivatives of order $|\alpha| \leq q$, $q < k - N/p$, can be constructed and the estimate

$$(2.3) \quad \|f - F_\pi f\|_{\ell,p} \leq K_\ell |\pi|^{k-\ell} |f|_{k,p}, \quad 0 \leq \ell \leq q$$

is valid. The only restriction on $\Omega$ for (2.3) to hold is that there exist a continuous extension operator from $W^{k,p}(\Omega)$ to $W^{k,p}(\Gamma)$, where $\Gamma$ is a rectangular polygon containing $\overline{\Omega}$.

Some discussion is in order about the historical antecedants of this quasi-interpolation method. In [2.4] Birkhoff had introduced univariate spline moment approximation for odd degree splines. De Boor subsequently refined this method in [2.11] to include even degree splines and more general interval partitions. In [2.12], de Boor introduced a univariate alternate of [2.13]. Using tensor product methods, Schultz [2.30] subsequently generalized the results of [2.12] to rectangular parallelepipeds. The equivalence of the approaches of [2.11] and [2.13] is demonstrated in [2.13] in the univariate case. It is of interest whether moment methods are equivalent in the multivariate case. Some evidence to support this has been found in certain moment matching results of section one.

The great strength of the quasi-interpolation method of

de Boor and Fix is the wide variety of regions $\Omega$ for which the method is applicable. Only function values in $\Omega$ are required for the construction of the quasi-interpolant. We wish to discuss now other methods of approximation for special regions. We begin with rectangular parallelepipeds.

Let $I = \prod_{\ell=1}^{N} I_\ell$, $I_\ell$ a compact interval and suppose that separate partitionings $\pi_\ell$, inclusive of endpoints of each interval $I_\ell$, are prescribed. Thus a partition $\pi = \prod_{\ell=1}^{N} \pi_\ell$ of points in $I$ is induced. For fixed positive integers $m_1,\ldots,m_N$ let $S \subset C^{m_1-1,\ldots,m_N-1}$ be a finite dimensional space of piecewise polynomials of degree $2m_\ell - 1$ in each variable $x_\ell \in I_\ell$ which is specified by a tensor product scheme of univariate cardinal spline functions on each interval $I_\ell$. We briefly discuss these cardinal splines for certain Hermite-Birkhoff interpolation schemes on an interval.

Let $a = t_0<t_1<\ldots<t_n = b$ be a partition of $[a,b]$ and let an $(n+1)$-row by m column incidence matrix $(e(i,\nu))$, $0\le i\le n$, $0\le\nu\le m-1$, be given where $e(i,\nu) = 1$ or $0$ depending upon whether the $\nu^{th}$ derivative is specified at $t_i$ or not. We shall assume $n\ge m-1$ and $e(i,0) = 1$ for each $i=0,\ldots,n$. Corresponding to each nonzero entry with indices $(i_*,\nu_*)$ of the incidence matrix there is a uniquely determined cardinal spline function satisfying

(2.4)

$$\text{(i)}\quad s^{(2m)}(t) = 0,\ t \ne t_i,\ i = 0,\ldots,n,$$

$$\text{(ii)}\quad s \in C^{m-1}[a,b],$$

$$\text{(iii)}\quad s^{(\nu)}(t_i) = \begin{cases} 1 \text{ if } (i,\nu) = (i_*,\nu_*) \\ 0 \text{ if } e(i,\nu) = 1 \text{ and } (i,\nu) \ne (i_*,\nu_*) \end{cases},$$

$$\text{(iv)}\quad s^{(2m-1-\nu)}(t_i) = 0 \text{ if } e(i,\nu) = 0.$$

Of considerable interest is the range of cases when $e(i,\nu) = 0$

for $\nu > 0$ and $i=1,\ldots,n-1$. In these cases, $s \in C^{2m-2}[a,b]$. If, in addition, $e(0,\nu) = 0 = e(n,\nu)$ for $\nu > 0$, then we have the so-called natural cardinal spline functions $C_0,\ldots,C_n$ associated with the given partition. In his dissertation, Richards [2.26] proved that the projections, in $C[a,b]$, onto the natural spline subspaces associated with a countable family of uniform partitions of $[a,b]$ are uniformly bounded in norm. Known convergence results for dense families of $C[a,b]$, together with the Banach-Steinhaus theorem, imply the uniform convergence of natural interpolating spline functions to arbitrary continuous functions on $[a,b]$, provided the partitions are uniform. This has the following consequence for the natural cardinal spline functions. Given a family of uniform partitions indexed by $\mu$, then

$$(2.5) \qquad \sum_i |C_i^{(\mu)}(t)| \leq K$$

for some positive constant K, independent of $t \in [a,b]$ and $\mu$. The implications of (2.5) for multidimensional approximation will be seen subsequently. Before proceeding to these, we discuss inequalities similar to (2.5), in special cases, for boundary conditions other than those leading to the natural splines. Starting with periodic boundary conditions, in the cubic case ($m=2$), Cheney and Schurer [2.18] have initiated a vigorous area of research into the precise properties of locally bounded mesh ratios which give rise to uniformly bounded projections. In the case $m=2$, Birkhoff and de Boor [2.6] have investigated the exponential decay of the cardinal functions away from the point of nonzero interpolation when the incidence matrix has an initial column of ones and its further entries satisfy

$$(2.6) \qquad \begin{array}{ll} \text{(i)} & e(0,1) = 1 = e(n,1) \\ \text{(ii)} & e(i,1) = 0, \quad i = 1,\ldots,n-1. \end{array}$$

They have observed that only the cardinal splines associated with the entries e(i,0) need be estimated. For these cardinals, inequality (2.5) holds for families of quasi-uniform partitions.

Suppose now that, for each interval $I_\ell$ with partition $\pi_\ell$, a family of cardinal spline functions, piecewise polynomials of degree $2m_\ell-1$ is determined according to an Hermite-Birkhoff interpolation scheme satisfying card $\pi_\ell \geq m_\ell - 1$, with incidence matrices having all nonzero first (zero$^{th}$) column entries. The class $S$ may be taken to be the linear span of basis functions obtained by tensor products of the univariate cardinals. We are not suggesting that such a basis is computationally useful; indeed, it leads to full representation matrices. We have chosen this approach since we wish to illustrate inequality (2.5) in the sequel.

There has been a continuing search for linear operators mapping smooth function classes into $S$ which require minimal smoothness and yield maximal computability and error estimation. Perhaps the pioneering paper in this development was that of Birkhoff and Garabedian [2.8] who succeeded in constructing $C^1$ bicubic spline approximations on rectangles. De Boor [2.10], using a space $S$ generated by incidence matrices satisfying (2.6), constructed convergent $C^2$ bicubic spline approximations on rectangles and gave an efficient algorithm for the determination of the coefficients of the bicubic polynomials in each subrectangle. Ahlberg, Nilson and Walsh [2.1, 2.2] extended de Boor's results to include periodic and natural boundary conditions and obtained orthogonality and minimization results, the latter for the functional

$$(2.7) \qquad F(f) = \left\{ \int_I \left[ \left(\frac{\partial}{\partial x_1}\right)^{m_1} \cdots \left(\frac{\partial}{\partial x_N}\right)^{m_N} f \right]^2 \right\}^{1/2}$$

for $N=m_1=m_2=2$ over the class $C^{2,2}(I)$ with constrained interpolation conditions. It was observed by these authors [2.2,

p. 249] that the variational class $C^{2,2}(I)$ could be replaced by a larger class in which it is dense, viz., the direct product of the Hilbert spaces $W^{2,2}(I_1)$ and $W^{2,2}(I_2)$, which is the completion in the semi-norm (2.7) of the tensor product of these two Hilbert spaces. This device, dating back to Von Neumann, is elaborated by Sard [2.29, p. 354] and is used by him in a variety of applications. Mansfield [2.23, 2.24] has explicitly described the completion in the general situation of the direct product of $W^{m_1,2}(I)$ and $W^{m_2,2}(I_2)$ using a norm which dominates the semi-norm (2.7). Ritter [2.27], using a class only slightly more restrictive, has obtained a general minimization principle where Hermite-Birkhoff interpolation equality constraints are relaxed to inequality constraints. Haussmann [2.22] and Tippenhauer [2.33] have given abstract formulations of tensor product variational principles.

As will be familiar from the one-dimensional theory, no discrimination of boundary conditions is necessary for a convergence theory for the approximation of the completion of $C^{m_1,\ldots,m_N}(I)$ functions by the interpolating members of $\mathcal{S}$, nor are mesh restrictions required. We refer, for $N=m_1=m_2=2$ to [2.2, pp. 247-248]. Only the analogue of the first integral relation is required here and this is satisfied by the interpolating member of $\mathcal{S}$. Alternative procedures, utilizing tensor products of reproducing kernels in Hilbert spaces and corresponding projections S have been employed by Mansfield [2.23] to obtain the estimate

$$(2.8) \qquad \|D^\alpha(f-Sf)\|_{L^\infty(J)} \le K|\pi|^{m-1/2-\max(\alpha_1,\alpha_2)}[f,f]^{1/2}, \qquad 0 \le \alpha_1,\alpha_2 \le m-1$$

for functions f in the direct product of $W^{m,2}(I_1)$ and $W^{m,2}(I_2)$ where $[f,f]^{1/2}$ is a certain semi-norm and S is a generalized interpolation operator.

It is advantageous to consider larger classes than $C^{m_1,\ldots,m_N}(I)$ or its completion. Schultz [2.31] has shown that, by using the $L^2(I)$ projection $\Pi$ onto $S$, estimates of the form

$$(2.9)\quad \|f-\Pi f\|_{0,2} \leq K|\pi|^m \sum_{i=1}^{N} \left\|\left(\frac{\partial}{\partial x_i}\right)^m f\right\|_{0,2}$$

hold for smooth f. A variety of other $L^2$ estimates are obtained, including derivative estimates and estimates for Sobolev classes. The derivative estimates require a mesh restriction since $\Pi f$, in general, is not interpolating and $L^2$ Markov-type inequalities are employed. Some of the results are stated for the case where $S$ is a tensor product of generalized spline subspaces. The incidence matrices considered have the property that, if $e(i,\nu) = 1$ then $e(i,\nu') = 1$ for $0 \leq \nu' \leq \nu$.

We wish now to obtain estimates for the approximation of a dense linear subspace of $C(I)$ by the unique interpolants of $S$ for certain partitions $\pi$ and certain incidence matrices, using inequality (2.5). For concreteness, suppose that univariate Hermite-Birkhoff interpolation schemes and corresponding families of partitions are prescribed for each interval $I_\ell$ so that (2.5) holds with constant $K_\ell$ on $I_\ell$. Specifically, we consider the dense class $C^{m_1,\ldots,m_N}(I)$ and estimate the difference $f-Sf$ in $L^\infty(I)$, where $Sf \in S$ interpolates f on $\pi$ in the Hermite-Birkhoff sense:

$$(2.10)\quad Sf(x_1,\ldots,x_N) = \sum_j f_{j_1,\ldots,j_N} C_{j_1}(x_1) \cdots C_{j_N}(x_N).$$

Here $C_{j_\nu}(x_\nu)$, for fixed $1 \leq \nu \leq N$, represent the cardinal spline functions on $I_\nu$; $f_{j_1,\ldots,j_N}$ the corresponding data yielded by f for the tensor product interpolation conditions. Let $S_\nu$ represent the univariate spline interpolation operator on $I_\nu$; $S_\nu$ operates on functions in $C^{m_1,\ldots,m_N}(I)$ by holding all variables fixed except the $\nu^{th}$. Moreover, $S=S_N\ldots S_1$. Now, setting $S_0$ equal to the identity, the collapsing series

$$f - Sf = \sum_{\nu=1}^{N} [S_{\nu-1} \cdots S_0 f - S_\nu \cdots S_0 f],$$

together with (2.5) and well-known univariate estimates, imply

$$(2.11) \quad \|f - Sf\|_{0,\infty} \leq \sum_{\nu=1}^{N} K_1 \cdots K_{\nu-1} m_\nu! \, |\pi_\nu|^{m_\nu - 1/2} \, |I_\nu|^{1/2} \|f\|_{C^{m_1,\ldots,m_N}}$$

Thus, we conclude from the Banach-Steinhaus theorem and earlier remarks that the natural interpolating multivariate spline functions converge to arbitrary continuous functions on I if the family of partitions $\pi$ is uniform in each coordinate variable.

We turn our attention now to higher order estimates. When the approximants are interpolants derived from a minimum seminorm setting, such estimates require a generalized second integral relation to hold (cf. [2.2, p. 248]). Here is where boundary conditions enter. It is to be expected that the higher order estimates will hold on I only for incidence matrices whose first and last rows consist entirely of ones. It is of interest, however, that the $L^2$ projection results of Schultz [2.31] hold for higher order approximation without special boundary conditions. We cite now the result of Carlson and Hall [2.17], valid for $N=m_1=m_2=2$. Without any mesh restrictions, they have obtained the estimate for the interpolate Sf

$$(2.12) \quad \|(f-Sf)^{(\alpha_1,\alpha_2)}\|_{0,\infty} \leq \varepsilon_1 \|f^{(4-\alpha_2,\alpha_2)}\|_{0,\infty} |\pi_1|^{4-\alpha_1}$$
$$+ \varepsilon_{12} \|f^{(2,2)}\|_{0,\infty} |\pi_1|^{2-\alpha_1} |\pi_2|^{2-\alpha_2}$$
$$+ \varepsilon_2 \|f^{(\alpha_1,4-\alpha_1)}\|_{0,\infty} |\pi_2|^{4-\alpha_2}$$

for $f \in C^4(I)$, $0 \leq \alpha_1, \alpha_2 \leq 2$, and for the incidence matrices of (2.6).

We turn now to a brief description of blending methods.

Hermite (local) blending methods were introduced by Coons [2.19] and were analyzed further in [2.2] and by Birkhoff and Gordon [2.9]. The nonlocal or spline blending methods were developed by Gordon in [2.20]. As distinct from interpolation to discrete data at mesh points, these methods interpolate to continuous functions along mesh lines, i.e., they interpolate curve networks. Variational principles are satisfied by the spline blended surfaces and the convergence estimates involve the product of the mesh lengths $|\pi_\nu|$ rather than the sum as in (2.12), for example. Of course, the subspace of blending functions is infinite dimensional. The relation of the spline-blending projection to the tensor product projection is an interesting one. They may be viewed, respectively, as maximal and minimal projections in a distributive lattice of commutative projection operators [2.21], an idea attributable to Birkhoff. The search for intermediate projections in the distributive lattice, which lead to discretizations of the blending methods and hence to finite-dimensional problems is critical for their practical implementation.

We close the discussion of the approximation on rectangular parallelepipeds by recalling the papers of Rosen [2.28] who employs linear programming methods to minimize the error of multidimensional spline approximations at grid points and hence globally, Munteanu and Schumaker [2.25] who employ tensor product and blending methods of univariate variation diminishing methods and of Schultz [2.32] and Zavjalov [2.34].

The necessity of constructing spline approximations for non-rectangular regions is illustrated in very practical problems such as the fairing of a ship's hull [2.3]. Some precise information is now available for rectangular polygons such as L-shaped regions and other polygonal domains. Significant steps in constructing convergent bicubic spline interpolation schemes for L-shaped regions have been taken by Carlson and

Hall [2.14, 2.15] in obtaining fourth order schemes and by Mansfield [2.24] for bivariate splines. Of considerable interest is the role of blending in these approaches. In [2.24], extremal blending methods are used to obtain extensions to rectangles containing L-shaped regions and the interpolation operators are constructed explicitly from these. The estimate (2.8) is, in fact, valid for L-shaped regions. Convergent schemes for right triangular domains have been obtained by Carlson and Hall [2.16]. An excellent survey of some of these results for polygonal domains as well as related questions can be found in the survey papers of Birkhoff and de Boor [2.7] and Birkhoff [2.5].

## References

[2.1] Ahlberg, J. H., E. N. Nilson and J. L. Walsh, Extremal, orthogonality and convergence properties of multidimensional splines. J. Math. Appl. 12 (1965), 27-48.

[2.2] ________, The Theory of Splines and their Applications. Academic Press, New York, 1967.

[2.3] Berger, S. A., W. C. Webster, R. A. Tapia and D. A. Atkins, Mathematical ship lofting. J. Ship Research 10 (1966), 203-222.

[2.4] Birkhoff, G., Local spline approximation by moments. J. Math. Mech. 16 (1967), 987-990.

[2.5] ________, Piecewise bicubic interpolation and approximation in polygons. In Approximation, with Special Emphasis on Spline Functions, I. J. Schoenberg, ed., Academic Press, New York, 1969, pp. 185-221.

[2.6] Birkhoff, G. and C. R. de Boor, Error bounds for spline interpolation. J. Math. Mech. 13 (1964), 827-835.

[2.7] ________, Piecewise polynomial interpolation and approximation. In Approximation of Functions, H. L. Garabedian, ed., Elsevier Publishing Co., Amsterdam, 1965, pp. 164-190.

[2.8] Birkhoff, G. and H. L. Garabedian, Smooth surface interpolation. J. Math. Phys. 39 (1960), 258-268.

[2.9] Birkhoff, G. and W. J. Gordon, The draftsman's and

related equations. J. Approx. Th. 1 (1968), 199-208.

[2.10] de Boor, C., Bicubic spline interpolation. J. Math. Phys. 41 (1962), 212-218.

[2.11] ________, A note on local spline approximation by moments. J. Math. Mech. 17 (1968), 729-736.

[2.12] ________, On uniform approximation by splines, J. Approx. Th. 1 (1968), 219-262.

[2.13] de Boor, C. and G. J. Fix, Spline approximation by quasi-interpolants. J. Approx. Th., to appear.

[2.14] Carlson, R. E. and C. A. Hall, On piecewise polynomial interpolation in rectangular polygons. J. Approx. Th. 4 (1971), 37-53.

[2.15] ________, Bicubic spline interpolation in L-shaped domains. WAPD-T-2452, Bettis Atomic Power Laboratory, West Mifflin, Pa., 1971.

[2.16] ________, Bicubic spline interpolation and approximation in right triangles. WAPD-T-2488, 1971.

[2.17] ________, Error bounds for bicubic spline interpolation. J. Approx. Th., to appear.

[2.18] Cheney, E. W. and F. Schurer, A note on operators arising in spline approximation. J. Approx. Th. 1 (1968), 94-102.

[2.19] Coons, S. A., Surfaces for computer-aided design of space forms. Report MAC-TR-41 (revision of 1964 report) Project MAC, M.I.T., Cambridge, Mass., 1967.

[2.20] Gordon, W. J., Spline-blended surface interpolation through curve networks. J. Math. Mech. 18 (1969), 931-952.

[2.21] ________, Distributive lattices and the approximation of multivariate functions. In Approximation, with Special Emphasis on Spline Functions, I. J. Schoenberg, ed., Academic Press, New York, 1969, pp. 223-277.

[2.22] Haussmann, W., On multivariate spline systems. To appear.

[2.23] Mansfield, L. E., On the variational characterization and convergence of bivariate splines. Numer. Math. 20 (1973), 99-114.

[2.24] ________, On the variational approach to defining splines on L-shaped regions. J. Approx. Th., to appear.

[2.25] Munteanu, M. J. and L. L. Schumaker, Some multi-dimensional spline approximation methods. CNA-25, The

University of Texas, Austin, Texas, 1971.

[2.26] Richards, F. B., A generalized minimum norm property for spline functions and applications. Dissertation, University of Wisconsin, Madison, Wisc., 1970.

[2.27] Ritter, K., Two dimensional splines and their extremal properties. ZAMM 49 (1969), 597-608.

[2.28] Rosen, J. B., Minimum error bounds for multidimensional spline approximation. J. Comp. Sys. Sc. 5 (1971), 430-452.

[2.29] Sard, A., Linear Approximation. Mathematical Surveys 9, Amer. Math. Soc., Providence, R. I., 1963.

[2.30] Schultz, M. H., $L^\infty$-multivariate approximation theory. SIAM J. Num. Anal. 6 (1969), 161-183.

[2.31] ________, $L^2$-multivariate approximation theory. SIAM J. Num. Anal. 6 (1969), 184-209.

[2.32] ________, Multivariate L-spline interpolation. J. Approx. Th. 2 (1969), 127-135.

[2.33] Tippenhauer, U., Mehrdimensionale invariante Interpolationssysteme in Hilberträumen, ZAMM 52 (1972), 222-224.

[2.34] Zavjalov, Ju. S., Interpolation by bicubic splines, Vyčisl. Sistemy Vyp. 38 (1970), 74-101 (Russian).

## 3. Finite Elements and Local Approximation Theory

In his recent paper, [3.14], Strang distinguishes between two levels of development in the finite element method after its introduction by Courant in 1943. Progress in the first stage was due to engineers and there is a vast literature on the subject dating back to at least 1956. For a partial survey of this literature, we refer to the bibliographies and discussion of references [3.3, 3.8, 3.17, 3.18, 3.21]. The key idea, from which of course stems the name, was the generation of trial functions for the Ritz-Galerkin method through the decomposition of the Euclidean domain $\Omega$ into cells or elements with the dependence of the trial functions limited to a finite number of nodes in each element and with certain continuity requirements on the trial functions. The secondary stage referred to by Strang is that of numerical analysis and an extremely informative discussion of this can be found in [3.14] (see also the forthcoming book of Strang and Fix [3.15]). We shall now summarize certain portions of this discussion relating to multidimensional approximation theory, reserving mention of the Ritz-Galerkin method until the final section.

A domain $\Omega$, typically but not exclusively, a convex polyhedron, is prescribed in $R^N$ and a closed linear subspace V of $W^{m,2}(\Omega)$, dictated by boundary conditions or other external factors, is selected. The trial space $S$ is restricted to lie in V and to be of degree k-1, i.e., the restriction of $S$ to each element $e_i$ contains the polynomials of total degree k-1. Here k is any integer satisfying $1 \leq m \leq k$. The requirement $S \subset W^{m,2}(\Omega)$ introduces compatibility conditions at inter-element boundaries; a sufficient conforming condition is that the trial functions possess continuous derivatives of order less than m. The nodal finite element method is a convenient way of constructing a local basis for $S$. $\overline{\Omega}$ is partitioned into a union of closed elements $e_i$ overlapping only at element boundaries. Each element contains no more than some fixed number of prescribed nodes $z_\nu$, allowing multiplicities, and to each basis function $\phi_\nu$ there correspond such a node $z_\nu$ and a differential operator $D_\nu$, possibly the identity, of order less than k. The basis functions $\phi_\nu$ satisfy the property that they vanish in elements not containing $z_\nu$; moreover, they are interpolating:

(3.1) $D_i \phi_\nu(z_i) = \delta_{i\nu}$.

The problem we wish to consider is the accuracy of approximation of members of $W^{k,2}(\Omega)$ by functions in $S$ in terms of the quantities

(3.2) $h_i$ = diameter $e_i$.

We wish to obtain a mean square convergent process of order $h^k$,

(3.3) $h = \max h_i$.

It is essential, in extending the approximation methods to functions with singularities, to build up the golbal estimates from local estimates within the individual elements. To do this it is desirable to construct a continuous linear mapping T from $V \cap W^{k,2}(\Omega)$ into $S$ satisfying:

(3.4i) the restriction of T to each element preserves polynomials of degree less than k;

(3.4ii) $|v-Tv|_{\ell,2} \leq Ch^{k-\ell}|v|_{k,2}$, $v \in V \cap W^{k,2}(\Omega)$,

where C does not depend on h. Now the construction of such a mapping T is carried out in [3.14] via the composition map of smoothing (which preserves polynomials) and interpolation via the cardinal basis $\phi_\nu$, under a certain uniformity assumption on the basis, which enables one to obtain C in (3.4ii) independent of h and also to pass from local estimates, obtainable from (3.4i) and the generalized Peano kernel theorems of section one, to global estimates. The uniformity assumption on the basis is described as follows. Let a (directed) sequence of subspaces $S^h$, parametrized by h, be given. The basis functions $\phi_\nu = \phi_\nu^h$ are uniform to order q if there exist constants $c_\ell$ such that for all h, i and $\nu$,

$$(3.5) \qquad \max_{\substack{x \in e_i \\ |\alpha| = \ell}} |D^\alpha \phi_\nu(x)| \leq c_\ell h_i^{|D_\nu| - \ell}, \quad 0 \leq \ell \leq q \leq k.$$

Thus, if the basis is uniform to order q, (3.4ii) holds for $\ell \leq q$. It is possible, as Strang points out, to relate the uniformity assumption to the geometry of the $e_i$. Thus, when N=2 and the $e_i$ are obtained by triangulation, it suffices that the smallest angle of the triangulation be bounded from below independently of h and, in this case, the $\phi_\nu$'s may be generated from those associated with a single fixed triangle by linear transformations. More generally, as is shown in [1.5] by Ciarlet and Raviart, it suffices that each $e_i$ should contain a sphere of radius $\sigma h_i$, where $\sigma$ is fixed.

Two final remarks are in order concerning [3.14]. First, the intermediate smoothing process may be omitted when the interpolate is well-defined for $W^{k,2}(\Omega)$ functions, e.g., if k-N/2 exceeds the highest order of the derivatives $D_\nu$. Second, the error estimation in the smoothing process requires only that the elements locally be of comparable size, similar to locally bounded mesh ratio assumptions in univariate spline theory.

The papers of Birkhoff, Schultz and Varga [3.1] and Zlamal [3.19] gave considerable impetus to the mathematical development of local approximation theory, the former involving full Hermite interpolation on rectangles by $C^{m-1}$ piecewise polynomials of degree 2m-1 in each variable and the latter involving continuous piecewise quadratic approximations and $C^1$ fifth degree approximations determined on triangular elements. At about the same time, Goël [3.7], examining local basis functions for rectangular and triangular elements, discovered the <u>necessity</u> in certain cases of these functions producing polynomials through linear combinations. Now one direction of generalization of [3.1] was subsequently carried out by Nitsche [3.12] via multi-dimensional Fourier series. Another direction of generalization of [3.1], which could be viewed as the nodal finite element method with rectangular elements, was subsequently carried out in N dimensions by Bramble and Hilbert [1.4] via a refined Peano kernel theorem permitting, as conjectured by Birkhoff, the omission of certain derivatives of order 2m on the right hand side of the error estimates in the approximation of $W^{2m,p}(\Omega)$ functions for certain ranges of p. As pointed out by Varga [0.18], an intermediate smoothing technique, followed by Hermite interpolation, yields an estimate similar to (3.4ii) for $k=2m$ and $0 \le \ell \le m$ when $p=2$.

Ciarlet and Wagschal [3.6], using a coordinate free approach, were able to extend the setting of [3.19] to a simplicial complex in $R^N$. Somewhat earlier, Zenisek [3.16] constructed a hierarchy of polynomials on triangles of total degree $4m+\ell$, $\ell=1,2,3,4$, with $C^m$ conforming conditions. The case $m=1$ and $\ell=1$ had been the case treated by Zlamal in [3.19] and simultaneously by six others (see the discussion in [3.21]). Zenisek was able to obtain interpolation theorems and consequent convergence results up to the thirteenth order. It remained for Bramble and Zlamal [3.3] to unify this work initiated in [3.19]. Treating, explicitly, piecewise polynomials of degree 4m+1, they obtained interpolation and subsequent convergence results for $W^{k,2}(\Omega)$ functions, $2m+2 \le k \le 4m+2$, of the form

$$(3.6)\qquad \|u-Tu\|_{\ell} \leq \frac{K}{(\sin \lambda)^{m+\ell}} h^{k-\ell}|u|_{k}, \qquad 0 \leq \ell \leq k,$$

where $\Omega$ is triangulated into triangles with maximum diameter h and smallest internal angle $\lambda$ and T represents the interpolation operator. Subsequent extensions of [3.3] have been given by Zlamal and Zenisek [3.21] involving triangular, tetrahedral and rectangular elements and by Ciarlet and Raviart [1.5] in which coordinate free, Frechet calculus, Lagrange and Hermite interpolation schemes over simplices obtainable from a fixed model simplex, are employed to obtain estimates in $W^{m,p}(\Omega)$, $1 \leq p \leq \infty$. Results of Nicolaides [3.11], who has considered principal lattices of order k for interpolation by polynomials of degree k on simplices in $R^N$, are particularly useful here Babuska and Rosenzweig [3.2], finally, have discussed the effect of possible singular points.

It is frequently desirable to concentrate the maximal number of element parameters at element vertices in order to reduce the dimension of the trial subspace. Sometimes, for computational reasons, it is even desirable to reduce the number of parameters and thus the accuracy of approximation. A discussion of these points, with references, may be found in [3.4], [3.5], [3.20] and [3.21].

Local interpolation schemes by bicubic polynomials for polygons which can be partitioned into rectangular and triangular elements with $C^1$ conforming conditions have been considered by Hall [3.9] and Hulme [3.10]. These may be viewed as bicubic generalized Hermite interpolation schemes, the latter especially suited for use in the solution of Neumann problems.

We close this section with a brief discussion of the use of elements with curved boundaries. Such elements are, of course, permitted in the approach of Strang outlined earlier and are becoming increasingly useful in so-called isoparametric methods for the solution of boundary value problems. We mention first the paper of Shah [3.13] in which derivatives are interpolated at element vertices and moments are matched along element boundaries. Shah's approach is especially well-suited for the use of curvilinear coordinate systems. Convergence results of optimal order are obtained for the Sobolev classes. Ciarlet and Raviart [3.5] have utilized elements obtained via smooth diffeomorphisms from model elements and have carefully estimated the constants in convergence expressions in terms of these diffeomorphisms. The reader is also referred to the expository article of Ciarlet [3.4]. Gordon and Hall [3.8] have combined discretized blending methods with isoparametric methods. The bibliography of [3.8] is a useful source of the appropriate engineering literature for the isoparametric methods. Finally, the reader is also referred to the discussion in [3.15].

## References

[3.1] Birkhoff, G., M. H. Schultz and R. S. Varga, Piecewise Hermite interpolation in one and two variables with applications to partial differential equations. Numer. Math. 11 (1968), 232-256.

[3.2] Babuska, I. and M. B. Rosenzweig, A finite element scheme for domains with corners. Numer. Math. 20 (1972), 1-21.

[3.3] Bramble, J. H. and M. Zlamal, Triangular elements in the finite element method. Math. Comp. 24 (1970), 809-820.

[3.4] Ciarlet, P. G., Orders of convergence in finite element methods. Conf. on Math. of Finite Elements, Brunel University, 1972.

[3.5] Ciarlet, P. G. and P.-A. Raviart, Interpolation de Lagrange sur des éléments finis courbes dans $R^n$. C. R. Acad. Sc., Paris 274 (1974), 640-643.

[3.6] Ciarlet, P. G. and C. Wagschal, Multipoint Taylor formulas and applications to the finite element method. Numer. Math. 17 (1971), 84-100.

[3.7] Goël, J. J., Construction of basic functions for numerical utilization of Ritz's method. Numer. Math.

12 (1968), 435-447.

[3.8] Gordon, W. J. and C. A. Hall, Discretization error bounds for transfinite elements, GMR-1196. General Motors Research Laboratories, Warren, Michigan, 1972.

[3.9] Hall, C. A., Bicubic interpolation over triangles. J. Math. Mech. 19 (1968), 1-11.

[3.10] Hulme, B. L., A new bicubic interpolation over right triangles. J. Approx. Th. 5 (1972), 66-73.

[3.11] Nicolaides, R. A., On the class of finite elements generated by Lagrange interpolation. SIAM J. Numer. Anal. 9 (1972), 435-445.

[3.12] Nitsche, J., Interpolation in Sobolevschen Funktionenräumen. Numer. Math. 13 (1969), 334-343.

[3.13] Shah, J. M., Two-dimensional polynomial splines. Numer. Math. 15 (1970), 1-14.

[3.14] Strang, G., Approximation in the finite element method. Numer. Math. 19 (1972), 81-98.

[3.15] Strang, G., and G. Fix, An Analysis of the Finite Element Method. Prentice-Hall, Englewood Cliffs, N. J. (to appear), Ch. 3.

[3.16] Zenisek, A., Interpolation polynomials on the triangle. Numer. Math. 15 (1970), 283-296.

[3.17] Zienkiewicz, O. C., The finite element method: from intuition to generality. Appl. Mech. Rev. 23 (1970), 249-256.

[3.18] ________, The Finite Element Method in Engineering Sciences. McGraw-Hill, New York, 1971.

[3.19] Zlamal, M., On the finite element method. Numer. Math. 12 (1968), 394-409.

[3.20] ________, A finite element procedure of the second order of accuracy. Numer. Math. 14 (1970), 394-402.

[3.21] Zlamal, M., and A. Zenisek, Mathematical aspects of the finite element method. In Technical, Physical and Mathematical Principles of the Finite Element Method, Academia, Prague, 1971, pp. 15-39.

## 4. Approximation in $R^N$ and Extrapolation

The Fourier transform,

$$\hat{f}(\xi) = \int_{R^N} f(x)e^{-i<x,\xi>}\,dx, \quad <x,\xi> = \sum_{\nu=1}^{N} x_\nu \xi_\nu,$$

is defined for functions $f \in S$ by the above and its inverse on $S$ is given by

$$\check{f}(\xi) = (2\pi)^{-N} \int_{R^N} f(x)e^{-i<x,\xi>}dx.$$

As usual, it may be extended by continuity to $L^2(R^N)$. For $\phi \in C_0^\infty(R^N)$ define the discrete Fourier transform, for $h > 0$,

$$(4.1) \qquad \tilde{\phi}(\xi) = h^N \sum_{x\in G_h} \phi(x)e^{-i<x,\xi>} \quad ,$$

where $G_h = \{hj = h(j_1,\ldots,j_N):j_i \text{ an integer}\}$. $\tilde{\phi}$ is a periodic function in each argument of period $2\pi/h$. Now define

$$E_h = \{x \in R^N: \ |x_\nu| \leq \pi/h, \ 1 \leq \nu \leq N\}$$

and let $\chi_h$ be the characteristic function of $E_h$. Bramble and Hilbert [1.3] have shown that, for $k > N/2$, there exists an extension of the discrete Fourier transform (4.1) to all of $W^{k,2}(R^N)$. If $u \in W^{k,2}(R^N)$, then $\tilde{u} \in L^2(E_h)$ and is extended by periodicity. Moreover, there exists a positive constant $C$ such that

$$(4.2) \qquad \|(\chi_h\tilde{u})^{\vee} - u\|_{\ell,2} \leq Ch^{k-\ell}|u|_{k,2}, \ 0 \leq \ell < k.$$

We remark that (4.2) is a generalized cardinal series: if $\phi \in C_0^\infty(R^N)$, we have

$$(\chi_h\tilde{\phi})^{\vee}(\xi) = \frac{h^N}{(2\pi)^N}\int_{E_h}\left\{\sum_{x\in G_h}\phi(x)e^{-i<x,y>}\right\}e^{i<y,\xi>}dy$$

$$= \sum_{x\in G_h}\phi(x)\prod_{\nu=1}^{N}\left\{\frac{\sin[(\xi_\nu - x_\nu)\pi/h]}{(\xi_\nu - x_\nu)\pi/h}\right\}.$$

We now wish to generate interpolating spline functions on $R^N$. Let $\psi_\nu(x_\nu)$ be given by

$$\psi_\nu(x_\nu) = \begin{cases} 1 & \text{if } -\frac{1}{2} \le x_\nu \le \frac{1}{2} \\ 0 & \text{otherwise} \end{cases}$$

and set

$$\psi(x) = \prod_{\nu=1}^{N}\psi_\nu(x_\nu).$$

Now define $\psi = \psi^{(1)}$ and, for $\ell > 1$, $\psi^{(\ell)}$ by the convolution

$$(4.3)\qquad \psi^{(\ell)}(x) = \psi*\psi^{(\ell-1)}(x) = \int_{R^N}\psi(y)\psi^{(\ell-1)}(x-y)dy.$$

$\psi^{(\ell)}$ is a piecewise polynomial of degree $N(\ell-1)$ with continuous partial derivatives of order $\ell-2$ and is of compact support. It was demonstrated in [1.3] that there exists a linear interpolation operator T, mapping $W^{k,2}(R^N)$ into $W^{k-1,2}(R^N)$ for $k > N/2$, of the form,

$$Tu(x) = h^{-N/2}\sum_{y\in G_h} w_y\psi^{(k)}\left(\frac{x-y}{h}\right),$$

with the property that

$$\sum_{y\in G_h}|w_y|^2 \le C'\|u\|_{0,2}^2,\quad C' > 0,$$

and such that

$$(4.4)\qquad \|u-Tu\|_{\ell,2} \le Ch^{k-\ell}\|u\|_{k,2},\quad 0 \le \ell < k$$

for some constant $C > 0$. Note that Tu agrees with u on $G_h$.

Now the generalized cardinal series and the interpolating splines are connected in a remarkable way. Specifically, if h is fixed, $k > N/2$ and $u \in W^{k,2}(R^N)$ then, for all $m \geq 1$, there is a piecewise polynomial interpolating u on $G_h$ of the form

$$P_m(x) = \sum_{y \in G_h} w_y^{(m)} \psi^{(m)} \left(\frac{x-y}{h}\right) .$$

Then Bramble and Hilbert [1.3] have shown that $P_m$ converges uniformly to $(\chi_h \tilde{u})^\vee$ on $R^N$ as $m \to \infty$.

We have seen that the function $\phi = \psi^{(k)}$ enjoys the property that any function $u \in W^{k,2}(R^N)$ can be approximated by (infinite) linear combinations of the functions

$$\phi_j^h(x) = h^{-N/2} \phi \left(\frac{x-jh}{h}\right)$$

with order of accuracy $h^k$ in the mean square norm. We shall now characterize all such functions $\phi$ through results of Fix and Strang [4.10]. (The univariate case is due to Schoenberg [4.16]).

<u>Theorem</u> (Fix-Strang). For any integers $k \geq m \geq 0$ the following conditions are equivalent:

(4.5i) $\phi \in W_c^{m,2}(R^N)$, $\hat{\phi}(0) \neq 0$ but $\hat{\phi}$ has zeros of order at least k+1 at the other points of $2\pi Z^N$:

$$D^\alpha \hat{\phi}(2\pi j) = 0 \text{ if } 0 \neq j \in Z^N, \quad 0 \leq |\alpha| \leq k;$$

(4.5ii) $\phi \in W_c^{m,2}(R^N)$ and, for $0 \leq |\alpha| \leq k$, the function $\sum_{j \in Z^N} j^\alpha \phi(t-j)$ is a polynomial in $t_1, \ldots, t_N$ with leading term $Ct^\alpha$, $C \neq 0$;

(4.5iii) $\phi$ is a distribution with compact support and for each $u \in W^{k+1,2}(R^N)$ there are weights $w_j^h$ such that, as $h \to 0$,

$$\|u - \sum_{j \in Z^N} w_j^h \phi_j^h\|_{r,2} \leq c_r h^{k+1-r} \|u\|_{k+1,2}, \quad 0 \leq r \leq m,$$

with

$$\sum_{j \in Z^N} |w_j^h|^2 \leq K \|u\|_{0,2}^2 .$$

The constants $c_r$ and K are independent of u.

We remark that if the basis elements $\phi_j^h$ satisfy a stability criterion of uniform linear independence as $h \to 0$ then the conditions of the theorem may be modified somewhat. Specifically, we assume

$$(4.6) \qquad \| \sum_{j\in Z^N} v_j \phi_j^h \|_{0,2}^2 \geq \sigma \sum_j |v_j|^2, \quad \sigma > 0,$$

holds for all sufficiently small h. It is known that (4.6) is equivalent to the condition that there is no real $\xi_0$ such that

$$(4.7) \qquad \hat{\phi}(\xi_0 + 2\pi j) = 0, \text{ for all } j \in Z^N.$$

Under this stability hypothesis, the condition in (4.5iii) on the weights is automatically satisfied. Moreover, (4.5ii) can be generalized to assert that

$$t^\alpha = \sum_{j\in Z^N} \lambda_j^\alpha \phi(t-j), \quad 0 \leq |\alpha| \leq k.$$

The exponent k+1-r is best possible for every $0\leq r\leq m$ if k is the largest integer for which (4.5i) holds. A thorough analysis of the constants $c_r$ is given in [4.10], by Strang in [4.15] and by Strang and Fix in [3.15]. It is shown that the infimum of these constants can be determined through the approximation of polynomials of degree k+1 by polynomials of lower degree, illustrating the local polynomial behavior of smooth functions.

If we set k=m in the theorem, then the function $\psi^{(k+1)}$ of (4.3) satisfies (4.5i):

$$(4.8) \qquad \hat{\psi}^{(k+1)}(\xi) = \prod_{\nu=1}^{N} \left\{ \frac{\sin(\xi_\nu/2)}{(\xi_\nu/2)} \right\}^{k+1} .$$

Aubin [4.1, 4.2, 4.3] had earlier used the function $\psi^{(k+1)}$ to obtain estimates of optimal order in Sobolev spaces. Employed were prolongation and restriction operators, $p_h$ and $r_h$

respectively, which had been introduced earlier by Cea [4.7]:

(4.9i)
$$p_h : \ell^2 \to W^{k,2}(R^N),$$
$$p_h v_h = \sum_{j \in Z^N} v_j \mu \left(\frac{x}{h} - j\right),$$

where $u \in W^{k,2}(R^N)$ has compact support and $\int_{R^N} \mu = 1$;

(4.9ii)
$$r_h : W^{k,2}(R^N) \to \ell^2,$$
$$r_h u = \left\{ h^{-N} \int_{R^N} \lambda\left(\frac{x}{h} - j\right) u(x)dx \right\}_{j \in Z^N},$$

where $\lambda \in L^\infty(R^N)$ has compact support and $\int_{R^N} \lambda = 1$. Aubin was able to show, for the choice $\mu = \psi^{(k+1)}$, that, corresponding to the prolongation operator $p_h$ of (4.9i) thus determined, there is a restriction operator (4.9ii) such that, for $u \in W^{k+1,2}(R^N)$,

(4.10) $$\|u - p_h r_h u\|_{\ell,2} \leq Ch^{k+1-\ell}|u|_{k+1,2}, \quad 0 \leq \ell \leq k.$$

In fact, Aubin discusses the more general problem of the approximation of $W^{k+1,p}(R^N)$ functions.

Another example of a function $\phi$ satisfying (4.5i) was considered by di Guglielmo [4.13]:

(4.11) $$\phi(x) = \int_{-1/2}^{1/2} \psi^{(k)}(x_1 - \tau, \ldots, x_N - \tau)d\tau.$$

In this case,

$$\hat{\phi}(\xi) = \prod_{\nu=1}^{N} \left\{\frac{\sin(\xi_\nu/2)}{\xi_\nu/2}\right\}^k \left\{\frac{\sin\left(\sum_{\nu=1}^{N} \xi_\nu/2\right)}{\sum_{\nu=1}^{N} \xi_\nu/2}\right\}.$$

The choice of $\mu = \phi$ in (4.9i) leads to simplicial spline functions in this case, since the support of $\phi$, defined by (4.11), lies in the convex envelope of the two hypercubes whose main diagonal vertices include the origin (common) and $(\frac{k}{2},\ldots,\frac{k}{2})$, $(-\frac{k}{2},\ldots,-\frac{k}{2})$ respectively. It was shown in [4.13] that estimates of the form (4.10) hold for suitable restriction operators. Di Guglielmo also computed the maximal number of multi-integers j such that $\phi(x)$ and $\phi(x-j)$ are nonzero for any x, viz., $[2(2k)^N-(2k-1)^N]$, and obtained a minimality property for this number with respect to other compact support functions $\phi$.

One of the more remarkable observations made by di Guglielmo and Aubin [4.4, Theorem 1-1, p. 141] is that, for functions $\phi = \mu$ satisfying, say (4.5i) and the stability condition (4.7), there exist restriction operators $r_h$ such that (4.10) is an estimate locally of optimal order, i.e., with respect to the n-widths of the injection of the unit ball of $W^{k+1,2}(\Omega)$ into $W^{\ell,2}(\Omega)$ where $\Omega$ is bounded. This is deducible, without the actual computation of the widths, from a study of the stability function, studied by Raviart, Aubin and others (see [4.4] and references therein).

It is possible to generalize the previous results by admitting translates of more than one function. This was observed by di Guglielmo [4.14], Babuska [4.5] and Fix and Strang [4.10] and is essential in the use of less smooth trial spaces in which more than one basis function is associated with a given nodal point. If $\phi_1,\ldots,\phi_n$ are given functions in $W_c^{m,2}(R^N)$, $m\leq k$, and functions $\phi^h_{-,j}$ are given by

(4.12) $\phi^h_{ij}(x) = h^{-N/2}\phi_i\left(\frac{x}{h} - j\right),\ 1 \leq i \leq n,\ j \in Z^N,$

then a generalization of the earlier theorem is afforded by the equivalence of:

(4.13i) there are linear combinations $\psi_\alpha$ of the $\phi_i$ which satisfy,

$$\hat{\psi}_0(0) = 1,\ \hat{\psi}_0(2\pi j) = 0 \text{ for } j \neq 0,$$

$$\sum_{\beta\leq\alpha} \frac{D^{\beta}\hat{\psi}_{\alpha-\beta}(2\pi j)}{\beta! i^{|\beta|}} = 0 \text{ for all } j \in Z^N,\ 1 \leq \alpha \leq k;$$

(4.13ii) $$\frac{t^{\alpha}}{\alpha!} = \sum_{\beta\leq\alpha} \sum_{j\in Z^N} \frac{j^{\beta}\psi_{\alpha-\beta}(t-j)}{\beta!},\quad |\alpha| \leq k;$$

(4.13iii) for each $u \in W^{k+1,2}(R^N)$ there are weights $w_{ij}^h$ such that for $0 \leq \ell \leq m$,

$$\|u - \sum_{i,j} w_{ij}^h \phi_{ij}^h\|_{\ell,2} \leq c_{\ell} h^{k+1-\ell} \|u\|_{k+1,2},$$

with

$$\sum_{i,j} |w_{ij}^h|^2 \leq K\|u\|_{0,2}^2 .$$

This fundamental result, due to Fix and Strang, of the abstract finite element method, together with the precise analysis of the constants $c_{\ell}$, represent the culmination of the work initiated by Aubin and carried on by his student di Guglielmo and by Babuska. The latter [4.5] had used the Fourier transform of a certain linear combination of $\phi_1,\ldots,\phi_n$ and, under certain growth assumptions on the transform, had constructed an explicit linear operator T, in terms of the $\phi_{ij}^h$, for which optimal order convergence holds in fractional order Sobolev spaces. It is likely that Babuska's conditions or modifications thereof furnish an additional equivalence in (4.13). (For additional development of the hill function method we cite [4.6].) Babuska did, in fact, suggest to Fix and Strang still another equivalence in (4.13): there exists a function $\Phi$ which is a linear combination of $\phi_1,\ldots,\phi_n$ and their translates such that (4.5) holds for this $\Phi$. This fourth equivalence is verified in [4.10].

It is reasonable to inquire whether a natural linear operator exists for the computation of the weights $w_j^h$ or $w_{ij}^h$. This question is resolved in [4.10] (see also [4.15]) through the construction of the quasi-interpolate of u on the functions $\psi_\alpha$ via the use of the values of $D^\alpha u$ at nodal points when they are well-defined. In particular, a careful error analysis is carried out in the $\|\cdot\|_{\ell,\infty}$ norm for $W^{k+1,\infty}(R^N)$ functions, $0\leq\ell\leq k$, via the quasi-interpolation process (see also Theorem 5.3 of Varga's monograph [0.18]). It is clear that the smoothing process suggested by Strang [3.14] is applicable when the above quasi-interpolant or the interpolant of Bramble and Hilbert [1.3] fail to be defined.

We turn now to a brief discussion of extrapolation methods. The Calderon extension theorem and its improvement as presented in Morrey [1.10, p. 74] make possible the application of results known for $R^N$ to bounded regular regions. This device was used by the writer in his 1966 doctoral dissertation and by many others, e.g., Aubin [4.4], di Guglielmo [4.13], Nitsche [3.12] and Schultz [2.31]. An extremely subtle version, in which continuity with respect to a Sobolev semi-norm (rather than norm) is obtained, is presented by Strang [3.14]. It should be emphasized, however, that these extension operators are not explicit and are hence primarily of theoretical usefulness. Coatmelec [4.8, 4.9] has constructed linear extension operators for functions in $C^m(E)$, E a compact subset of $R^N$, thus giving linear analogues of the nonlinear Whitney extension results and has estimated the extension norms. Frederickson [4.12] using spaces of triangular splines on $R^2$ developed in [4.11] has constructed computationally practical linear extensions of $C^m(\Omega)$ functions where $\Omega \subset R^2$ satisfies a cone condition.

## References

[4.1] Aubin, J. P., Approximation des espaces de distributions et des operateurs differentiels. Bull. Soc.

Math. France Mem. 12 (1967), 1-139.

[4.2] ________, Behavior of the error of the approximate solutions of boundary-value problems for linear elliptic operators by Galerkin's and finite-difference methods. Ann. Sci. Norm. Pisa 21 (1967), 599-637.

[4.3] ________, Evaluation des erreurs de troncature des approximations des espaces des Sobolev. J. Math. Anal. Appl. 21 (1968), 356-368.

[4.4] ________, Approximation of Elliptic Boundary Value Problems. Wiley-Interscience, New York, 1972.

[4.5] Babuska, I., Approximation by hill functions. Comm. Math. Univ. Carol. 11 (1970), 787-811.

[4.6] ________, Approximation by hill functions II. Comm. Math. Univ. Car. 13 (1972), 1-22.

[4.7] Céa, J., Approximation variationnelle des problèmes aux limites. Ann. Inst. Fourier 14 (1964), 345-444.

[4.8] Coatmelec, C., Approximation et interpolation des fonctions différentiables de plusieurs variables. Ann. Sc. L'Ecole Norm. Sup. 83 (1966), 271-341.

[4.9] ________, Prolongement d'une fonction en une fonction différentiable. Diverses majorations sur le prolongement. In Approximation with Special Emphasis on Spline Functions, I. J. Schoenberg, ed., Academic Press, New York, 1969, pp. 29-49.

[4.10] Fix, G. and G. Strang, Fourier analysis of the finite element method in Ritz-Galerkin theory. Proceedings CIME (Italy) Summer School, to appear; in summary form, Studies in Appl. Math. 48 (1969), 265-273.

[4.11] Frederickson, P. O., Generalized triangular splines. Rep. 7 (1971), Lakehead University, Ontario, Canada.

[4.12] ________, Quasi-interpolation, extrapolation, and approximation on the plane. Proceedings of the Manitoba Conference on Numerical Mathematics, University of Manitoba, Canada, 1971.

[4.13] di Guglielmo, F., Construction d'approximations des espaces de Sobolev sur des réseaux en simplexes, Calcolo 6 (1969), 279-331.

[4.14] ________, Méthode des éléments finis: une famille d'approximations des espaces de Sobolev par les translatés de p fonctions. Calcolo 7 (1970), 185-234.

[4.15] Strang, G., The finite element method in approximation theory. In Numerical Solutions of Partial Differential Equations II, SYNSPADE, B. Hubbard, ed., Academic Press, New York, 1971.

[4.16] Schoenberg, I. J., Contributions to the problem of approximation of equidistant data by analytic functions, Quart. Appl. Math. 4 (1946), 45-99, 112-141.

## 5. Widths and Entropies in Sobolev Spaces

Lorentz [5.9, 5.11] and Sprecher [5.17] have given comprehensive surveys on the subject of n-width and entropy and the corresponding relations with the complexity and structure of functions. It is not our intention to reproduce these surveys here nor the extensive bibliographies, particularly of the Russian literature. We simply mention briefly that the concept of metric entropy has been used successfully to answer, in the negative, conjectures on the representation of functions of smoothness index k/N by functions with smaller index. Here k represents the order of differentiability and N the number of Euclidean dimensions. Entropy was also utilized by Vituskin [5.18] as an effective lower bound for a measure of the number of parameters required in $\varepsilon$-accurate algorithms (see also the paper of Lorentz [5.10]). An excellent account of the subject, with proofs, is available in the book of Lorentz [5.12] and in [5.11].

We wish to discuss here recent results on the exact asymptotic order of the n-widths and entropies in $W^{\lambda,p}(\Omega)$ of the unit ball of $W^{\mu,p}(\Omega)$. Here $\lambda$ and $\mu$ are arbitrary real numbers, possibly negative, satisfying $\lambda < \mu$. The majoration results, for both widths and entropies, are stated and proved by Birman and Solomjak [1.2] for very generally weighted fractional order Sobolev spaces. The minoration results, for widths, are due to El Kolli [5.4] and, for entropies, to Mostefai [5.15]. The method of proof involves first a consideration of the case $\lambda = 0$ and then the general case via the interpolation theory of Peetrie (cf. Lions and Peetrie [5.8]) and the fundamental isomorphism theorem of Marcinkiewicz [5.13] (see also the book of Mikhlin [5.14]). $\Omega$ is a bounded regular domain in what follows.

Theorem. The n-widths $d_n$ of the unit ball $SW^{\mu,p}(\Omega)$ in $W^{\lambda,p}(\Omega)$, $\lambda<\mu$, $1<p\leq\infty$, satisfy the asymptotic estimate

$$(5.1i) \quad C_1 n^{-(\mu-\lambda)/N} \leq d_n \leq C_2 n^{-(\mu-\lambda)/N}$$

as $n \to \infty$; the $\varepsilon$-entropy, $H_\varepsilon$, for $1<p<\infty$, satisfies

$$(5.1ii) \quad C_1 \varepsilon^{-N/(\mu-\lambda)} \leq H_\varepsilon \leq C_2 \varepsilon^{-N/(\mu-\lambda)}$$

as $\varepsilon \to 0$.

(5.1i) is proved in [5.4]; minoration estimates depend upon the theorem of Gohberg and Krein [5.12, p. 137]. Babuska [5.2] obtained the result earlier for p=2. The estimate for p=1 is proved in [5.4] for $\lambda=0$ and $\mu$ an integer. (5.1ii) is proved in [5.15]. In the general majoration estimates of entropy in [1.2], partitionings of $\Omega$, dependent on the approximated function, are employed, a generalization of the univariate splines with variable knots. Such majorations include the case p=1. Special cases of (5.1i), for p=2, $\lambda=0$ and $\mu$ a positive integer, were obtained by the writer in [5.6] and subsequent papers, the most recent being [5.7]. Fundamental results of Golomb [5.5] on spectral characterizations of ellipsoidal n-widths were employed in [5.6]. The results of [5.6] gave precise constants in (5.1i). n-widths of $L^2$ classes have also been discussed by Nitsche [5.16]. Clements [5.3] has considered $L^1$ entropy results for certain classes of smooth functions and Aubin [5.1] has considered spectral characterizations of n-width.

## References

[5.1] Aubin, J. P., Approximation of nonhomogeneous Neumann problems, regularity of the convergence and estimates of error in terms of n-width. MRC Tech. Summary Report 924, Madison, Wisc., 1968.

[5.2] Babuska, I., The rate of convergence for the finite element method. SIAM J. Numer. Anal. 8 (1971), 304-315.

[5.3] Clements, G. F., Entropies of several sets of real-valued functions. Pacific J. Math. 13 (1963), 1085-1095.

[5.4] El Kolli, A., $n^{ieme}$ epaisseur dans les espaces de Sobolev. J. Approx. Th., to appear

[5.5] Golomb, M., Optimal approximating manifolds in $L_2$-spaces. J. Math. Anal. Appl. 12 (1965), 505-512.

[5.6] Jerome, J., On the $L_2$ n-width of certain classes of functions of several variables, dissertation, Purdue University, Indiana, 1966.

[5.7] ________, Asymptotic estimates of the n-widths in Hilbert space. Proc. Amer. Math. Soc. 33 (1972), 367-372.

[5.8] Lions, J. L. and J. Peetre, Sur une classe d'espaces d'interpolation. Inst. Hautes Etudes Sci. Publ. Math. 19 (1964), 5-68.

[5.9] Lorentz, G. G., Metric entropy, widths and superpositions of functions. Amer. Math. Monthly 69 (1962), 469-485.

[5.10] ________, Entropy and its applications. J. Soc. Indust. Appl. Math. (Ser. B) Numer. Anal. 1 (1964), 97-103.

[5.11] ________, Metric entropy and approximation. Bull. Amer. Math. Soc. 72 (1966), 903-937.

[5.12] ________, Approximation of Functions. Holt, Rinehart and Winston, New York, 1966.

[5.13] Marcinkiewicz, J., Sur les multiplicateurs des series de Fourier. Studia Mathematica 8 (1939).

[5.14] Mikhlin, S. G., Multidimensional singular integrals and integral equations. Moscow, 1962.

[5.15] Mostefai, A., Évaluations de l'$\varepsilon$-entropy dans les espaces de Sobolev. Dissertation, University of Algier, 1970.

[5.16] Nitsche, J., Zur Frage optimaler Fehlerschranken bei Differenzenverfahren I. Circolo Matematico series 2, 16 (1967), 69-80.

[5.17] Sprecher, D. A., A survey of solved and unsolved problems on superpositions of functions. J. Approx. Th. 6 (1972), 123-134.

[5.18] Vitushkin, A. G., Estimation of complexity of the tabulation problem. Gosizdat, Moscow, 1959. Translation, Pergamon Press, New York, 1961.

## 6. Variational Problems and Elliptic Spline Functions

It is our purpose in this section to state, briefly, results on the convergence of Galerkin projections to solutions of boundary value problems and to eigenfunctions and to present, in addition, certain generalizations of univariate spline functions. We begin with a result contained in the book of Strang and Fix [3.15, Theorem III.7]. If $B(v,v)$ is a quadratic form of order m satisfying the ellipticity condition

$$(6.1)\qquad \|v\|_{m,2}^2 \le C_1 B(v,v) \le C_2 \|v\|_{m,2}^2 , \quad v \in W^{m,2}(\Omega),$$

for positive constants $C_1$ and $C_2$ and if $u_h$ is a finite element approximation to the solution u of the variational boundary value problem obtained by Galerkin projection onto a trial space, exact of degree $k-1 \ge m-1$, then the estimates

$$(6.2)\qquad \begin{aligned} &(i) \quad \|u-u^h\|_{s,2} \le Ch^{k-s}\|u\|_{k,2}, \quad s \ge 2m-k, \\ &(ii) \quad \|u-u^h\|_{s,2} \le Ch^{2(k-m)}\|u\|_{k,2}, \quad s \le 2m-k \end{aligned}$$

hold for all values of $s \le m$.

The estimate (6.2) is also discussed in somewhat less general form by Strang in [3.14, 4.15]. In the latter reference he indicates that, with stability assumptions, the estimates also hold pointwise. In his book [4.4] Aubin systematically develops estimates of the form (6.2) (see also Schultz [6.23]). Special cases of (6.2) in the univariate case were considered by Nitsche [6.13] and in the multivariate case by Oganesjan and Rukhovets [6.16] and by Nitsche [6.14] (see also Kellogg [6.12]). The variational technique in [6.14], whereby an energy projection is shown to be quasi-optimal in the least-squares sense, is sometimes referred to as Nitsche's trick. Convergence for $s>m$ in [6.2], including convergence of residuals, appears to require stability hypotheses, similar to those in earlier sections, going back at least to Polskii [6.19]. One of the interesting facets of (6.2) is the validity for $s<0$; in particular, for $s=-1$, we obtain the average norm which is a measure of oscillation. Frequently such negative norms are required to obtain optimal possible convergence. This has been observed by Bramble and Osborn [6.5] who have shown that such negative norms resolve the discrepancy between the order of convergence of approximate eigenvalues and that of approximate eigenfunctions, which arises when the latter convergence is estimated in the least squares sense (see the papers of Pierce and Varga [6.17, 6.18] for a thorough background in this area). In the approach of [6.5], both rates of convergence are given by the exponent of (6.2ii) (provided $s<0$ is properly chosen).

Conjugate problems, giving lower bounds for estimation, are studied by Aubin and Burchard [6.3]. Babuska [6.4] studies general quadratic forms, not necessarily satisfying (6.1). Least-squares methods, involving trial functions which do not satisfy essential or forced boundary conditions, have been studied by Bramble and Schatz [6.7, 6.8], Bramble and Nitsche [6.6] and by Nitsche [6.15]. Schultz [6.20, 6.21, 6.22] has studied the applications of multivariate approximation theory to Rayleight-Ritz-Galerkin methods while Herbold and Varga [6.11] have studied the effect of quadrature approximations in these methods. Strang and Berger [6.25] and Thomee [6.26] have studied the effect of domain approximations while Ciarlet and Raviart [6.9] have utilized isoparametric methods and quadrature approximations. Finally, Strang [6.24] has executed strategies for various deviations from the usual finite element approach.

Atteia [6.1] presented one of the earliest characterizations of multidimensional spline functions in utilizing a Cartesian product T of partial differential operators with range a closed subspace of $\Pi L^2(\Omega)$ and in [6.2] he considered linear (Hilbert) subspaces of locally convex linear topological spaces together with reproducing kernels. Weinberger [6.27] considered generalizations of the univariate hypercircle methods [1.3] which lead to univariate splines.

In [6.10], S. D. Fisher and the writer consider multidimensional elliptic spline functions which arise through constrained minimization in the essential suprememum norm and the energy norm. In the $L^\infty$ minimization, it was shown that there exist open subsets $\Omega_+$ and $\Omega_-$ of bounded open domains $\Omega$, with $\Omega_+ \cup \Omega_-$ dense in $\Omega$, such that $Ls = \lambda$ on $\Omega_+$ and $Ls = -\lambda$ on $\Omega_-$, where s is a solution of the minimization problem, subject to interpolation constraints $s \in U \subset W^{2m}(\Omega) \cap W_0^m(\Omega)$,

$$\|Ls\|_{0,\infty} = \lambda = \inf_{f\in U} \|Lf\|_{0,\infty}$$

and L is an elliptic differential operator of order 2m mapping $W^{2m,2}(\Omega) \cap W_0^{m,2}(\Omega)$ onto $L^2(\Omega)$. This result extends the notion of one-dimensional perfect spline function. In the $L^2$ minimization, it was shown that, for any symmetric bilinear form $B(u,v)$, satisfying a Garding inequality on $W_0^m(\Omega)$, there exists a grid parameter $h_0$ such that a solution $s \in U \subset W_0^m(\Omega)$, satisfying,

$$B(s,s) = \inf_{u\in U} B(u,u), \tag{6.3}$$

exists for minimization with constrained interpolation on grids of mesh length $0<h\leq h_0$. Furthermore, if $u_0$ is a fixed function and U is defined in terms of agreement with $u_0$ at grid points, then the estimate

$$\|u_0-s\|_{\ell,2} \leq Ch^{m-\ell}\|u_0\|_{m,2}, \quad 0 < h \leq h_0, \quad 0 \leq \ell < m,$$

was obtained, where s solves (6.3). It is assumed $m > N/4$

and $m > N/2$ in the essential supremum and energy minimizations respectively.

## References

[6.1] Atteia, M., Existence et détermination des fonctions "Spline" à plusieurs variables. C. R. Acad. Sc., Paris 262 (1966), 575-578.

[6.2] _______, Fonctions "Spline" et noyaux reproduisants d'Aronszajn-Bergman. R.I.R.O. 4, R-3, (1970), 31-43.

[6.3] Aubin, J. P. and H. G. Burchard, Some aspects of the method of the hypercircle applied to elliptic variational problems. In Numerical Solutions of Partial Differential Equations II, SYNSPADE, B. Hubbard, ed., Academic Press, New York, 1971.

[6.4] Babuska, I., Error bounds for the finite element method. Numer. Math. 16 (1971), 322-329.

[6.5] Bramble, J. H. and J. E. Osborn, Rate of convergence estimates for nonselfadjoint eigenvalue approximations. MRC technical Summary Report 1232, University of Wisconsin, Madison, Wisc., 1972.

[6.6] Bramble, J. H. and J. A. Nitsche, A generalized Ritz-least-squares method for Dirichlet problems. manuscript.

[6.7] Bramble, J. H. and A. H. Schatz, Rayleight-Ritz-Galerkin methods for Dirichlet's problem using subspaces without boundary conditions. Comm. Pure Appl. Math. 23 (1970), 653-675.

[6.8] _______, Least squares methods for $2m^{th}$ order elliptic boundary value problems. Math. Comp., to appear.

[6.9] Ciarlet, P. G. and P. A. Raviart, The combined effect of curved boundaries and numerical integration in isoparametric finite element methods. Proceedings, Symposium on The Mathematical Foundations of the Finite Element Method, University of Maryland, Baltimore, 1972.

[6.10] Fisher, S. D. and J. W. Jerome, Elliptic variational problems in $L^2$ and $L^\infty$, submitted to Indiana J. Math.

[6.11] Herbond, R. J. and R. S. Varga, The effect of quadrature errors in the numerical solution of two dimensional boundary value problems by variational techniques. Aequationes Math. 7 (1972), 36-58.

[6.12] Kellogg, R. B., Difference equations on a mesh arising from a general triangulation. Math. Comp. 18 (1964), 203-210.

[6.13] Nitsche, J., Ein Kriterium für die Quasi-Optimalität des Ritzchen Verfahrens. Numer. Math. 11 (1968), 346-348.

[6.14] ________, Lineare Spline-Funktionen und die Methoden von Ritz für elliptische Randwertprobleme. Arch. Rat. Mech. Anal. 36 (1970), 348-355.

[6.15] ________, Über ein Variationsprinzip zur Lösung von Dirichlet-Problemen bei Verwendung von Teilräumen, die keinen Randbedingungen unterworfen sind. Abh. Math. Sem. Univ. Hamburg 36 (1970/71).

[6.16] Oganesjan, L. A. and P. A. Rukhovets, Investigation of the convergence rate of variational-difference schemes for elliptic second order equations in a two-dimensional domain with a smooth boundary. Ž. Vyčisl. Mat. i Mat. Fiz. 9 (1969), 1102-1120 (Russian).

[6.17] Pierce, J. G. and R. S. Varga, Higher order convergence results for the Rayleigh-Ritz method applied to eigenvalue problems. I: Estimates relating Raleigh-Ritz and Galerkin approximations to eigenfunctions. SIAM J. Numer. Anal. 9 (1972), 137-151.

[6.18] ________, Higher Order convergence results for the Rayleigh-Ritz method applied to eigenvalue problems, II: Improved error bounds for eigenfunctions. Numer. Math. 19 (1972), 155-169.

[6.19] Polskii, N. J., On the convergence of certain approximation processes in analysis. Ukr. Math. J. 7 (1955), 56-70 (Russian).

[6.20] Schultz, M. H., Rayleigh-Ritz-Galerkin methods for multidimensional problems. SIAM J. Numer. Anal. 6 (1969), 523-538.

[6.21] ________, Approximation theory of multivariate spline functions in Sobolev spaces. Ibid., 6 (1969), 570-582.

[6.22] ________, Multivariate spline functions and elliptic problems. In Approximation with Special Emphasis on Spline Functions, I. J. Schoenberg, ed., Academic Press, New York, 1969, pp. 279-347.

[6.23] ________, $L^2$ error bounds for the Rayleigh-Ritz-Galerkin method. SIAM J. Numer. Anal. 8 (1971), 737-748.

[6.24] Strang, G., Variational crimes in the finite element method. Proceedings, Symposium on The Mathematical Foundations of the Finite Element Method, University of Maryland, Baltimore, 1972.

[6.25] Strang, G. and A. Berger, The change in solution due to change in domain. Proceedings, A.M.S. Symposium on Partial Differential Equations, University of California, Berkeley, 1971.

[6.26] Thomee, V., Approximate solution of Dirichlet's problem using approximating polygonal domains. Proceedings, Conference on Numerical Analysis, Royal Irish Academy, Dublin, 1972.

[6.27] Weinberger, H. F., On optimal numerical solution of partial differential equations. SIAM J. Numer. Anal. 9 (1972), 182-198.

---

[1] Research supported in part by National Science Foundation Grant GP 32116.

Department of Mathematics
Northwestern University
Evanston, Illinois 60201

# A REVIEW OF MÜNTZ-JACKSON THEOREMS

D. J. Newman[1]

If two theorems were ever "made for each other" then surely the Müntz and Jackson theorems were.

Müntz' theorem tells us that if $0 < \lambda_1 < \lambda_2 < \ldots$ then the combinations $c_0 + \Sigma_{i=1}^{n} c_i x^{\lambda_i}$ span all of $C[0,1]$ if and only if $\Sigma_{i=1}^{\infty} (1/\lambda_i) = \infty$. Jackson's theorem, on the other hand, says that the (ordinary) polynomials $c_0 + \Sigma_{i=1}^{n} c_i x^i$ span $C[0,1]$ <u>at the exact rate</u> $1/n$. What could be more natural than to try to combine these results and determine the exact rate of spanning by the generalized polynomials of Müntz.

Such a wedding did take place in 1965 [7] but it was a rather lowly $L^2$ ceremony and it wasn't until fairly recently that the results were elevated to the attractive supremum norm setting. We wish to report on some of these developments.

We have said above that ordinary polynomials span $C[0,1]$ at the rate $1/n$. This is a quick way of saying that to each $f(x)$ in $C[0,1]$ and each n there exists an $n^{th}$ degree polynomial $P(x)$ such that

(1) $\|f(x) - P(x)\| \leq A\omega_f(1/n)$.

Here $\|\cdot\|$ means the supremum norm on $[0,1]$, $\omega_f$ is the modulus of continuity of f and A is an absolute constant. The word <u>exact</u> in our phrase indicated further that the above statement becomes false if $1/n$ is replaced by anything substantially smaller than $1/n$. More precisely this means the existence of another (smaller) positive constant $A_1$ with the property that for each n the relation $\|f - P\| \leq A_1\omega_f(1/n)$ is false for some f and all possible P of degree n.

## 1. Guessing the Exact Rate; $L^2$ Approximation

If we try to guess what to replace Jackson's 1/n in (1) by, then we are faced with the following. We desire an expression in $\lambda_1, \lambda_2, \ldots \lambda_n$ such that

(a) It tends to 0 if and only if $\Sigma_{i=1}^{n} (1/\lambda_i)$ tends to $\infty$ (Müntz' theorem).

(b) It is (of order) 1/n when $\lambda_i = i$ (Jackson's theorem).

The first inspired guess is $\exp(-\Sigma_{i=1}^{n} (1/\lambda_i))$, and of course, it is wrong.

There is, namely, a third clue which negates this guess. For by extending f(x) to [-1,1] so as to be even there and by applying Jackson's theorem on [-1,1] we find we are using only the even monomials and doing just as well as with all of them! Thus we have

(c) It is of order 1/n when $\lambda_i = 2i$.

The picture that emerges is that, when discussing integer $\lambda_i$, nothing new happens until the sequence becomes "thinner" than the even integers. We are led to hypothesize $\lambda_{i+1} - \lambda_i \geq 2$ and then to guess the rate

(2) $\varepsilon = e^{-2\Sigma_{i=1}^{n}(1/\lambda_i)}$ .

This turns out to be correct as far as has been stated (within the restriction $\lambda_{i+1} - \lambda_i \geq 2$) and leads to such lovely special cases as $\lambda_i = p_i = i^{th}$ prime. Here one predicts an <u>exact</u> spanning rate of $\log^{-2} n$. The formula (2) for the rate was originally proved for the $L^2$ version of the Müntz-Jackson theorem under the above separation restriction in [7].

Overcoming only slight difficulties, the separation restriction can be shed in the $L^2$ setup, and a new formula for the rate can be obtained which applies to general $\lambda_i$ [8]. Essentially this is

$$(3) \qquad \varepsilon = \varepsilon(n) = \max_{\operatorname{Re} z=1} \left| \frac{B_2(z)}{z} \right|,$$

where $B_2(z)$ is the Blaschke product with zeros at the $\lambda_i + \frac{1}{2}$, namely

$$B_2(z) = \prod_{i=1}^{n} \frac{z - \lambda_i - \frac{1}{2}}{z + \lambda_i + \frac{1}{2}}$$

In the separated case treated earlier the maximum in (3) is taken at $z = 1$, and the result is

$$\varepsilon = \prod_{i=1}^{n} \frac{\lambda_i - \frac{1}{2}}{\lambda_i + \frac{3}{2}}$$

which is of course equivalent to (2).

Again with $\lambda_i = i$, $i=1,2,\ldots$ we have, on $\operatorname{Re} z = 1$, $|z-i| = |z + i - 2|$, so that

$$|B(z)| = \prod_{i=1}^{n} \left|\frac{z - i}{z + i}\right| = \frac{|z|\ |z - 1|}{|z + n - 1|\ |z + n|}.$$

Thus we see that the expression $|B(z)/z|$ is maximized at $z = 1 + i\sqrt{n(n+1)}$ and the maximum value is $(2n+1)^{-1}$ in agreement with the ordinary Jackson theorem.

This final formula or "guess" for the rate, $\operatorname{Max}_{\operatorname{Re} z=1}|B(z)/z|$, which will simplify to $\exp(-2\Sigma_{i=1}^{n} (1/\lambda_i))$ in the separated case ($\lambda_{i+1} - \lambda_i \geq 2$), also simplifies in the exact opposite case. If we call $\lambda_i$ <u>unseparated</u> when $\lambda_{i+1} - \lambda_i \leq 2$ for all $i$, then it is possible to prove that our formula simplifies to

$$\varepsilon = \frac{1}{\sqrt{\Sigma_{i=1}^{n}\lambda_i}}.$$

Since a mere finite number of $\lambda_i$ do not affect the spanning rate by any more than a constant factor, we may delight in the fact that our two cases (separated case where $\lambda_{i+1} - \lambda_i \geq 2$ for all i and unseparated case where $\lambda_{i+1} - \lambda_i \leq 2$ for all i) cover all sequences of "ordinary" growth. Thus all "formula" sequences such as $n^\alpha \log^\beta n \log^\gamma \log n$, ... are eventually separated or eventually unseparated and so succumb to one of our simplified formulas.

One obtains a rather interesting wrinkle if one pushes the unseparated case to its extreme. Imagine, then, that the $\lambda_1, \lambda_2, \ldots, \lambda_n$ are equal to $1+\varepsilon, 1+2\varepsilon, \ldots, 1+n\varepsilon$ where $\varepsilon$ is infinitesimal. Müntz monomials become $x$, $x \log x$, $x \log^2 x, \ldots,$ $x \log^{n-1} x$ and it follows easily from our previously cited formula that the spanning rate becomes $n^{-1/2}$. If we then write $t = -\log x$ the setting changes to the interval $(0,\infty)$ and we find that we are doing approximation with $e^{-t}P(t)$, $P(t)$ an ordinary polynomial of degree less than n. In short we have the Bernstein problem with weight $e^{-t}$. The slight change in the modulus of continuities' meaning when we switch from x to t can easily be absorbed and we obtain a Jackson-Bernstein theorem: The spanning rate of (ordinary) polynomials on $[0,\infty)$ with weight $e^{-t}$ is $n^{-1/2}$.

## 2. Upper Bounds: The Other $L^p$ Norms

Here we treat the positive problem of finding good approximations to arbitrary functions by hook or crook! Of course we attempt to obtain best possible results but our success can only be measured later when we obtain the lower bounds and see how well they match.

It is perhaps not surprising that this aspect of the problem has received the most attention from researchers. Here, after all, is the job of work to be done, an approximators task being to approximate! (Still, it is nice to know that no Martian will ever approach you and say, "Oh you have $\omega(n^{-1/2})$, well we had that 180 years ago. We now have $\omega(n^{-2/3})$.")

We first revisit the $L^2$ variant of the problem if only to see the genesis of the expression

$$\varepsilon = \varepsilon_\Lambda = \max_{\operatorname{Re} z=1} \left|\frac{B_2(z)}{z}\right|$$

The technique for $L^2$ is to consider the dual problem and then to take Fourier transforms. The "spanning rate" is thereby realized as the norm of a certain operator. Indeed, $\varepsilon$ is easily identified with the supremum of the ratio of $\inf_{c_i} \|\psi(z) - \Sigma c_i x^{\lambda_i}\|$ and $\|\psi'(z)\|$ and the usual duality argument (as given for example in [7]) equates this with the infimum of the ratio of $\|\varphi'(x)\|$ and $\|\varphi(x)\|$, for all $\varphi$ satisfying

$$\int_0^1 \varphi'(x)x^{\lambda_i}dx = 0, \qquad i=1,\ldots,n.$$

If we introduce the "Fourier transform"

$$f(z) = \int_0^1 x^{z+\frac{1}{2}} \varphi'(x)dx,$$

then integration by parts gives

$$\frac{f(z)}{z+\frac{3}{2}} = -\int_0^1 x^{z+\frac{1}{2}} \varphi(x)dx$$

and we are led to considerations of the map $f(z) \to f(z+1)/(z+\frac{3}{2})$, or, equivalently, $f(z) \to f(z+1)/(z+1)$, ranging over the Paley-Wiener class and vanishing at $\lambda_1 + \frac{1}{2}, \lambda_2 + \frac{1}{2},\ldots,\lambda_n + \frac{1}{2}$.

Introducing

$$B_p(z) = \prod_{i=1}^{n} \frac{z - (\lambda_i + 1/p)}{z + (\lambda_i + 1/p)}$$

and writing $f(z) = B_2(z)g(z)$ we have, then,

$$\|f(z)\| = \|g(z)\|, \quad \left\|\frac{f(z+1)}{z+1}\right\| = \left\|\frac{B_2(z+1)}{z+1}\, g(z+1)\right\|.$$

Using the elementary fact that $\int|FG|^2 \leq \text{Max}\ |F|^2. \quad \int|G|^2$ this becomes

$$\left\|\frac{f(z+1)}{z+1}\right\| \leq \max_{\text{Re } z=1} \left|\frac{B_2(z)}{z}\right| \|g(z+1)\| \leq \max_{\text{Re } z=1} \left|\frac{B_2(z)}{z}\right| \cdot \|g(z)\|.$$

This establishes $\varepsilon_2 = \text{Max}_{\text{Re } z=1}|B_2(z)/z|$ as an upper bound for the operator norm in question and hence for the spanning rate.

This same approach, dualizing and Fourier transforming, may be used for general $L^p$ norms (as well as the supremum norm) and the problem again reduces to determining the norm of the operator $f(z) \to \frac{f(z+1)}{z+1}$ where the $f(z)$ range through the appropriate space of analytic functions vanishing at $\lambda_i + 1/p$.

What is lacking in this general setup is the simple $L^2$ norm which allowed one in (3) to pull out the Blaschke product and so directly estimate its effect on the norm of this operator. Indeed in all but the $L^2$ case the calculation involves an $L^q$ norm of an inverse Fourier transform. No one in his "right mind" would attempt the necessary Fourier inversion and expect anything decent to result. And so in 1970 Ganelius and his student Westlund [4] performed the bold contour integration involved in the evaluation of the inverse Fourier transform. They had (in the supremum norm) perfect success when $\lambda_{i+1} - \lambda_i \geq s > 2$ and were a mere logarithm off when s=2!

A short time later this technique was improved somewhat and it was found to be successful for all the $L^p$ spaces, $p \geq 2$, while the onerous logarithm was removed for s=2. The general result then read:

If $\lambda_{i+1} - \lambda_i \geq 2$ then $\varepsilon_p$, the spanning rate in $L^p$, is given by

$$\varepsilon_p = \max_{\text{Re } z=1} \left|\frac{B_p(z)}{z}\right|, \text{ for } p \geq 2,$$

where, again, $B_p(z)$ is the Blaschke product with the zeros $\lambda_i + 1/p$, the $L^p$-norm for $p = \infty$ being interpreted as the supremum norm [1].

At that time the case $1 \leq p < 2$ did not succumb perfectly but was off by the customary extra logarithms while the case where the $\lambda_i$ were not separated was quite hazy and imperfect. If, however, one told oneself that only supremum norm results were important and that only integer $\lambda_i$ thinner than the even integers were interesting, then one could feel total bliss.

Next came a startling development! A whole new strategy for approximation was developed by von Golitschek [5]. This consisted in first approximating the given function by an ordinary polynomial $\Sigma_{k=0}^{N} c_k x^k$ and then approximating each of the $x^k$ which occur by Müntz polynomials.

Now, to the sophisticated among us, this approach seems clearly doomed. All of us (in this charmed inner circle) realize the enormity of the coefficients $c_k$ necessary to produce good polynomial approximations. We also know (?) that we cannot expect to approximate $x^k$ by Müntz polynomials very closely. After all $x^k$ has a modulus of continuity like $k \cdot \varepsilon$ and this isn't very good at all. Well, we are dead wrong! Think of the <u>higher smoothness</u> theorems. Our monomial $x^k$ has many (fairly small) derivatives, it might very well be enormously well approximable--and it is.

Following von Golitschek's lead Leviatan [6] refined the technique and found many more cases where this elementary approach led to the same results as the previous Fourier inversion method.

Finally the elementary approach was pushed all the way and was found to give the previous general result and even more [9]. No longer was the separation of the $\lambda_i$ necessary. The new result reads [3]:

For $2 \leq p \leq +\infty$ the spanning rate $\varepsilon_p$ is given by

$$\varepsilon_p = \operatorname*{Max}_{\mathrm{Re}\, z=1} \left| \frac{B_p(z)}{z} \right|,$$

(the $B_p(z)$ defined as before). The germ of the proof is as follows: (supremum norm from now on).

1. $x^k$ can be approximated within $|B(k)|$ by Müntz polynomials.

Let $d\mu(x)$ be any measure for which

$$\int_0^1 |d\mu(x)| \leq 1 \text{ and } \int_0^1 x^{\lambda_i} d\mu(x) = 0, \; i=1,2,\ldots,n.$$

Define $f(z)$ by the relation

$$\int_0^1 x^z d\mu(x) = f(z) \cdot B(z).$$

Since $\int_0^1 x^z d\mu(x)$ is clearly bounded by 1 on the imaginary axis and since $|B(z)| \equiv 1$ there, we conclude that $f(z)$ is bounded by 1 on the imaginary axis. We may apply the maximum modulus theorem to the $H^\infty$ function $f(z)$ and deduce thereby that $|f(k)| \leq 1$. Hence $|\int_0^1 x^k d\mu(x)| \leq |B(k)|$ and the usual duality argument yields the result.

2. $|B(k)| \leq (\varepsilon k)^k$.

Indeed, $|B(k)| \leq |B(1 + i\sqrt{k^2 - 1})|^k$ by elementary estimates and by definition,

$$\left| \frac{B(1+i\sqrt{k^2-1})}{1+i\sqrt{k^2-1}} \right| \leq \varepsilon.$$

Together these estimates tell us that $x^k$ can be approximated within $(k\varepsilon)^k$ and this gives enough to cancel the huge $c_k$ and still have an $\varepsilon$ left for the upper bound.

In attacking the remaining $L^p$ cases ($1 \leq p < 2$) a different technique had to be employed. This was quite amusing and also seemed to expose a kind of duality in the problem. It turned out, namely, that the lower bound technique for $L^q (1/p+1/q=1)$ gave the upper bound result for $L^p$. We illustrate with $p=1$.

We assume that $\lambda_{i+1} - \lambda_i \geq 2$. The lower bound technique will show that the $L^1$ norm of the inverse Fourier transform of $B_\infty(z)/(z+k+1)$ is bounded by 3. But, by partial fraction decomposition,

$$\frac{B_\infty(z)}{z+k+1} = \frac{B_\infty(-k-1)}{z+k+1} + \sum_{i=1}^{n} \frac{c_i}{z+\lambda_i+1} .$$

Hence the inverse Fourier transform is exactly $B_\infty(-k-1)x^k + \Sigma_{i=1}^n c_i x^{\lambda_i}$. Thus

$$\int_0^1 |B_\infty(-k-1)x^k + \sum_{i=1}^{n} c_i x^{\lambda_i}| dx \leq 3$$

and, multiplying by $B_\infty(k+1) = B_\infty^{-1}(-k-1)$, we have an explicit $L^1$ approximation to $x^k$ within $3|B_\infty(k+1)| \leq 3[(k+1)\varepsilon]^{k+1}$.

This time $\varepsilon^{k+1}$ is what is needed to cancel the coefficients and have a single $\varepsilon$ left and we are again swept to success.

An examination of the two apparently diverse approaches (contour integrals vs. monomial approximation) shows something strangely common. The breakthrough in the elementary approach is the inequality $|B(k)| \leq (\varepsilon k)^k$, while the crucial estimate for the Ganelius-Westlund contour integral method was the inequality $|B(z)| \leq A|(\varepsilon|z|)^z|$. Virtually the same. What seems to be behind these theorems is a kind of Stirling formula!

## 3. Lower Bounds

We are able to produce lower bounds for the spanning rates which are virtually the same as the previously obtained upper bounds. Thus we can rejoice at having found the right and correct answers to the problems.

Here the original Fourier analysis method seems indispensible and the scene will always be the z-plane. For starters then, let us go back to the easy $L^2$ case. We are comparing the norms of $g(z)B_2(z)$ and $g(z+1)B_2(z+1)(z+1)^{-1}$. If $z=1+i\tau$ is the place where $\varepsilon_2 = \max_{\text{Re } z=1} |B_2(z)/z|$ is taken on, then we simply choose $g(z) = (z+1-i\tau)^{-1}$. This has norm of size 1 and we need only establish that

$$\int_{1-i\infty}^{1+i\infty} \frac{1}{|z+1-i\tau|^2} \left|\frac{B_2(z)}{z}\right|^2 |dz|$$

has size larger than $\varepsilon_2^2$. Thus we desire that $B_2(z)/z$ not fall off too quickly from its maximum value and this is established by noting that its logarithmic derivative exceeds -2 (again see [8]).

As always $L^2$ is especially simple but it does point the way for the general case. We must seek an f(z) (divisible by $B_p(z)$) such that

$$\left\|\frac{f(z+1)}{z+1}\right\| \geq c\varepsilon_p \|f(z)\|,$$

c being an absolute constant and $\|\cdot\|$ the $L^q$ norm of the inverse Fourier transform.

For $p \geq 2$ this can be achieved by the (rather complicated) choice

$$f(z) = \frac{z^2 s^2}{(z+s)^4} \frac{B_p(z)}{z+1-i\tau}.$$

where $\tau$ is, as before, such that

$$\varepsilon_p = \left|\frac{B_p(1+i\tau)}{1+i\tau}\right| \text{ and } s = \sqrt{1+\tau^2} .$$

The proof is complicated and uses Carlsons estimate for $L^1$ norms of Fourier transforms and Young's theorem for $L^p$ norms [3].

The case $p < 2$ is again different, incomplete, and displays the same amusing property that it did for the upper bound analysis, the upper bound theorem for $L^q$ giving the lower bound for $L^p$. Again we must assume $\lambda_{i+1} - \lambda_i \geq 2$ and we will illustrate with $L^1$.

Using our upper bound technique for $L^1$ we approximate 1 with the terms $x^{\lambda_i - 1}$. The result is, namely,

$$\int_0^1 \left|1 - \sum_{i=1}^{n} c_i x^{\lambda_i - 1}\right| dx \leq 3\varepsilon.$$

It follows that the integrated function $x - \Sigma_{i=1}^{n}(c_i/\lambda_i)x^{\lambda_i}$ has its $L^1$ modulus of continuity, $\omega(\delta)$, bounded by $3\varepsilon\delta$. It is this function which cannot be approximated too well. If it could be gotten much within $\omega(\varepsilon) \leq 3\varepsilon^2$ then this means that x itself could be gotten much within $\varepsilon^2$ and we disprove this by a duality argument.

Simply define

$$\varphi(x) = \int_{-i\infty}^{i\infty} x^{-z} \frac{B_1(z)}{(z+1)^2} dz.$$

Observe that

$$|\varphi(x)| \leq \int_{-\infty}^{\infty} \frac{dy}{|1+iy|^2} = \pi,$$

while

$$\int_0^1 x^{\lambda_i}\varphi(x)dx = \int_{-i\infty}^{i\infty} \frac{1}{\lambda_i+1-z} \frac{B_1(z)}{(z+1)^2} dz = 0$$

by contour integration. Also we obtain

$$\int_0^1 x\varphi(x)dx = \int_{-i\infty}^{i\infty} \frac{1}{2-z} \frac{B_1(z)}{(z+1)^2} dz = -\frac{2\pi i}{9} B_1(2).$$

Thus x cannot be approximated any better than $\frac{2}{9} B_1(2)$ which is indeed not much smaller than $\varepsilon^2$.

We shall allow ourselves the sloppiness of not distinguishing between $\varepsilon, \varepsilon_1, \varepsilon_p$ etc., since it can be shown that they are all equivalent in size. Summarizing the present results, then, the story reads as follows:

For $2 \leq p \leq \infty$, no restriction on the $\lambda_i$, the upper bound and lower bound match at $\varepsilon$.

For $1 \leq p < 2$, the $\lambda_i$ separated, i.e. $\lambda_{i+1} - \lambda_i \geq 2$, the upper bound and lower bound match at $\varepsilon$.

For $1 \leq p < 2$, no restriction on the $\lambda_i$, the upper bound and lower bound are off from $\varepsilon$ by factors of a few logarithms of $1/\varepsilon$.

The "higher smoothness" situation is also treated in [3] and the expected results are obtained.

Concluding, we remark that there seems to be little hope in obtaining decent lower bound estimates by means of the standard Bernstein's method of obtaining upper bounds for the derivatives of Müntz polynomials. Indeed, if we add $x^\lambda$ with a very large $\lambda$ to the $x^k$, $k=0,1,\ldots,n$, then the Bernstein bound must be very large ($(x^\lambda)' = \lambda x^{\lambda-1}$), while the addition of $x^\lambda$ cannot very much affect the approximation index.

## 4. Open Questions

It is not difficult to spot gaps in the present theory, the most embarrassing being perhaps the missing $L^p$ cases, namely $1 \leq p < 2$ when the $\lambda_i$ are not assumed to be separated. Another annoying fact is that Jackson's original theorem is an input and does not emerge as a corollary. This could be remedied by replacing the elementary arguments by the contour integral arguments but the price is a high one in terms of complexity.

Also if one traces some of the special cases one finds interesting results which beg for generalization. We already spoke of the quantitative Bernstein's theorem with weight $e^{-t}$. What of other weights? Another special case is obtained by choosing the $\lambda_i$ to be $1,2,3,\ldots,N$ each with multiplicity N. The monomials are then $x^i(\log x)^j$, the rate being $N^{-3/2}$. We may interpret this as saying that on the curve $y = e^x$, $0 \leq x \leq 1$, one can approximate by polynomials $P(x,y)$ of degree N to within $\omega_f(N^{-3/2})$ [10]. What is the result for other curves?

It seems, however, that the paramount problem is whether this hard work was necessary at all! It is just conceivable that the $L^2$ spanning rate for a set of functions automatically serves as the $L^p$ spanning rate and even as the supremum norm spanning rate. Such a result would be a real boon, allowing us to work only in the cozy comfort of $L^2$. A plausible conjecture along these lines might be the following: Let $(\varphi_1(x),\ldots,\varphi_n(x))$ have $L^2$ spanning rate $\varepsilon$; if $I^{\frac{1}{2}-\frac{1}{p}}$ denotes the operator of fractional integration, then

$$(I^{\frac{1}{2}-\frac{1}{p}}\varphi_1(x),\ldots,I^{\frac{1}{2}-\frac{1}{p}}\varphi_n(x))$$

has $L^p$ spanning rate $\varepsilon$. (This would explain the irksome $1/p$ which must be added to the $\lambda_i$). We would then recognize the Müntz monomical case as simply one where fractional integration takes an especially simple form. Again, Fourier series, basically an $L^2$ invention would be explained in $L^p$!

---

[1]Research supported in part by Grant AFOSR 72380.

References

[1] Bak, J. and D. J. Newman, Müntz-Jackson theorems in $L^p[0,1]$ and $C[0,1]$. Amer. J. Math. 94 (1972), 437-457.

[2] Bak, J. and D. J. Newman, Müntz-Jackson theorems in $L^p$, $p<2$. To appear, J. Approx. Theory.

[3] Bak, J., D. Leviatan and D. J. Newman, Generalized polynomial approximation. To appear in Israel J. Math.

[4] Ganelius, T. and S. Westlund, The degree of approximation in Müntz's theorem. To appear in Proceedings of the Conference on Math. Analysis, Jyväskylä, Finland, August 1970.

[5] von Golitschek, M., Erweiterung der Approximationssätze von Jackson im Sinne von Ch. Müntz II. J. Approx. Theory 3 (1970), 72-86.

[6] Leviatan, D., On the Müntz-Jackson theorem. To appear in J. Approx. Theory.

[7] Newman, D. J., A Müntz-Jackson theorem. American J. Math. 87 (1965), 940-944.

[8] __________, Müntz-Jackson theorem in $L^2$. To appear, J. Approx. Theory.

[9] __________, A general Müntz-Jackson theorem. To appear, Amer. J. Math.

[10] Newman, D. J. and L. Raymon, Quantitative polynomial approximation on certain planar sets. Trans. A.M.S. 136 (1965), 247-259.

Department of Mathematics
Yeshiva University
New York, N.Y. 10033

## ADDITIONAL REMARKS TO THE PAPER OF D. J. NEWMAN

M. v. Golitschek

D. J. Newman was the first in 1965 to formulate and prove a Jackson-Müntz theorem in the $L^2$-norm. Since then, the interest in theorems of this type has grown considerably. As an addition to D. J. Newman's talk I will summarize some results of may paper [2], which is to appear in the Journal of Approximation Theory in 1973. Some of the theorems of D. J. Newman are confirmed and extended in this paper, and perhaps some questions previously considered open are answered. I have proved my subsequent results by elementary methods: By approximating the given function by ordinary polynomials, and then by substituting appropriate Müntz polynomials for the monomials $x^q$.

### 1. Supremum-norm

Let $0 = \lambda_0 < \lambda_1 < \dots$ be a sequence of real numbers; we define

$$E_s(f;\lambda_i) = \inf_{a_i} \|f(x) - \sum_{i=0}^{s} a_i x^{\lambda_i}\|_{C[0,1]} \ .$$

If there exists a real number $\rho > 0$ such that $\lambda_i \geq \rho i$ for $i \geq i_0$, then for each $f \in C[0,1]$

(1) $E_s(f;\lambda_i) \leq K\ \omega(f; \{\varphi(s)\}^{-d})$,

where $\omega(f;\cdot)$ is the modulus of continuity, $d = \min\{2;\rho\}$ and $\varphi(s) = \exp(\Sigma_{i=1}^{s} (1/\lambda_i))$. The function $\varphi$ plays the same important role as in D. J. Newman's talk for the separated case ($\lambda_{i+1} - \lambda_i \geq 2$). In an earlier paper [1] I have proved that (1) is best possible for the sequence $\lambda_i = \rho i$, $i=0,1,\dots$, $0<\rho<\infty$.

In the following example the sequence $\{\lambda_i\}$ does not tend to $\infty$ too fast. Let $\lambda_i = i^\beta$, $i=0,1,\ldots$, $0<\beta<1$. Then

(2) $E_s(f;i^\beta) \leq K\ \omega(f;s^{-(1+\beta)/2})$.

Moreover, if the sequence $\{\lambda_i\}$ converges, i.e. $\lim_{i\to\infty} \lambda_i = \lambda$, $0<\lambda<\infty$, then we obtain

(3) $E_s(f;\lambda_i) \leq K\ \omega(f;s^{-1/2})$.

Finally, if $\lim_{i\to\infty} \lambda_i = 0$, $\lambda_0 = 0$, $\lambda_i > 0$, then

(4) $E_s(f;\lambda_i) \leq K\ \omega(f;\ \{\sum_{i=1}^{s} \lambda_i\}^{-1/2})$.

For the unseparated case ($\lambda_{i+1} - \lambda_i \leq 2$ for all i) D. J. Newman obtains a "spanning rate of $\{\Sigma_{i=1}^{s}\ \lambda_i\}^{-1/2}$." At first glance this result seems to differ much from the inequalities (1) - (3). Actually, however, D. J. Newman's and my results in (2) - (4) are essentially the same, because

$$\sum_{i=1}^{s} \lambda_i \approx s^{1+\beta} \text{ for } \lambda_i = i^\beta, \quad \sum_{i=1}^{s} \lambda_i \approx s \text{ for } \lim_{i\to\infty} \lambda_i = \lambda,\ \lambda>0.$$

For the separated case my inequality (1) agrees with Newman's result. But our formulas differ for the unseparated case. Moreover, many sequences are neither separated nor unseparated, but satisfy the condition $\lambda_i \geq \rho i$ for some $\rho > 0$ and all $i \geq i_0$.

## 2. $L^p$-norms, $1 \leq p < \infty$

Let $-1/p < \lambda_1 < \lambda_2 < \ldots$. Using again the "substitution method" I obtain upper bounds for the best approximation in the $L^p$-norms. Here, instead of the integral modulus of continuity, I use the following concept.

Let $f \in L^p(0,1)$. Let there exist a function $F \in L^p(-1,1)$ and a sequence of polynomials $P_n(x) = \Sigma_{i=0}^{n}\ a_{in}x^i$ such that

$$f(x) = F(x) \text{ for } 0 \leq x \leq 1, \quad \|F - P_n\|_{L^p(-1,1)} \leq M\, n^{-\gamma},$$

$$n=1,2,\ldots,$$

where M and $\gamma$ are two positive constants.

If $\lambda_i \geq \rho i$ for $i \geq i_o$, and $d = \min\{2;\rho\}$, then we obtain for arbitrary $\epsilon > 0$

$$E_s(f;\lambda_i)_p \leq C_\epsilon\{\varphi(s)\}^{\epsilon-\gamma d}$$

$$+ K_\epsilon \sum_{q=0,q\neq\lambda_i}^{k} c_q|f^{(q)}(+0)|\{\varphi(s)\}^{\varepsilon-2q-2/p},$$

where the second sum vanishes for $\gamma \leq 1/p$ and where $k = [\gamma - 1/p]$, if $\gamma - 1/p$ is not an integer, and $k = \gamma - 1/p - 1$ otherwise.

If $\lim_{i\to\infty} \lambda_i = \lambda$, $-1/p < \lambda < \infty$, then

$$E_s(f;\lambda_i)_p \leq Cs^{-\gamma/2}$$

$$+ 2^{1/p} \sum_{q=0,q\neq\lambda_i}^{k} \frac{1}{q!}|f^{(q)}(+0)|\exp(-cs/(q+1)),$$

where $c > 0$ depends only on $\{\lambda_i\}$.

Many other examples are calculated in my paper. I believe that for arbitrary positive sequences $\{\lambda_i\}$ the "substitution method" yields good upper bounds for the best approximation $E_s(f;\lambda_i)_p$.

## References

[1] v. Golitschek, M., Die Sätze von Jackson für Polynome $\Sigma_{i=0}^{s} a_i x^{p_i}$. Dissertation, Würzburg, W. Germany, 1969, 89 pp.

[2] __________, Jackson-Müntz Sätze in der $L^p$-Norm. J. Approx. Theory 7 (1973), 87-106.

Institut für Angewandte Mathematik
Universität Würzburg
8700 Würzburg
West Germany

# APPROXIMATION BY $C^k$-FUNCTIONS

Daniel Wulbert

Let M be a finite-dimensional $C^k$-manifold, and let p be a positive continuous function on M. If f is a continuous real function on M, there is a $C^k$-function g such that $|f(m) - g(m)| < p(m)$ for all $m \in M$. If M is infinite dimensional, the theory is more delicate. However in the last few years a series of remarkable papers have finally settled some of the original conjectures. This article is a survey of the state of the theory of $C^k$-approximation of functions. We will outline a series of results leading to the identification of the currently known Banach manifolds which admit $C^k$-approximations. Although we will not prove all the results presented here, a surprising number of results are accessible by simple and elegant arguments. Our outline favors such results over results of greater generality.

The final section contains historical notes, indications of related results, and the statements of open problems.

Notation. Throughout this paper M will denote a metrizable $C^k$-manifold modelled on a Banach space E ($k=0,1,2,\ldots,\infty$). The dual of E and the closed unit ball of E are written $E^*$ and S(E).

Let F be a Banach space. The set of F-valued functions on M which have k continuous derivatives will be denoted by $C^k(M,F)$. When $k=0$, or F is the real line, R, the k or R respectively is deleted. For $f \in C^k(M)$, carr $f = \{x \in M : f(x) \neq 0\}$, is the carrier of f.

If $A \subseteq M$, card A is the cardinality of A, and dens A $= \min \{\text{card } B : B \text{ is a dense subset of } A\}$. The closure of A is written cl A, and for x in M, $d(x,A) = \min \{d(x,a) : a \in A\}$.

The composite of two functions f and g is written $f \circ g$. The k-linear functional which is the $k^{th}$ derivative at x of a function f defined on E is written $D_x^k f$. However we will write $D_x f$ for $D_x^1 f$.

For an arbitrary set A, and a Banach space F

$$\ell_\infty(A,F) = \{f : f \text{ is an } F\text{-valued function on } A \text{ such that } \|f\| \equiv \max \|f(a)\| < \infty\}$$

$$c_o(A,F) = \{f \in \ell_\infty(A,F) : \{a \in A : \|f(a)\| > r\} \text{ is finite for } r > 0\}, \text{ and}$$

$$\ell_2(A,F) = \{f \in c_o(A,F) : \|f\|_{\ell_2}^2 \equiv \sum_{a \in A} [f(a)]^2 < \infty\}.$$

When $F = \mathbb{R}$ the symbol $\mathbb{R}$ is deleted. If in addition A is countable the corresponding spaces are written $\ell_\infty$, $c_o$, and $\ell_2$. When the context permits, we will delete the subscript, $\ell_2$, on the norm for $\ell_2$.

Let $\{U_i\}$ be an open cover of M. A collection of nonnegative functions $\{f_j\} \subset C^k(M)$ is a $C^k$-partition of unity subordinate to $\{U_i\}$, if for each j there is an i such that carr $f_j \subseteq U_i$, and if $\Sigma f_j \equiv 1$. The manifold M is said to admit $C^k$-partitions of unity if for each open cover $\mathcal{U}$ of M there is a $C^k$-partition of unity subordinate to $\mathcal{U}$.

The space $C(M,F)$ admits $C^k$-approximations if for each $f \in C(M,F)$ and each positive continuous function p on M there is a $g \in C^k(M,F)$ such that $\|f(m)-g(m)\| \leq p(m)$.

## 1. $C^k$-Partitions of Unity

Before exhibiting manifolds M for which C(M) admits $C^k$-approximations, we reformulate the approximation problem. The first step is to show that C(M) admits $C^k$-approximations if and only if E admits $C^k$-partitions of unity. The proof of this reformulation uses an argument on paracompactness. For example, we will use the following elementary fact.

Let A and B be closed subsets of M. We will say that a real function f separates A and B if $f(a) > 0$ for all $a \in A$, and $f(b) = 0$ for $b \in B$. Since M is paracompact, M admits $C^k$-partitions of unity if and only if every two closed subsets of M can be separated by a nonnegative $C^k$-function.

Proposition 1.1. Let F be a normed linear space. C(M,F) admits $C^k$-approximations if and only if M admits $C^k$-partitions of unity.

Proof: Suppose M admits $C^k$-partitions of unity, f is a continuous function of M into F, and g is a positive real function on M. There is an open cover $\{U_n\}$ of M which has the property that for y and z in $U_n$,

$$\|f(y) - f(z)\| \leq \inf \{g(w) : w \in U_n\}$$

Let $\{\psi_n\}$ be a $C^k$-partition of unity subordinate to $\{U_n\}$. Let $x_n \in \operatorname{carr} \psi_n$. Then $\|f(x) - \Sigma f(x_n) \psi_n(x)\| \leq g(x)$.

Now suppose C(M,F) admits $C^k$-approximations. Let A and B be closed subsets of M. Let f be a continuous real function on M such that $f(A) = 3$, and $f(B) = 0$. Let k be a norm-one vector in F. Let p be a linear projection of F onto the span of k. Let $h \in C^k(M,F)$ satisfy $\|f(x)k - h(x)\| < 1/\|p\|$. Then there is q in $C^k(M)$ such that $q(x)k = p \circ h(x)$. Let $r \in C^\infty(R)$ and carry $(-\infty,1)$ into zero and $(2,\infty)$ into 1. One can now verify that $r \circ q$ is a $C^k$-function that separates A and B. Hence M admits $C^k$-partitions of unity.

Lemma 1.2. If M admits $C^k$-partitions of unity then every open submanifold of M admits $C^k$-partitions of unity. Conversely if each point of M is in an open submanifold of M which admits $C^k$-partitions of unity, then M also admits $C^k$-partitions of unity.

Proof: Let $\mathcal{U}$ be an open cover of an open submanifold W. A short argument using paracompactness shows that if there is a refinement $\{V_i\}$ of $\mathcal{U}$ such that $\mathrm{cl}V_i \subseteq W$, for each i, then W also admits a $C^k$-partition of unity. Since any common refinement of $\mathcal{U}$ and

$$\{\{w \in W : d(w,M-W) > 1/n\}\}_{n=1}^{\infty}$$

has this property, W admits $C^k$-partitions of unity.

For the converse, let $\mathcal{U}$ be an open cover of M. From the hypothesis, there is a locally finite refinement of $\mathcal{U}$, $\{V_i\}$, such that for each i, $\mathrm{cl}V_i$ is contained in an open submanifold which admits partitions of unity. Using again paracompactness, one constructs, from this, a partition of unity of M subordinate to $\mathcal{U}$.

The following is now immediate.

Proposition 1.3. M admits $C^k$-partitions of unity if and only if E admits $C^k$-partitions of unity.

The next theorem is presented partly as an example of an application of partitions of unity.

Proposition 1.4. Let M be a paracompact $C^k$-manifold modelled on E. If M admits $C^k$-partitions of unity then M can be $C^k$-embedded onto a closed submanifold of $\ell_2(A, R \times E)$, where card A = dens M.

Proof: We can find $\{U_a, \phi_a, q_a\}_{a\varepsilon A}$ such that $\{U_a, \phi_a\}_{a\varepsilon A}$ is an atlas of M, $q_a$ is a $C^k$-partition of unity on M, and $U_a$ = carr $q_a$. For each m $\varepsilon$ M let,

$$P_a(m) = q_a(m)/[\sum_{a\varepsilon A} q_a^2(m)]^{1/2}, \text{ and}$$

$$f(m)_a = (P_a(m), P_a(m)\phi_a(m))$$

The corresponding map $f: m \to \{f(m)_a\}_{a\varepsilon A}$ is a $C^k$-mapping of M into $\ell_2(A, R \times E) - \{0\}$.

If $f(x_n) \to y$, there must be an $a \in A$ for which $P_a(y_n)$ converges to a nonzero number. Hence $\phi_a(y_n)$ converges, and $x_n$ converges to say x. Therefore $f(x) = y$. Therefore f is a homeomorphism, and also f(M) is closed.

Now let $(U,\phi)$ be a chart, and let $x \in \phi(U)$. To complete the proof it suffices to show that $D_x(f\circ\phi^{-1})$ is an isomorphism of E onto a split subspace of $\ell_2(A, R\times E)$ (the inverse function theorem).

We first show that $D_x(f\circ\phi^{-1})$ is an isomorphism. If $\phi^{-1}(x) \in U_a$, then

$$D_x(f\circ\phi^{-1}) = D_{\phi_a\circ\phi^{-1}(x)}(f\circ\phi_a^{-1}) \circ D_x(\phi_a\circ\phi^{-1}).$$

Since $D_x(\phi_a\circ\phi^{-1})$ is an isomorphism it suffices to show that $D_y(f\circ\phi_a^{-1})$ is an isomorphism for $y \in \phi_a(U_a)$. For $w \in E$

$$D_y(f_a\circ\phi_a^{-1})(w) = (D_y(P_a\circ\phi_a^{-1})(w), [D_y(P_a\circ\phi_a^{-1})(w)]y + P_a(\phi_a^{-1}(y))w).$$

Hence $D_y(f_a\circ\phi_a^{-1})(w_n) \to 0$ implies $w_n \to 0$, and $D_y(f_a\circ\phi_a^{-1})$ is an isomorphism of E into $(R\times E)$.

Furthermore if N is the null space of $D_y(P_a\circ\phi_a^{-1})$ then $(\{0\}\times N)$ is contained in the range of $D_y(f_a\circ\phi_a^{-1})$. Therefore $D_y(f_a\circ\phi_a^{-1})(E)$ has finite codimension in $R\times E$ and is complemented.

It follows from the closed graph theorem that $D_y(f\circ\phi_a^{-1})$ is also an isomorphism which has a complemented range. This completes the proof.

<u>Corollary 1.5</u>. A paracompact Banach manifold is metrizable.

<u>Corollary 1.6</u>. If E is a Hilbert space, M can be $C^k$-embedded onto a closed submanifold of $\ell_2(A)$ where card A = dens M (see Proposition 3.7).

If M is separable, the same proof shows that M can be $C^k$-embedded onto a closed submanifold of $c_0(I, R\times E)$, where I is the integers. The only modification required is changing the definition of $P_a$ to

$$p_a(m) = q_a(m)/\eta(q(m)),$$

where $\eta$ is an equivalent $C^\infty$-norm on $c_o$.

## 2. $C^k$-smoothness

An open set $U \subseteq E$ is a $C^k$-carrier if there is a nonnegative $C^k$-function f such that U = carr f. Obviously translations, dilations, unions, and intersections of $C^k$-carriers are $C^k$-carriers, also if f is a $C^k$-function $\{x : f(x) > a\}$ is a $C^k$-carrier.

Definition. E is $C^k$-smooth if E contains a bounded $C^k$-carrier.

Theorem 2.1. Let E be a separable Banach space. E admits $C^k$-partitions of unity if and only if E is $C^k$-smooth.

Proof: The necessity is obvious. To prove the sufficiency we show that any two disjoint closed sets A and B can be separated by a nonnegative $C^k$-function. Since E is second countable there is a countable family of nonnegative $C^k$-functions $\{f_i\}$ such that if $U_i$ = carr $f_i$, $\{U_i\}$ covers E, and $U_i \cap A \neq \emptyset$ implies $U_i \cap B = \emptyset$. Let

$$V_r = \{x : f_r(x) > 0,\ f_i(x) < 1/r,\ i < r\}.$$

Let $p_i$ be a nonnegative $C^k$-function with carrier $V_i$. Let $J = \{i : V_i \cap A \neq \emptyset\}$. Since $\{V_i\}$ is a locally finite cover of E, $\Sigma\{P_i : i \in J\}$ is a $C^k$-function which separates A and B.

Since we will be able to characterize the $C^1$-smooth Banach spaces, the last theorem identifies the separable spaces which admit $C^1$-partitions of unity. The first lemma below is an adaptation, to our setting, of a known result.

Definition. A submanifold H of a Banach space E is a hypersurface if for each $h \in H$ the tangent space at H, T(h), has codimension one in E.

Let

$$N(h) = \{L \in E^* : \|L\| = 1, T(h) = L^{-1}(0)\}, \text{ and}$$

$$N(H) = \cup\{N(h) : h \in H\}$$

<u>Lemma 2.2</u>. If H is a closed and bounded $C^1$-hypersurface in E, then N(H) is dense in $\{L \in E^* : \|L\| = 1\}$.

<u>Lemma 2.3</u>. If E contains a closed bounded $C^1$-hypersurface, H, then dens E = dens E*.

<u>Proof:</u> It is always true that dens E $\leq$ dens E*. Since N(h) is continuous we have,

$$\text{dens } E \geq \text{dens } M \geq \text{dens } N(H) = \text{dens } \{L \in E^* : \|L\| = 1\}$$

$$= \text{dens } E^*.$$

<u>Theorem 2.4</u>. If E is $C^1$-smooth then dens E = dens E*.

<u>Proof:</u> Let g be a nonzero $C^1$-function with carr g $\subseteq$ $\{x : \|x\| \leq 1\}$, and $0 \leq g \leq 1$. We define a new function on E - {0},

$$h(x) = \int_{-\infty}^{\infty} g(sx)ds.$$

Then h has the following properties: (1) h is $C^1$, (2) th(tx) = h(x) for t real, (3) $h^{-1}(1) \subseteq \{x : \|x\| \leq 1, x \neq 0\}$, and (4) for all $x \neq 0$, $D_x h \neq 0$. From the implicit function theorem $h^{-1}(1)$ is a closed and bounded $C^1$-hypersurface.

The next theorem will show that when E is separable, dens E = dens E* implies that E is $C^1$-smooth. In general, though, this is false. For example if A is uncountable, dens $(\ell_1 \times \ell_2(A))$ = dens $(\ell_\infty \times \ell_2(A))$. But $\ell_1 \times \ell_2(A)$ contains a copy of $\ell_1$. The above theorem shows that $\ell_1$ is not $C^1$-smooth.

<u>Definition</u>. A norm on E is a $C^k$-norm if it is a $C^k$-function on E - {0}.

If E has an equivalent $C^k$-norm, then E is $C^k$-smooth.

However, except for a few specific spaces, it is not generally known which spaces have $C^k$-norms. A recent powerful theorem by S. L. Trojanski clarifies the special case for k=1.

Theorem 2.5. There is an equivalent $C^1$-norm on E if either (a) E* is separable, or (b) (Trojanski) E is reflexive.

Putting together several previous results, we can characterize the separable Banach spaces E such that C(E) admits approximation by $C^1$-functions. The characterization is the paradigm on which the conjectures for $C^k$-approximation of functions on nonseparable spaces are based.

Corollary 2.6. Let E be a separable Banach space. The following are equivalent: (1) E admits $C^1$-partitions of unity, (2) E is $C^1$-smooth, (3) dens E* = dens E, and (4) there is an equivalent $C^1$-norm on E.

## 3. Specific Spaces

If X is a compact Hausdorff space, then dens C(X)* = card X. Hence

Proposition 3.1. If X is an uncountable, compact, metric space C(X) is not $C^1$-smooth.

Proposition 3.2. If A is any infinite set $\ell_\infty(A)$ is not $C^1$-smooth.

Proposition 3.3. $\ell_1(A)$ is not $C^1$-smooth.

Let $L^p (1 \leq p < \infty)$ denote the space $L^p(S,\Sigma,\mu)$ of real functions defined for the positive measure space $(S,\Sigma,\mu)$. The smoothness of the $L^p$ spaces depends on the specific number p. For example if p is an even integer, the function $\|f\| = \int |f|^p$ is the diagonal of the multilinear functional

$$m(f_1,f_2,\ldots,f_p) = \int \prod_{i=1}^{p} f_i .$$

Hence $\|f\|$ is $C^\infty$ and $L^p$ is $C^\infty$-smooth.

If p is not an even integer $L^p$ is not $C^\infty$-smooth. The precise smoothness is given by the following result. If p is not an even integer let a(p) be the largest integer strictly smaller than p. If p is an even integer let $a(p) = \infty$.

Proposition 3.4. $\|\cdot\|_p \in C^{a(p)}(L^p)$. If p is not an even integer $L^p$ is not $a(p) + 1$ smooth.

It is easy to show that $c_o(A)$ is $C^\infty$-smooth. Let g be a nonnegative $C^\infty$-real-valued function on the real line such that $g(x) = 1$ if $|x| \leq 1/2$ and $g(x) = 0$ if $|x| \geq 1$. Define,

$$h(x) = \prod_{a\varepsilon A} g(x_a) \qquad \text{for } x = \{x_a\}_{a\varepsilon A} \ \varepsilon\ c_o(A).$$

Since H is a $C^\infty$ function with bounded carrier; $c_o(A)$ is $C$-$^\infty$ smooth.

The following more difficult result was announced by H. Tarunczyk.

Theorem 3.5. $c_o(A)$ admits $C^\infty$-partitions of unity.

Tarunczyk's theorem is the fundamental result on partitions of unity for nonseparable spaces. For example it implies that all Hilbert spaces admit $C^\infty$-partitions of unity. The applications of the above theorem require the following lemma.

Lemma 3.6. If there is a homeomorphism h of a Banach space E into $c_o(A)$ such that h is a $C^k$-mapping, then E admits $C^k$-partitions of unity.

Proof: Let $\{U_i\}$ be an open cover of E. Let $V_i \subseteq c_o(A)$ be an open set such that $h^{-1}(V_i) = U_i$. Let $\{f_j\}$ be a $C^\infty$-partition of unity of $\cup V_i$ subordinate to $\{V_i\}$. Then $\{f_i \circ h\}$ is a $C^k$-partition of unity of E subordinate to $\{U_i\}$.

It is now immediate that a Hilbert space admits $C^\infty$-partitions of unity.

Proposition 3.7. Every Hilbert space admits $C^\infty$-partitions of unity.

Proof: For $\{x_a\} \in \ell_2(A)$, define

$$h(\{x_a\}) = \{\Sigma\, x_a^2, \{x_a\}\} \in c_o(\{0\} \cup A)$$

Then h is a $C^\infty$-mapping which is a homeomorphism of $\ell_2(A)$ into $c_o(\{0\} \cup A)$.

A refinement of the above proof for Hilbert space can be applied to the $L^p$ spaces. The homeomorphism needed depends on a construction of Lindenstrauss.

Lemma 3.8. If E is reflexive, there is a set A and a one-to-one bounded linear mapping, L, of E into $c_o(A)$.

Let a(p) be as defined before.

Proposition 3.9. $L^p$ admits $C^{a(p)}$-partitions of unity.

Proof: Let L be a bounded one-to-one linear mapping of $L^p$ into $c_o(A)$. Define

$$L(f) = \{\|f\|^p, L(f)\} \in c_o(\{0\} \cup A).$$

Clearly, h is a one-to-one, $C^{a(p)}$-mapping. If $h(f_n)$ converges to h(f), then $\|f_n\|$ converges to $\|f\|$. Moreover, since $L^p$ is reflexive and the kernel of L = {0}, $f_n$ converges weakly to f. Since $L^p$ is locally uniformly convex $f_n$ converges to f [Day, 1962, VII, 2].

From the results on the $L^p$ spaces, the best result possible for general reflexive spaces would be that they admit $C^1$-partitions of unity. This, in fact, is true. An averaging process of Asplund [1967] combined with Trojanski's renorming theorem shows that a reflexive space E has an equivalent norm $|||\ \ |||$ which is both $C^1$ on E - {0}, and which is locally uniformly convex. Hence, if, in the definition of the homeomorphism h is the last proof, we replace $\|f\|^p$ with $|||f|||^2$ we have proved the following.

Proposition 3.10. A reflexive space admits $C^1$-partitions of unity.

## 4. Higher Order Derivatives

Although continuous functions on $c_0$ can be approximated by $C^\infty$-functions, a theorem by J. Wells [1969] shows that they cannot be approximated by $C^2$-functions which have bounded second derivatives.

Proposition 4.1. A real-valued $C^1$-function defined on $c_0$ which has a uniformly continuous derivative has an unbounded carrier.

Proof: Suppose that there is a $C^1$ function f, and an integer N such that $f(0) = 1$, $f(x) = 0$ for $\|x\| \geq 1$ and $\|D_{x+h}f - D_x f\| < 1/2$ whenever $\|h\| \leq 1/N$.

Suppose there exist points $\{x_i\}_{i=0}^M \subseteq (c_0)$, $x_0 = 0$ such that

$$\text{(i)} \quad \|x_k - x_{k-1}\| \leq 1/N, \qquad \text{(ii)} \quad \|x_M\| \geq 1$$

$$\text{(iii)} \quad \sum_{k=1}^{M} \|x_k - x_{k-1}\| \leq 2, \qquad \text{(iv)} \quad D_{x_k} f\,(x_k - x_{k-1}) = 0.$$

Then we derive the following contradiction:

$$|f(x_M) - f(0)| \leq \sum_{k=1}^{M} |f(x_k) - f(x_{k-1}) - D_{x_{k-1}} f(x_k - x_{k-1})|$$

$$\leq \sum \frac{1}{2} \|x_k - x_{k-1}\| < 1.$$

The second inequality above follows by applying the mean value theorem to the function

$$g(h) = f(x_k + h) - D_{x_{k-1}} f(h)$$

Hence it remains to find points $x_k$ in $c_0$ satisfying (i) - (iv) above. However, this is not difficult. Let

$$A = \{x(i) \in c_0 : x(i) = 0,\ i > 2^N,\ |x(i)| \leq 1/N \text{ for } i \leq 2^N, \text{ and } |x(i)| = 1/N \text{ for all except possibly one } i \leq 2^N\}.$$

Let $h_0 = 0$. Suppose we have found points $\{h_j\}_{j=1}^{k+1} \in A$, such that $D_{s(k)}f(h_{k+1})=0$, where $s(k) = \Sigma_{j=0}^{k} h_j$. Also assume there exist $2^{N-k}$ coordinates, $B_k$, such that $i \varepsilon B_k$ implies $|\Sigma_{j=1}^{k+1} h_j(i)| = k/N$.

Since A is a connected symmetric set there is an $h_{k+2}$ in A for which $D_{s(k+1)}f(h_{k+2}) = 0$, where $s(k+1) = \Sigma_{j=0}^{k+1} h_j$, and $h_{k+2}$ agrees with $h_j$ $(j \leq k+1)$ on at least half of the coordinates in $B_k$. Let $B_{k+1}$ denote these coordinates. Hence $\{h_i\}_{i=0}^{k+2}$ also satisfies the two conditions above.

Let $x_k = \Sigma_{i=0}^{k} h_i$. Then $\{x_k\}_{k=1}^{N}$ satisfies conditions (i) - (iv) and the proof is complete.

Let p be a continuous positive real function on $c_o$ which vanishes at infinity. Wells' theorem implies that there does not exist a nontrivial $C^2$-function f with bounded second derivative such that $|f(x)| \leq p(x)$. Hence if f is a $C^2$-function which is not $C^3$, it is not possible simultaneously to approximate f and its second derivative.

N. Moulis [1971] has proved the following:

<u>Proposition 4.2</u>. Let f be a $C^1$-function defined on $c_o$. Let $p_1$ and $p_2$ be continuous positive functions on $c_o$. There is a $C^\infty$-function g on $c_o$, such that $\|f(x) - g(x)\| < p_1(x)$ and $\|D_xf - D_xg\| < p_2(x)$ for all $x \varepsilon c_o$.

<u>Proposition 4.3</u>. Let f be a $C^{2k-1}$-function on $\ell_2$. Let $p_i$ be continuous positive functions on $\ell_2$ for $i=0,1,2,\ldots,k$. There is a $C^\infty$-function g on $\ell_2$ for which $\|D_x^if - D_x^ig\| < p_i(x)$ for $i=0,1,\ldots,k$ and for all x in $\ell_2$.

## 5. Notes and Open Problems

Proposition 1.4 is a typical example of Banach manifold embedding theorems. McAlpin introduced embeddings of M into an infinite product of $R \times E$ in his original proof that a separable $\ell_2$-manifold can be embedded in $\ell_2$ [see, Abraham, 1963]. Some major refinements, and variants are in [Bonic and

Frampton, 1966] and [Bonic, Frampton and Tromba, 1969] also see [Colojora, 1965].

The general problem is still open.

<u>Problem.</u> Which E-manifolds can be split embedded in E?

To be more specific:

<u>Problem.</u> If E is reflexive, can an E-manifold be embedded in E?

These problems are related to the following Banach space problem [see Kuiper, 1966, Theorem 3.3].

<u>Problem.</u> If E is reflexive, is E isomorphic to $E \times E$?[1]

The most striking embedding theorem of course is the following well-known recent result:

<u>Theorem.</u> A separable $C^\infty$-manifold modelled on $\ell_2$ is $C^\infty$-diffeomorphic to an open subset of $\ell_2$.

For the strongest versions of this theorem see [Eells and Elworthy, 1970], [Kuiper and Burghelea] and [Moulis, 1971, p. 331].

Except for proposition 1.4, we always hypothesize that M is metrizable. However the paracompactness of M is the only essential property used in proofs. Corollary 1.5, of course, shows that the seemingly more general hypothesis of paracompactness is spurious. A different proof for Corollary 1.5, for $C^1$-manifolds is given in [Palais, 1965]. Lemma 1.3 is from [Palais, 1965].

Theorem 2.1 is from [Bonic and Frampton, 1966]. The general theorem is still open.

<u>Problem.</u> Does every $C^k$-smooth Banach space admit $C^k$-partitions of unity? k=1?

<u>Problem.</u> Does a $C^k$-smooth reflexive space admit $C^k$-partitions of unity?

Proofs for Theorem 2.5(a) are in [Klee, 1960], [Kadec, 1958, 1965], [Rainwater, 1966] and [McKessock and Whitfield].

Trojanski's theorem (2.5(b)) [1971] answered a long-standing conjecture. It was known that for each reflexive space, E, there exists an equivalent Gateaux differentiable norm on E - {0} [Lindenstrauss, 1966] and [Amir and Lindenstrauss, 1968]. Trojanski's proof is a delicate construction which uses techniques from [Amir and Lindenstrauss, 1968]. Other significant studies are [Cudia, 1964], [Day, 1955, 1962], [Clarkson, 1936], [Kadec 1955] [Smulian, 1940, 1954], and [Lovaglia, 1955].

Restrepo [1964] proved that if a separable space E has an equivalent $C^1$-norm, then E* is also separable (the converse followed from Kadec's theorem mentioned above). Leach and Whitfield [1972] generalized Restrepo's theorem to show that dens E* = dens E if there is a differentiable function on E which has a bounded carrier. Leduc [1970] found the elegant proof for the special case of the Leach-Whitfield theorem presented as Theorem 2.4. Restrepo's proof was a direct application of the Bishop-Phelps theorem. Similarly the idea of Leduc's proof is to apply a modification of the Bishop-Phelps theorem. For example, Lemma 2.2 above can be derived from the Bishop-Phelps constructions (see [Phelps, 1973] for a similar proof of Leduc's theorem, and for the ideas needed to prove Lemma 2.2].

Complete characterizations of $C^1$-smooth (nonseparable) spaces are still not known.

Problem. If every separable subspace of E has a separable dual is E $C^1$-smooth?

Problem. Does there exist a $C^k$-smooth space which does not have a $C^k$-norm?

Problem. Characterize $C^k$-smooth spaces.

The Restrepo-Leduc-Leach-Whitfield theorem, is one in a series of results which show that only very restricted classes of spaces can satisfy smoothness conditions. The following is a folk theorem, probably due to R. R. Phelps. (Also see

[Dixmier, 1948]).

Proposition. If the norm on E*** is $C^1$-smooth, E is reflexive.

Proof: Let $Q_i$ be the natural embedding of the $i^{th}$ conjugate of E into the $2 + i^{th}$ conjugate. Suppose E is not reflexive. By the Bishop-Phelps theorem [Phelps, 1973] we can choose $h \varepsilon S(E^{**}) - Q_0(E)$ such that $h(y) = 1$ for some $y \varepsilon S(E^*)$. From the Hahn-Banach theorem $Q_0^{**}(h) \neq Q_2(h)$ in E*****. But both attain their norm at $Q_1(y) \varepsilon$ E***. Hence E*** is not smooth.

Restrepo has shown [1966]:

Proposition. If the norms of both E and E* are $C^1$-norms, then E is reflexive.

Proof: If $x \varepsilon E$ has norm one, $d(x) = \|x\|'$ is the unique norm one functional in E* that attains its norm at x. Hence $D = \{d(x) : \|x\| = 1\}$ is the set of all norm-one functionals in E* which attain their norm on S(E). The Bishop-Phelps theorem states that D is dense in $\{y \varepsilon E^* : \|y\| = 1\}$. Similarly $\{d^*(y) = \|y\|' : y \varepsilon E^*, \|y\| = 1\}$ is dense in $S = \{z \varepsilon E^{**} : \|z\| = 1\}$. Since the norms have continuous derivatives, $T = \{d^* \circ d(x) : x \varepsilon E, \|x\| = 1\}$ is dense is S. But $d^* \circ d$ is the natural isometric injection of E into E**, so T is complete, T = S, and E is reflexive.

The following theorem appears to have been proved simultaneously by [Bonic and Reis, 1966], [Sundaresan, 1967] and [Rao, 1967].

Proposition. If the norms of E and E* are both $C^2$-norms, E is isomorphic to Hilbert space.

A similar result was found by Veic [1971]. He showed that if the norm of E is a $C^2$-norm, then E is isomorphic to a Hilbert space if and only if the second derivative of $\|\cdot\|^2$ is uniformly continuous on a neighborhood of zero.

The properties of the James space, J, are relevant to the

above results [James 1951]. J is a separable space isomorphic to J**. Hence Kadec's theorems shows that all the higher conjugates of J have equivalent $C^1$-norms. Yet J is not reflexive.

Problem. If the norms on E and E* are Gateaux differentiable, is E reflexive?

Problem. If both E and E* are $C^2$-smooth, is E isomorphic to a Hilbert space

Kurzweil [1954] proved proposition 3.4 in an important paper that preceded the interest in smooth partitions of unity, stimulated by Banach manifold theory. A special case of his work shows that if p is an even integer, $C(\ell_p)$ admits $C^\infty$-approximations. Hence from proposition 1.1, $\ell_p$ admits $C^\infty$-partitions of unity. Using a different method, Eells also constructed $C^\infty$-partitions of unity for $\ell_2$ [see Lang, 1962, p. 30].

Proposition 3.1 follows easily from the fact that card $\beta A = 2^{2^{\text{card } A}}$ where $\beta A$ is the Stone-Čech compactification of the discrete space A [Gillman and Jerison 1960, 9.2]. It is known, in fact, that $\ell_\infty(A)$ does not have an equivalent Gateaux differentiable norm [Day, 1955].

In 1971 Wells showed that every Hilbert space admits $C^1$-partitions of unity. This was the first such result for non-separable spaces. The remarkable theorem by Tarunczyk [1972] is more general than the statement of Theorem 3.5. It can be used, for example, to show that if E and F admit $C^k$-partitions of unity then $E \times F$ admits $C^k$-partitions of unity.

Problem. Do $c_o$ (or Hilbert) products of spaces $E_i$ admit $C^k$-partitions of unity when each $E_i$ admits $C^k$-partitions of unity

Problem. Construct an explicit $C^\infty$-norm for $c_o(I,\ell_2)$ or $\ell_2(I,c_o)$ where I is the integers.

Problem. Is the James space $C^\infty$-smooth

Much of the current research involves discriminating smoothness classes of finer delineation than the $C^k$-classes of Banach spaces (see [Palais, 1965] [Wells, 1972], [Bonic and Frampton, 1966], [Goodman, 1971] and [Whitfield, 1973]). Since smoothness is a Banach space invariant, such results also have applications to Banach space theory [Bonic and Frampton, 1966].

If K is a compact subset of E, and f is a continuous mapping of K into a Banach space F, then f can be approximated by the restriction to K of polynomials on E [Prenter, 1970] and [Germanov, 1970]. (If F is finite dimensional the Stone-Weierstrass theorem applies.) If F is strictly convex, if M is a finite dimensional subspace of F, and if $P_M(y)$ is the best approximation from M to $y \in F$, then $P_M \circ f$ and hence f can be approximated by the restriction to K of polynomials on E. The general case follows from the fact that the span of the range of f is separable, and therefore admits an equivalent strictly convex norm.

Let f be a $C^1$-mapping of an open subset of a Banach space E into a Banach space F. A point $x \in E$ is a <u>critical</u> point of f if $D_x f(E) \neq F$. Let C be the set of critical points of f. Then f is a <u>Sard</u> map if f(c) has no interior in F. The classic theorem implies that if $\dim E = n < \infty$, $\dim F = m < \infty$, $k = 1 + \max(n-m,0)$, and if f is a $C^k$-function, then f is a Sard map. The infinite dimensional results are in [Smale, 1965], [Kupka, 1965] and [Sard, 1965]. The Banach space results are for restricted classes of functions. The following result however shows that arbitrary maps can be approximated by Sard maps [Moulis, 1971] and [Eells and McAlpin, 1967].

<u>Theorem</u>. Let p and q be positive functions on an open set $U \subseteq \ell_2$. For each $C^1$-function f of U into $\ell_2$ there is a $C^\infty$ Sard function $g : U \to \ell_2$ such that for $u \in U$ $\|f(u) - g(u)\| < p(u)$ and $\|D_u f - D_u g\| < g(u)$.

Problem. For which Banach spaces E and F can the function in C(E,F) be approximated by Sard mappings?

---

[1]After preparing this manuscript, I learned that this problem has recently been solved by T. Figiel: "An example of an infinite dimensional reflexive Banach space non-isomorphic to its cartesian square," Studia Math 42 (1972), 295-305.

References


[1] Abraham, R., Lectures of Smale on Differential Topology. Columbia University, 1963.

[2] Abraham, R., and Joel Robbin, Transversal Mappings and Flows. Benjamin, Inc., New York, 1967.

[3] Amir, D., and J. Lindenstrauss, The Structure of Weakly Compact Sets in Banach Spaces. Annals of Mathematics 8 (1968), 35-46.

[4] Anderson, R. D., T. A. Chapman and R. M. Schori, Problems in the Topology of Infinite-Dimensional Spaces and Manifolds. ZW-reports, Mathematical Centre, Amsterdam, 1971.

[5] Asplund, E., Averaged Norms. Israel J. Math., 5 (1967), 227-233.

[6] Bonic, R. A., A Note on Sard's Theorem in Banach Spaces. Proc. A.M.S. 17 (1966), 1218.

[7] Bonic, R. A., Four Brief Examples Concerning Polynomials on Certain Banach Spaces. J. Diff. Geometry, 2 (1968), 391-392.

[8] Bonic, R., and J. Frampton, Smooth Functions on Banach Manifolds, J. Math. and Mech. 15 (1966), 872-898.

[9] Bonic, R., J. Frampton and A. Tromba, Λ-Manifolds. J. of Functional Analysis, 3 (1969), 310-320.

[10] Bonic, R. and F. Reis, A Characterization of Hilbert Space. Anais Da Acad. Brasil De Ciencias, 33 (1966), 239-241.

[11] Clarkson, J., Uniformly Convex Spaces. Trans. A.M.S., 40 (1936), 396-414.

[12] Colojoara, I., On Whitney's Imbedding Theorem. Rev. Roum. Math. Pures et Appl., 10 (1965), 291-296 and 1051-1052.

[13] Cudia, D. F., The Geometry of Banach Spaces, Smoothness. Trans. A.M.S. 110 (1964), 283-314.

[14] Day, M., Strict Convexity and Smoothness. Trans. A.M.S. 73 (1955), 516-528.

[15] Day, M., Every L-Space is Isomorphic to a Strictly Convex Space. Proc. A.M.S. 8 (1957), 415-417.

[16] Day, M., Normed Linear Spaces. Springer Verlag, Berlin, 1962.

[17] Dixmier, J., Sur un Théorème de Banach. Duke Math J., 15 (1948), 1057-1071.

[18] Dold, A., Partitions of Unity in the Theory of Fibrations. Ann. Math., 78 (1963), 223-255.

[19] Eells, J., A Setting for Global Analysis. Bull. A.M.S., 72 (1966), 751-807.

[20] Eells, J. and K. D. Elworthy, On the Differential Topology of Hilbert Manifolds. Proc. Summer Institute on Global Analysis, 1968.

[21] Eells, J. and K. D. Elworthy, Open Embeddings of Certain Banach Manifolds. Ann. of Math., 91 (1970), 465-485.

[22] Eells, J., and Mac Alpin, An Approximate Morse-Sard Theorem. J. Math. and Mech., 17 (1967), 1055-1064.

[23] Germanov, L. L., Uniform Approximation to Continuous Functions with Values in Locally Convex Spaces. Proc. of Conference on Constructive Function Theory, Varna (1970), 19-25.

[24] Gillman, L. and M. Jerison, Rings of Continuous Functions. Van Nostrand, Princeton, 1960.

[25] Goncar, A. A., Uniform Approximation of Continuous Functions by Harmonic Functions. Izv. 27 (1963), 1239-1250.

[26] Goncar, A. A., Approximation of Continuous Functions by Harmonic Functions. Dokl. 154 (1964), 503-6.

[27] Goodman, V., A Divergence Theorem for Hilbert Space, Trans. A.M.S., 164 (1972), 411-426.

[28] Goodman, V., Quasi-Differentiable Functions on Banach Spaces. Proc. A.M S., 30 (1971), 367-370.

[29] Graves, L. M., Some General Approximation Theorems. Ann. Math., 42 (1941), 281-293.

[30] Harvey, F. R. and R. O. Wells, Jr., Holomorphic Approximation & Hyperfunction Theory on a $C^1$ Totally Real Submanifold or a Complex Manifold. Math.Ann., 192 (1972), 287-319.

[31] Holmes, R., and B. Kripke, Smoothness of Approximation.

Mich. Math. J., 15 (1968) 225-248.

[32] James, R. C., A non-Reflexive Banach Space Isometric with its Second Conjugate Space. Proc. Nat. Acad. Sci., 37 (1951), 174-177.

[33] Kadec, M. I., On Weak and Norm Convergence. Dokl. Akad. Nauk SSS , 112 (1958), 13-16.

[34] Kadec, M. I., Spaces Isomorphic to a Locally Uniformly Convex Space (Russian) Izv. Vyss. Ucebn. Zaved., Seria Mat. 13 (1959) 51-57.

[35] Kadec, M. I., Conditions for Differentiable Norms on Banach Spaces. Uspehi Mat. Nauk, 20 (1965) 183-187.

[36] Klee, V., Mappings into Normed Linear Spaces. Fund. Math. 49 (1960), 25-34.

[37] Kuiper, N., Variétés Hilbertiennes Aspects Géométriques. University of Montreal Press, Montreal, 1966.

[38] Kuiper, N. H. and D. Burghelea, Hilbert Manifolds. Ann. of Math. 90 (1969), 389-417.

[39] Kuiper, N. and Terpstra-Keppler, Differentiable Closed Embeddings of Banach Manifolds. Symposium in Honour of Prof. G. deRham (Springer-Verlag) 1970, 118-125.

[40] Kupka, I., Counterexample to the Morse-Sard Theorem in the Case of Infinite Dimensions. Proc. A.M.S., 16 (1965), 954-957.

[41] Kurzweil, J., On Approximation in Real Banach Spaces. Studia Math. 14 (1954), 214-231.

[42] Lang, S., Introduction to Differentiable Manifolds. Interscience, New York, 1962.

[43] Leach, E. B. and J. H. M. Whitfield, Differentiable Functions and Rough Norms on Banach Spaces. Proc A.M.S., 33 (1972), 120-126.

[44] Leduc, M., Jauges Différentiables et Partitions de l'Unité. Seminarie Choquet, 4e année, 1964/1965, No. 12, Paris.

[45] Leduc, Michel, Densité de Certaines Familles d'Hyperplans Tangents. C. R. Acad. Sci., Paris, Ser. A-B, 270 (1970), A326-A328.

[46] Lindenstrauss, J., On Non-Separable Reflexive Banach Spaces. Bull. A.M.S., 72 (1966), 967-970.

[47] Lovaglia, A. R., Locally Uniformly Convex Banach Spaces. Trans. A.M.S. 78 (1955), 225-238.

[48] McKessock, D. J. R. and J. H. M. Whitfield, A Differentiable Norm on a Separable Banach Space. Proc. 25th Meeting Canadian Math. Congress (1971), 485-491.

[49] Mitjagin, B. S., Approximation of Functions in $L^p$- and C-spaces on the Torus. Mat. Sb., 58 (1962), 397-414, MR 27-2772.

[50] Moulis, N., Approximation de Functions Differentiables sur Certains espaces de Banach. Ann. Instit. Fourier, 21 (1971), 293-345.

[51] Neugebauer, C. J., Smoothness and Differentiability in $L_p$, Studia Math. 25 (1964), 81-91.

[52] Palais, R. S., Lectures on the Differential Topolocy of Infinite Dimensional Manifolds. (Notes by S. Greenfield) Brandeis University, 1965.

[53] Palais, R. S., Foundations of Global Non-Linear Analysis. Benjamin, Inc., New York, 1968.

[54] Pelczynski, A., On Almost Diffeomorphic Banach Spaces. Koninkl. Ned. Akad. Wetensch, Proc. Ser. A, 71 (1968), 202-208.

[55] Phelps, R. R., Support Cones in Banach Spaces and Their Applications (to appear).

[56] Prenter, P., A Weierstrass Theorem for Real Separable Hilbert Spaces. J. Apprx. Theory, 3 (1970), 341-351.

[57] Quinn, Frank, Transversal Approximation on Banach Manifolds. Proc. Symp. in Pure Math., 15 (1970), 213-223.

[58] Rainwater, J., On a Renorming Theorem of Klee. Bull. A.M.S., 72 (1966), 145-146.

[59] Rainwater, J., Local Uniform Convexity of Day's Norm on $c_o(\Gamma)$. Proc. A.M.S., 22 (1969), 335-339.

[60] Rao, M. M., Characterizing Hilbert Spaces by Smoothness. Nederl. Akad. Wetensch, Proc. Ser A, 70 (1967), 132-135.

[61] Renz, P., Smooth Extensions in Infinite Dimensional Banach Spaces. Trans. A.M.S., 168 (1972), 121-132.

[62] Restrepo, Guillermo, Differentiable Norms in Banach Spaces. Bull. A.M.S., 70 (1964), 413-414.

[63] Restrepo, G., Differentiable Norms. Bol. Soc. Mat. Mexicana, 10 (1965), 47-55.

[64] Restrepo, Guillermo, An Infinite Dimensional Version of a Theorem of Bernstein. Proc. Amer. Math. Soc., 23 (1969), 193-198.

[65] Sard, A., The Measure of the Critical Values of Differentiable Maps. Bull. A.M.S., 48 (1942), 883-890.

[66] Sard, A., Images of Critical Sets. Annals of Math., 68 (1958), 247-259.

[67] Sard, A., Hausdorff Measure of Critical Images on Banach Manifolds. Amer. J. Math., 87 (1965), 158-174.

[68] Smale, S., An Infinite Dimensional Version of Sard's Theorem. Amer. J. Math., 87 (1965), 861-866.

[69] Sundaresan, K., Smooth Banach Spaces. Math Ann., 173 (1967), 191-199.

[70] Smulian, V., Sur la derivabilité de la norme dans l'espace de Banach. Dokl. Akad. Nauk SSSR, 27 (1940), 643-648.

[71] Smulian, V L., Sur la Structure de la Sphere Unitaire dans l'Espace de Banach. Math. Sb., 9 (51) (1954), 545-561.

[72] Tarunczyk, H., Smooth Partitions of Unity on Some Non-Separable Banach Spaces. Preprint 31, Polish Acad. Sci. Institute of Math., 1972, to appear in Studia Math.

[73] Trojanski, S. L., On Locally Convex and Differentiable Norms in Certain Non-Separable Banach Spaces. Studia Math., 37 (1971), 173-180.

[74] Veic, B. E., Criteria for the Hilbertness of a Banach Space in Terms of Smoothness (Russian). Mat Zametki, 9 (1971) 385-390, Math Notes, 9 (1971), 221-224.

[75] Vituskin, A. G., Some Theorems on the Possibility of Uniform Approximation of Continuous Functions by Analytic Functions. Dokl., 123 (1958), 959-962.

[76] Wells, J., Differentiable Functions on $c_o$. Bull. A.M.S. 75 (1969), 117-118.

[77] Wells, J., Differentiable Functions on Banach Spaces with Lipschitz Derivatives. J. Diff. Geo. (to appear).

[78] Wells, J., Smooth Partitions of Unity and Extension Theorem. Proc. Summer College on Global Analysis, Trieste, 1972 (to appear).

[79] White, H. E., Jr , The Approximation of One-to-One Measurable Transformations by Diffeomorphisms. Trans. A.M.S. 141 (1969), 305-321.

[80] Whitfield, J. H. M., Rough Norms on Banach Spaces. Proc. 25th Meeting Canadian Math. Congress (1971), 595-601.

[81] Whitfield, J. H. M., Quasi-Differentiable Norms. Notices A.M.S., 20 (1973), A144.

[82] Whitney, H., On Totally Differentiable and Smooth Functions. Pacific J. Math, 1 (1951), 143-159.

Department of Mathematics
University of Washington
Seattle, Washington 98105

and

University of California San Diego
La Jolla, California 92037

Note added in proof: E. Leach has communicated a simple proof that E is reflexive if the norm on E* is $C^1$ -smooth.

J. Whitfield has informed us that he and Davis, Johnson, and Figiel have constructed counterexamples to the conjecture that E must be reflexive when the norms in E and E* are Gateaux differentiable.

The following reference should be noted: Lindenstrauss, J., Weakly Compact Sets - Their Topological Properties and the Banach Spaces They Generate. Symposium on Infinite Dimensional Topology. Annals of Math. Studies 69. Princeton University Press, 1972.

# CHARACTERIZATIONS OF GENERALIZED CONVEX FUNCTIONS BY THEIR BEST $L^p$-APPROXIMATIONS

Dan Amir and Zvi Ziegler

Let $C[a,b]$ denote the linear space of continuous real-valued functions on the closed interval $[a,b]$. Let $\mathcal{U} = \{u_j; j = 0,1,\ldots\}$ be a Tchebycheff system on $[a,b]$ (for the terminology and notation, as well as standard results concerning Tchebycheff systems and generalized convex functions, see [3]). We denote by $C(u_0,\ldots,u_n) = C(u_0,\ldots,u_n;a,b)$ the cone of generalized convex functions with respect to $\{u_0,\ldots,u_n\}$ in $C[a,b]$. $W_n$ denotes the linear subspace spanned by $\{u_0,\ldots,u_n\}$. For $1 \leq p \leq \infty$, denote by $T_n^p(f) = T_n^p(f;a,b) = \Sigma_{j=0}^n a_{n,j}^p(f;a,b)u_j$ the "polynomial" in $W_n$ nearest to f in the $L^p$-norm over $[a,b]$, $E_n^p(f) = E_n^p(f;a,b) = \|f-T_n^p f\|_p$.

Some characteristic properties of the Tchebycheff best approximation to generalized convex functions obtained by us in [1] can be summarized in the following:

<u>Theorem A.</u> If $f \in C(u_0,\ldots,u_{n-1})\setminus W_{n-1}$ then: (i) $a_{n,n}^\infty(f) > 0$. (ii) $f-T_{n-1}^\infty f$ has n sign changes and the last sign is +. Consequently, if either f or -f belongs to $C(u_0,\ldots,u_{n-1})\setminus W_{n-1}$, then (iii) $E_n^\infty f < E_{n-1}^\infty f$. In particular, if f is regularly $\mathcal{U}$-monotone (i.e., if for each n either f or -f is in $C(u_0,\ldots,u_n)$) and if $f \notin \text{span}\,\mathcal{U}$, then $\{E_n^\infty f;\ n=0,1,\ldots\}$ is a strictly decreasing sequence.

The conclusions of Theorem A obviously hold also for every subinterval $[\alpha,\beta]$ of $[a,b]$. The converse theorems proved in [1] show that the validity of any of the conclusions (i), (ii), or (iii) on <u>every</u> subinterval $[\alpha,\beta] \subset [a,b]$ implies that

$f \in C(u_0,\dots,u_{n-1})\backslash W_{n-1}$ or (in the case of (iii)) that either f or -f is in this cone. This can be summarized in the following:

Theorem B. (i) If $a^{\infty}_{n,n}(f;\alpha,\beta) \geq 0$ for every subinterval $[\alpha,\beta] \subset [a,b]$, then $f \in C(u_0,\dots,u_{n-1};a,b)$. (ii) If $f-T^{\infty}_{n-1}(f;\alpha,\beta)$ has n sign-changes for every $[\alpha,\beta] \subset [a,b]$, then either f or -f is in $C(u_0,\dots,u_{n-1};a,b)\backslash W_{n-1}$. (iii) If $E^{\infty}_n(f;\alpha,\beta) < E^{\infty}_{n-1}(f,\alpha,\beta)$ for every $[\alpha,\beta] \subset [a,b]$, then either f or -f is in $C(u_0,\dots,u_{n-1};a,b)\backslash W_{n-1}$. In particular, if $\{E^{\infty}_n(f;\alpha,\beta);\ n=0,1,\dots\}$ is strictly decreasing for every $[\alpha,\beta] \subset [a,b]$, then f is regularly $\mathcal{U}$-monotone.

It was also shown, by a category argument, that a countable number of restrictions: $E^{\infty}_{n_j}(f;\alpha_i,\beta_i) < E^{\infty}_{n_j-1}(f;\alpha_i,\beta_i)$, $i=1,2,\dots;\ j=1,2,\dots$, will not suffice to guarantee that $f \in C(u_0,\dots,u_{n-1})$ for any $n \geq 1$.

In [2], the exactly analogous results are proved for the weighted $L^2$-norm approximations (only that in the analogue of Theorem B, $\mathcal{U}$ is assumed to be an Extended Complete Tchebycheff system, of the form $u_n(t) = w_0(t)\int_a^t w_1(x_1)\int_a^{x_1} w_2(x_2)\dots\int_a^{x_{n-1}} w_n(x_n)dx_n\dots dx_1$, where $w_i(t)$ are positive $C^{\infty}[a,b]$-functions). The methods used in [1] and [2] are quite different, due to the difference between the characterizations of best approximants in the sup-norm and the $L^2$-norm. The main tool in the proof of the analogue of Theorem B in this case is the following

Lemma: If $g \in C^{(n)}[a,b]\backslash C(u_0,\dots,u_{2n-1})$, then there are $Q \in W_{2n-1}$ and $t_1 < t_2$ in (a,b) such that: (i) $(-1)^n(g-Q) < 0$ on $(t_1,t_2)$ and (ii) $(g-Q)(t_i) = D_0(g-Q)(t_i) = \dots = D_{n-1}D_{n-2}\dots D_0(g-Q)(t_i)$ for $i = 1,2$, where $D_jf(t) = (d/dt)(f(t)/w_j(t))$. Part (i) of the theorem is proved from the lemma by a suitable choice of g and integration by parts. The other parts of the theorem follow from (i) by standard arguments.

We wish to add a simple corollary to Theorem B:

Theorem 1. Let $u_n = t^n/n!$, $n=0,1,\ldots$ and $p=2$ or $p=\infty$. If $f \in C[a,b]$ is not analytic on $(a,b)$, then there are a sequence of intervals $[\alpha_k,\beta_k] \subset [a,b]$ and a sequence $1 \le n_1 < n_2 < n_3 < \ldots$ of integers such that $E^p_{n_k}(f;\alpha_k,\beta_k) = E^p_{n_k-1}(f;\alpha_k,\beta_k)$ for $k=1,2,\ldots$

Proof: If not, then by Theorem B there exists some $N_0$ with $f \in C(u_0,\ldots,u_n) \cup -C(u_0,\ldots,u_n)$, for all $n \ge N_0$. By a theorem of Bernstein (see [4], p. 141) this implies that f is analytic on (a,b).

The question which arises naturally is whether the analogues of Theorems A and B hold for all $L^p$-norms, $1 \le p \le \infty$. For Theorem A the answer is in the affirmative:

Theorem 2. If $f \in C(u_0,\ldots,u_{n-1})\setminus W_{n-1}$ then, for every $1 < p \le \infty$: (i) $a^p_{n,n}(f) > 0$. (ii) $f - T^p_{n-1}f$ has n sign changes and the last sign is +. (iii) $E^p_n f < E^p_{n-1} f$. For p=1, the same conclusions hold provided that f does not coincide with a u-polynomial on any subinterval of [a,b]).

Proof: For $1 < p < \infty$, the best approximant $g = T^p_{n-1}f$ is characterized by: $\int_a^b h(x)|f(x)-g(x)|^{p-1}\operatorname{sgn}\{f(x)-g(x)\}dx = 0$ for all $h \in W_{n-1}$. For p=1 this condition is sufficient, and it is necessary if $f \ne g$ a.e. (see [4], p. 64). Since $f-g \in C(u_0,\ldots,u_{n-1})$, it cannot have more than n zeros. If it has less than n sign changes, there is a polynomial $h \in W_{n-1}$ with $\operatorname{sgn} h(x) = \operatorname{sgn}\{f(x)-g(x)\}$ on (a,b), which contradicts the orthogonality condition. Thus f-g has exactly n sign changes and no other zeros in (a,b). Similarly, $f - T^p_n f$ has at least n+1 sign changes, hence $T^p_{n-1}f \ne T^p_n f$ and $E^p_n f < E^p_{n-1} f$. If $a \le t_0 < t_1 < \ldots < t_n \le b$ and $(f-g)(t_i)$ are of alternating signs, then the sign of the determinant

$$U\begin{pmatrix} u_0,\ldots,u_{n-1},f-g \\ t_0,\ldots,t_{n-1},t_n \end{pmatrix}$$

is the same as $\operatorname{sgn}(f-g)(t_n)$, and since $f-g \in C(u_0,\ldots,u_{n-1})$, it must be +. The fact that $a^p_{n,n} f > 0$ follows by continuity of the nonvanishing functional $a^p_{n,n} f$ on the convex (hence connected) set $C(u_0,\ldots,u_{n-1})\setminus W_{n-1}$, since $a^p_{n,n}(u_n) = 1$.

A counterexample which shows the necessity of the additional restriction in the $L^1$-case is:

$$f(x) = \begin{cases} 0, & 0 \leq x \leq 0.9, \\ x-0.9, & 0.9 \leq x \leq 1. \end{cases}$$

A simple geometric argument shows that $E^1_1 f = E^1_0 f$, $T^1_1 f = T^1_0 f \equiv 0$, although $f \in C(1)\setminus W_0$.

The analogue of Theorem B follows easily from Theorem 2 in the case that f is sufficiently smooth:

<u>Theorem 3</u>. Let $\{u_0,\ldots,u_n\}$ be an Extended Complete Tchebycheff system and let $f \in C^{(n)}[a,b]$. If $f \notin C(u_0,\ldots,u_{n-1})$ then there exists an $[\alpha,\beta] \subset [a,b]$ such that for each p, $1 \leq p \leq \infty$, $a^p_{n,n}(f;\alpha,\beta) < 0$. If neither f nor -f is in $C(u_0,\ldots,u_{n-1})\setminus W_{n-1}$ then there exists an interval $[\alpha,\beta] \subset [a,b]$ such that $a^p_{n,n}(f;\alpha,\beta) = 0$, $E^p_n(f;\alpha,\beta) < E^p_{n-1}(f;\alpha,\beta)$ and $f - T^p_{n-1}(f;\alpha,\beta)$ has fewer than n sign changes.

<u>Proof:</u> The first statement follows from Theorem 2 and the fact that in this case $f \in C(u_0,\ldots,u_{n-1};\alpha,\beta)$ iff $D_{n-1}\cdots D_0 f \geq 0$ on $[\alpha,\beta]$. The second statement follows from the first by a continuity argument.

<u>Remark</u>. The points $\alpha,\beta$ in the conclusions of the theorem can be chosen to be rational. On the other hand, the next theorem shows that the smoothness of generalized convex functions implies that if f is only assumed to be continuous, then no countable number of conditions of the type $E^p_n(f;\alpha,\beta) < E^p_{n-1}(\alpha,\beta)$ will suffice.

<u>Theorem 4</u>. The set $\{f\colon f \in C[a,b];\ a^p_{n,n}(f;\alpha,\beta) = 0\}$ is closed and nowhere dense (in the sup-norm). Therefore, if $\{u_0,\ldots,$

$u_n, \ldots\}$ is an Extended Complete Tchebycheff system, no countable set of conditions of the type $a^F_{r_j,n_j}(f;\alpha_i,\beta_i) \neq 0$, $i=1,2,\ldots$, $j=1,2,\ldots$, can characterize generalized convex functions.

Proof: Immediate, since the elements of $C(u_0,\ldots,u_{n-1})$ are differentiable for $n\geq 2$ (and those of $C(u_0)$ - differentiable almost everywhere), and the set of such functions is of the first category in $C[a,b]$.

Theorem 4 shows that the application of standard smoothing procedures to Theorem 3 cannot be expected to yield the general analogue of Theorem B for $p \neq 2,\infty$.

## References

[1] Amir, D. and Z. Ziegler, Functions with Strictly Decreasing Distances from Increasing Tchebycheff Subspaces. J. Approximation Theory 6 (1972), 332-344.

[2] Amir, D. and Z. Ziegler, Characterization of Generalized Convex Functions by Best $L^2$-Approximations. To appear.

[3] Karlin, S. and W. Studden, Tchebycheff Systems: With Applications in Analysis and Statistics. Interscience, New York, 1966.

[4] Timan, A. F., Theory of Approximation of Functions of a Real Variable. Pergamon, New York, 1963.

Department of Mathematical Sciences
Tel Aviv University
Tel Aviv, Israel

Faculty of Mathematics
Israel Institute of Technology
Haifa, Israel

THE SEPARABLE PROJECTION PROPERTY

John Warren Baker

## 1. Introduction

A separable Banach space X has the separable projection property if for every separable Banach space Y and every isometric embedding u: X → Y, there is a projection $\Pi$ of Y onto u(X). If u can always be selected with $\|\Pi\| \leq \lambda$, X is a $P'_\lambda$ space (denoted $X \in P'_\lambda$). In [5], D. Dean showed that if X has the separable projection property, it is a $P'_\lambda$ space for some finite $\lambda$. D. Amir (see [1] and [2]) has shown that if S is a compact metric space, then C(S) has the separable projection property if and only if S is homeomorphic to the set $\Gamma(\omega^n k)$ of ordinals not exceeding $\omega^n k$ for some positive integers n and k.

If a Banach space X has the separable projection property, the number $p_s(X) = \inf\{\lambda \mid X \in P'_\lambda\}$ will be called the (separable) projection constant. In [10], A. Sobczyk established that $p_s(c_o) = 2$ and McWilliams in [3] showed that $p_s(c) = 3$. Recently, A Pełczyński [9, p. 74] indicated it would be interesting to know the values of $p_s(C(\omega^n))$ for $1 \leq n < \omega$. Here we show $p_s(C(\omega^n k)) = 2n + 1$ for $1 \leq n,k < \omega$. This establishes the value of the projection constant for all continuous function spaces with the separable projection property and includes McWilliam's result.

## 2. Preliminaries

If X is a topological space, a decomposition D of X is a disjoint collection of closed subsets of X with $X = \cup\{a : a \in D\}$. The notation X/D denotes the set D with its quotient topology.

A set $A \in D$ is called plural if it contains at least two elements. Also, a set $A \in D$ is called a limit set if each open set containing A has nontrivial intersection with a plural set in $D - \{A\}$. The $n^{th}$ derived decomposition $D^{(n)}$ of X is defined as follows: $D^{(1)}$ is the decomposition of X consisting of the plural limit sets in D and singleton sets. Inductively, $D^{(n+1)} = (D^{(n)})^{(1)}$. If $D^{(n)}$ contains no plural sets, we write $D^{(n)} = 0$. The concept of the $n^{th}$ derived set is due to R. Arens [3]. For additional terminology and basic properties of decompositions, see [7].

If X and Y are compact Hausdorff spaces and $\phi$ is a (continuous) map of X onto Y, then $\phi^\circ$ denotes the isometric isomorphism from $C(X)$ to $C(Y)$ that takes f to $f \circ \phi$. If $Y = X/D$ and $\phi$ is the quotient map of D, then $\phi^\circ[C(X/D)]$ is identified with $C(X/D)$ and consists of all functions in $C(X)$ which are constant on each set in D. If X' is the derived set of X, the topological derivative $X^{(n)}$ is defined by $X^{(1)} = X'$ and, inductively, $X^{(n+1)} = (X^{(n)})'$.

## 3. Results

Lemma. Let X be a compact metric space. If H is a subspace of $C(X)$ isometrically isomorphic to $C(\omega^n k)$, then there is a projection $\Pi: C(X) \to H$ with $\|\Pi\| \leq 2n + 1$.

Outline of Proof: Let u be an isometric isomorphism from $C(\omega^n k)$ onto H. By a theorem of W. Holsztyński [6], there is a closed subset Q of X, a map $\phi$ of Q onto $\Gamma(\omega^n k)$ and $\varepsilon$ in $C(Q)$ with $|\varepsilon(q)| = 1$ and $\varepsilon(q)\ (ug)(q) = \phi^\circ g(q)$ for all $q \in Q$ and $g \in C(\omega^n k)$. Let D be the upper semicontinuous decomposition $\{\phi^{-1}(\alpha) | \alpha \in \Gamma(\omega^n k)\}$ of Q induced by the closed map $\phi$. Define $Q_1 = \phi^{-1}[0, \omega^n]$ and $Q_i = \phi^{-1}(\omega^n(i-1), \omega^n i]$ for $1<i\leq k$, and let $D_i$ be the restriction of D to $Q_i$. Then $D_i$ is an u.s.c. decomposition of the compact set $Q_i$. If $H_i$ is the u.s.c. decomposition

$D_i^{(1)}$ of $Q_i$, then $(\omega^n(i-1), \omega^n i]^{(n+1)} = \emptyset$ implies $H_i^{(n)} = 0$. By Theorem 1.9 in [4], there is a projection $P_i$ of $C(Q_i)$ onto $C(Q_i/H_i)$ with $\|P_i\| \leq 2n+1$. Let $Y_i$ be the union of the plural sets in $D_i - H_i$. Each plural set S in $D_i - H_i$ is open and closed in $Q_i$. Let x(S) be a point in S and for $f \in C(Q_i)$, define $P_i^* f(x) = P_i f(x)$ for $x \in Q_i - Y_i$ and $P_i^* f(x) = P_i f(x(S))$ for $x \in S \subset Y_i$. It can be shown that $P_i^*$ is a projection of $C(Q_i)$ onto $C(Q_i/D_i)$ with $\|P_i^*\| \leq \|P_i\| \leq 2n+1$.

Since $C(Q) = C(Q_1) \oplus \ldots \oplus C(Q_k)$, the operator $P = P_1 \oplus \ldots \oplus P_k$ is a projection of C(Q) onto C(Q/D) with $\|P\| \leq 2n+1$. If $R\colon C(X) \to C(Q)$ is the restriction operator and $T_\varepsilon\colon C(Q) \to C(Q)$ is defined by $T_\varepsilon(f) = \varepsilon f$, then $u(\emptyset^\circ)^{-1} P T_\varepsilon R$ is the desired projection.

<u>Theorem</u>. $P_s(C(\omega^n k)) = 2n+1$.

<u>Proof</u>: Let E be a separable Banach space, $S = \Gamma(\omega^n k)$, and let $u\colon C(S) \to E$ be an isometric embedding. By the Banach-Mazur theorem, we may assume E is a subspace of C([0,1]). By the preceding lemma, there is a projection $\Pi$ of C([0,1]) onto u[C(S)] with $\|\Pi\| \leq 2n+1$. The restriction P of $\Pi$ to E is a projection of E onto u[C(X)]; hence, $C(S) \in P'_{2n+1}$. The fact that $C(S) \notin P'_\lambda$ for $\lambda < 2n+1$ is established by Amir in the proof of the theorem in [1].

## References


[1] Amir, D., Continuous function spaces with the separable projection property. Bull. Res. Counc. Israel, <u>10F</u> (1962), 163-164.

[2] Amir, D., Projections onto continuous function spaces. Proc. Amer. Math. Soc. <u>15</u> (1964), 396-402.

[3] Arens, R., Projections on continuous function spaces. Duke Math. J. <u>32</u> (1965), 469-478.

[4] Baker, J., Some uncomplemented subspaces of C(X) of the type C(Y). Studia Math. <u>36</u> (1970), 85-103.

[5] Dean, D., Projections in certain continuous function spaces. Canad. J. Math. 14 (1962), 395-401.

[6] Holsztyński, W., Continuous mappings induced by isometries of spaces of continuous functions. Studia Math. 26 (1966), 133-136.

[7] Kelley, J., General Topology. D. Van Nostrand, Princeton, 1955.

[8] McWilliams, R., On projections of separable subspaces of (m) onto (c). Proc. Amer. Math. Soc. 10 (1959), 872-876.

[9] Pełczyński, A., Linear extensions, linear averagings, and their applications to linear topological classification of spaces of continuous functions. Dissertationes Math. (Rosprawy Mat.) 58 (1968), 1-92.

[10] Sobczyk, A., Projections of the space (m) on its subspace ($c_0$). Bull. Amer. Math. Soc. 47 (1941), 938-947.

Department of Mathematics
Florida State University
Tallahassee, Florida 32306

# APPROXIMATION AND INTEGRAL REPRESENTATION FOR OPERATORS ON BOUNDED ANALYTIC FUNCTIONS

M. W. Bartelt

## Introduction

Much work has been done on B, the algebra of bounded analytic functions on the open unit disc D in the complex plane. It has often been studied in the topology, $\sigma$, of uniform convergence on D, given by the norm $\|f\| = \sup \{|f(z)|: z \in D\}$. Recent research indicates that B may be more easily studied in the strict topology, $\beta$, introduced by R. C. Buck [3]. (See the references for related results.)

The strict topology on B is the locally convex topology defined by the collection of semi-norms $|f|_{\phi} = \|f\phi\|$ for $\phi$ a continuous function on D which vanishes at infinity. One of the principal advantages of the strict topology is that the polynomials are strictly dense in B, whereas the norm closure of the polynomials is only C, the algebra of functions in B which are continuous on the closure of D. In general, $\beta$ combines some of the better features of the norm topology and of $\kappa$, the topology of uniform convergence on compact subsets of D.

This note is concerned in general with $[\beta:\beta]$, the algebra of all continuous linear operators from B, endowed with $\beta$, into itself. In particular it involves the approximation and integral representation of operators in $[\beta:\beta]$. Analogous to previous studies of the strict topology, $[\beta:\beta]$ is more easily understood than the larger algebra of all bounded linear operators from B into B.

Let $(B,\tau)$ denote B endowed with the topology $\tau$, where $\tau$ is one of the topologies $\kappa,\beta$ or $\sigma$. Also letting $\tau_1$ and $\tau_2$ be one of the topologies $\kappa,\beta$, or $\sigma$, we denote by $[\tau_1:\tau_2]$ the algebra of continuous linear operators from $(B,\tau_1)$ into $(B,\tau_2)$. Thus $[\sigma:\sigma]$ is the algebra of all bounded linear operators from B into B. The only distinct operator algebras $[\tau_1:\tau_2]$ and all their proper inclusions are given [1] by $[\kappa:\sigma] \subset [\beta:\sigma] \subset [\beta:\beta] \subset [\sigma:\sigma]$ and $[\kappa:\sigma] \subset [\kappa:\kappa] \subset [\beta:\beta]$.

Two appropriate topologies for the study of $[\beta:\beta]$ are the operator norm topology inherited from $[\sigma:\sigma]$ and the topology of uniform convergence on norm bounded subsets of B. The latter is equivalent to the compact open topology.

## Results

On $[\beta:\beta]$, the relationship between the operator norm topology and the topology of uniform convergence on bounded subsets (written u.b.) is similar to the relationship between $\kappa,\beta$ and $\sigma$ on B. It is known [4] that in B, a sequence of functions converges strictly if and only if it is uniformly bounded and pointwise (or $\kappa$) convergent, the $\beta$ and $\sigma$ bounded subsets of B coincide and on a $\beta$ bounded subset of B, the topology $\beta$ is metrizable. The proofs of the corresponding statements for $[\beta:\beta]$ and the operator norm and u.b. topologies follow from the definitions (see [2]) and Theorem 4.

<u>Theorem 1</u>. A subset of $[\beta:\beta]$ is u.b. bounded if and only if it is norm bounded. On norm bounded subsets of $[\beta:\beta]$ the u.b. topology is metrizable. A sequence of operators in $[\beta:\beta]$ is u.b. convergent if and only if it is pointwise convergent and the operator norms of the operators in the sequence are uniformly bounded.

An integral representation for the operators in $[\kappa:\sigma]$ can be employed to study approximation in $[\beta:\beta]$.

<u>Theorem 2</u>. If T is in $[\kappa:\sigma]$, then there exists a function

$k(z,w)$ analytic for $|z| < 1$, $|w| > r_o$ for some $r_o < 1$ and such that if $1 > r_1 > r_o$, then there exists an M such that $|k(z,w)| \leq M$ for all $|z| < 1$, $|w| \geq r_1$, and such that

$$Tf(z) = \int_{|w|=r_1} f(w)\, k(z,w)\, dw, \qquad \text{all } f \in B.$$

Conversely, using this representation formula any such function $k(z,w)$ yields an operator T in $[\kappa:\sigma]$.

Proof: Showing that such a kernel $k(z,w)$ yields an operator T in $[\kappa:\sigma]$ is straightforward. If T is in $[\kappa:\sigma]$, let

$$k(z,w) = (2\pi i)^{-1} \sum_{k=0}^{\infty} T(z^k) w^{-k-1}$$

Then it follows that $k(z,w)$ satisfies the stated conditions and thus the operator S defined by

$$Sf(z) = \int_{|w|=r_1} f(w)\, k(z,w)\, dw$$

is in $[\kappa:\sigma]$. Since S and T agree on the polynomials which are $\beta$ dense in B, it follows that $S = T$.

It can be shown that this integral representation can be modified to

$$T f(z) = \int_{\Gamma} f(w)\, k(z,w)\, dw, \text{ where } \Gamma = \{z:|z| = 1\}$$

and $k(z,w)$ is modified slightly from the last theorem. A similar representation of all continuous linear functionals on $(B,\beta)$ has been obtained in [9].

Using the integral representation of Theorem 2 it can be shown that $[\kappa:\sigma] = \{T_r: T \in [\kappa:\sigma],\ 0 < r < 1\}$, where $T_r(f)(z) = T(f_r)(z)$ and $f_r(z) = f(rz)$. It is known [2] that if T is in $[\beta:\beta]$, then $\{T_r\}$ converges u.b. to T as $r \uparrow 1$ and $\{T_r\}$ converges in norm to T if and only if T is in $[\beta:\sigma]$. It should be pointed out that $[\beta:\beta]$ is a u.b. (and hence norm) closed subalgebra of $[\sigma:\sigma]$. Thus we obtain

Theorem 3. The algebra $[\kappa:\sigma]$ is u.b. dense in $[\beta:\beta]$ and the norm closure of $[\kappa:\sigma]$ is $[\beta:\sigma]$.

The convergence of operators in $[\beta:\beta]$ can be studied by using the kernel $k(z,w)$ associated with an operator by Theorem 2.

Theorem 4. Let $\{T_n\}$, $n=1,2,\ldots$, and $T$ be in $[\beta:\beta]$. Then $\{T_n\}$ converges u.b. to $T$ if and only if the corresponding kernels $\{k_n(z,w)\}$ converge uniformly on compact subsets of $D \times \{w: |w|>1\}$ to $k(z,w)$ and for any $\rho > 1$, there exists a number $M_\rho$ such that $|k_n(z,w)| \leq M_\rho$ for $|z| < 1$ and $|w| \geq \rho > 1$.

Proof: The result is established by studying the associated sequence of operators $[T_n]_r$ for $0 < r < 1$.

The last result can be applied to the multiplier operators in $[\beta:\beta]$. A multiplier is a linear operator $T$ defined on $B$ for which there exists a sequence $\{c_n\}$ such that $T(\Sigma a_n z^n) = \Sigma a_n c_n z^n$, for all functions $\Sigma a_n z^n$ in $B$. A linear operator $T$ is a multiplier from $B$ into $B$ if and only if the sequence $\{c_n\}$ is one side of the sequence of Fourier-Stieltjes coefficients of a bounded complex regular Borel measure $\mu$ on $\Gamma$. Let $\hat{\mu}$ denote the function defined on the non-negative integers $P$ whose value at the integer $k$ if $\hat{\mu}(k)$, the $k^{th}$ Fourier-Stieltjes coefficient. Then one can easily show that if $\{T_n\}$ is a sequence of multipliers converging in norm to a multiplier $T$, then $\{\hat{\mu}_n\}$ converges uniformly to $\hat{\mu}$ on $P$.

Theorem 5. Let $\{T_n\}$, $n=1,2,\ldots$, and $T$ be multipliers from $B$ into $B$ with associated measures $\{\mu_n\}$ and $\mu$. Then $\{T_n\}$ converges u.b. to $T$ if and only if $\{\hat{\mu}_n\}$ converges strictly to $\hat{\mu}$ on $P$.

## References

[1] Bartelt, M. W., Multipliers and operator algebras on bounded analytic functions. Pacific J. Math. 37 (1971), 575-584.

[2] Bartelt, M. W., Approximation in operator algebras on bounded analytic functions. Trans. Amer. Math. Soc. 170 (1972), 71-83.

[3] Buck, R. C., Operator algebras and dual spaces. Proc. Amer. Math. Soc. 3 (1952) 681-687.

[4] Buck, R. C., Algebraic properties of classes of analytic functions. Seminars on Analytic Functions, Vol. II, Princeton, 1957, 175-188.

[5] Buck, R. C., Bounded continuous functions on a locally compact space. Michigan Math. J. 5 (1958), 95-104.

[6] Piranian, G., A. L. Shields and J. H. Wells, Bounded analytic functions and absolutely continuous measures. Proc. Amer. Math. Soc. 18 (1967), 818-826.

[7] Rubel, L. A., Bounded convergence of analytic functions. Bull. Amer. Math. Soc. 77 (1971), 13-24.

[8] Rubel, L. A. and J. V. Ryff, The bounded weak-star topology and the bounded analytic functions. J. Functional Analysis 5 (1970), 167-183.

[9] Rubel, L. A. and A. L. Shields, The space of bounded analytic functions on a region. Ann. Inst. Fourier (Grenoble) 16 (1966) fasc. 1, 235-277.

[10] Sentilles, F. D., Semigroups of operators in C(S). Can. J. Math. 22 (1970), 47-54.

[11] Summers, W. H., The general complex bounded case of the strict weighted approximation problem. Math. Ann. 192 (1971), 90-98.

Mathematics Department
Rensselaer Polytechnic Institute
Troy, New York 12181

# INCLUSIONS, INTERPOSITION AND APPROXIMATION

J. Blatter and G. L. Seever

Let X be a set. An <u>inclusion</u> on X is any relation $\prec$ on $2^X$ (=set of all subsets of X) which satisfies

(1) $\emptyset \prec A$ and $A \prec X$ all A

(2) $A \prec B \Rightarrow A \subset B$

(3)
$$A \cup B \prec C \Longleftrightarrow A \prec C \text{ and } B \prec C$$
$$C \prec A \cap B \Longleftrightarrow C \prec A \text{ and } C \prec B$$

An inclusion $\prec$ which also satisfies

(4) $A \prec B \Rightarrow \exists C \ni A \prec C \prec B$

is called <u>idempotent</u>. For $\prec$ an inclusion on X, we define a relation, also denoted by $\prec$, on B(X) (= space of all bounded real-valued functions on X) by

$$u \prec v \Longleftrightarrow \forall r,s \in R, \text{ if } r<s, \text{ then } \{x: u(x) \geq s\} \prec \{x: v(x) \geq r\}.$$

Note that for $A,B \in 2^X$, $A \prec B \Longleftrightarrow 1_A \prec 1_B$ (= characteristic function of B). Denote by $M(X,\prec)$ the set of all $f \in B(X)$ such that $f \prec f$.

Here are some examples of inclusions. Let $\mathcal{C}$ be any sublattice of $2^X$ to which $\emptyset$ and X belong. Define $\prec_{\mathcal{C}}$ on $2^X$ by

$$A \prec_{\mathcal{C}} B \Longleftrightarrow \exists\, C \in \mathcal{C} \ni A \subset C \subset B.$$

$\prec_{\mathcal{C}}$ is an idempotent inclusion and $M(X,\prec_{\mathcal{C}})$ can be shown to be the closed convex cone generated by the constant functions and the functions $1_C$, $C \in \mathcal{C}$. In the case that X is a topological space and $\mathcal{C}$ is the set of open (resp., closed) subsets of X,

$M(X,\prec_C)$ is the set of bounded lower (resp., upper) semi-continuous functions on X. If $\prec_1$ and $\prec_2$ are inclusions on X, then the relation $\prec_1 \times \prec_2$ defined by

$$A \prec_1 \times \prec_2 B \Leftrightarrow \exists C \ni A \prec_1 C \text{ and } C \prec_2 B$$

is also an inclusion on X, $M(X,\prec_1\times\prec_2) = M(X,\prec_1) \cap M(X,\prec_2)$ and $u_i \in M(X,\prec_i)$, $i=1,2$, $u_1 \leq u_2 \Rightarrow u_1 \prec_1 \times \prec_2 u_2$. Note that an inclusion is idempotent iff $\prec\times\prec = \prec$; thus the term idempotent. If $C_1$ and $C_2$ are sublattices of $2^X$ to which $\emptyset$ and X belong, then $A \prec_{C_1} \times \prec_{C_2} B \Leftrightarrow \exists C_1 \in C_1$ and $C_2 \in C_2 \ni A \subset C_1 \subset C_2 \subset B$, and we see that $\prec_{C_1} \times \prec_{C_2}$ is idempotent iff whenever $C_1 \in C_1$, $C_2 \in C_2$ and $C_1 \subset C_2$, $\exists D_1 \in C_1$, $D_2 \in C_2 \ni C_1 \subset D_2 \subset D_1 \subset C_2$. In particular, if X is a topological space, $C_1$ the set of closed subsets of X and $C_2$ the set of open sets, then $\prec_{C_1} \times \prec_{C_2}$ is idempotent iff X is normal.

For reasons that shall shortly be apparent, idempotent inclusions are by far the most important ones and so an alternate description of them is of interest.

<u>Characterization Theorem</u>. Let X be a set. If $\prec$ is an idempotent inclusion on X, then

(i) $M(X,\prec)$ is a closed convex cone in $B(X)$ which contains the constant functions and is closed under the lattice operations.

(ii) For $A,B \subset X$, $A \prec B \Leftrightarrow \exists f \in M(X,\prec) \ni 1_A \leq f \leq 1_B$.

Conversely, if K is a closed convex cone in $B(X)$ which contains the constant functions and is closed under the lattice operations, then the relation $\prec_K$ on $2^X$ defined by

$$A \prec_K B \Leftrightarrow \exists f \in K \ni 1_A \leq f \leq 1_B$$

is an idempotent inclusion and $M(X,\prec_K) = K$.

Note that if X is a normal topological space and if $\prec = \prec_{C_1} \times \prec_{C_2}$, $C_1$ (resp., $C_2$) the set of closed (resp., open) subsets of X, then $M(X,\prec)$ is the set of bounded continuous functions on X and (ii) of the Characterization Theorem is just Urysohn's lemma. The Characterization Theorem also includes Nachbin's extension of the Urysohn lemma to topological spaces equipped with a normal pre-order [5, §2, Theorem 1]: let X be such a space, $C_1$ the set of closed increasing subsets of X, $C_2$ the set of open increasing sets, take $\prec$ to be $\prec_{C_1} \times \prec_{C_2}$, and now observe first that $M(X,\prec)$ is the set of bounded, increasing continuous functions and second that if A is closed and increasing, B open and increasing, and if $A \subset B$, then $A \prec B$ and so by (ii) there is an $f \in M(X,\prec)$ such that $1_A \leq f \leq 1_B$.

A result of Tong [6,7] and Katětov [3,4] states that for X a normal topological space, $u: X \to R$ upper semi-continuous, $\ell: X \to R$ lower semi-continuous, and $u \leq \ell$ there is a continuous $f: X \to R$ such that $u \leq f \leq \ell$. This result as well as its extension to topological spaces equipped with a normal pre-order [2; 6.16, 6.20] (cf. [5; Appendix, §5, Theorem 5]) has the following extension.

<u>Interposition Theorem</u>. Let X be a set, and let $\prec$ be an inclusion on X. Then the following conditions are equivalent.

(a) $\prec$ is idempotent.

(b) for $u,v \in B(X)$, $u \prec v \Rightarrow \exists f \ni M(X,\prec) \ni u \leq f \leq v$.

As an application of the above theorems we prove the following result. Further applications as well as the proofs of these theorems will appear elsewhere.

<u>Approximation Theorem</u>. Let X be a set, and let K be a closed convex cone in B(X) which contains the constant functions and is closed under the lattice operations. Let $X_0 \subset X$, and let $K_0 = \{f \in K: f \text{ vanishes on } X_0\}$. Then $K_0$ is an existence subset of B(X), i.e., for any $f \in B(X)$ there is a $g \in K_0$ such that for any $k \in K_0$, $\|f - g\| \leq \|f - k\|$.

Proof: Let $f \in B(X)$. Define $f_*, f^*$ as follows. Let d be the distance of f to $K_0$, i.e., $d = \inf\{\|f-k\|: k \in K_0\}$. For $x \in X$, set

$$f_*(x) = \inf\{k(x)+\varepsilon: k \in K_0, \varepsilon > 0, k+\varepsilon \geq f-d\}$$

$$f^*(x) = \sup\{k(x)-\varepsilon: k \in K_0, \varepsilon > 0, k-\varepsilon \leq f+d\}.$$

We claim that $f_* \prec_K f^*$, $\prec_K$ being defined as in the Characterization Theorem. Let $r,s \in \mathbb{R}$, $r > s$. Let $\varepsilon > 0$ be such that $r+2\varepsilon<s$. Let $k \in K_0$ be such that $\|f-k\| \leq d+\varepsilon$. $f_* \leq k+\varepsilon$ and $f^* \geq k-\varepsilon$ so that $\{x: f_*(x) \geq s\} \subset \{x: k(x)+\varepsilon \geq s\} \prec_K \{x: k(x)+\varepsilon \geq r+2\varepsilon\} \subset \{x: f^*(x) \geq r\}$ and it follows that $\{x: f_*(x) \geq s\} \prec_K \{x: f^*(x) \geq r\}$. Since by the Characterization Theorem $\prec_K$ is idempotent, there is by the Interposition Theorem a $g \in K$ such that $f_* \leq g \leq f^*$. $f_*$ and $f^*$ vanish on $X_0$ and so $g \in K_0$. Observe that $f - d \leq f_*$ and $f^* \leq f + d$ so that $\|f-g\| \leq d$.

If for example, A is a closed subalgebra of $B(X)$, then A is an existence subset of $B(X)$: if we take for K the set $A + \mathbb{R}$ and for $X_0$ the set of common zeros of the functions in A, then $A = K_0$ (cf. [1; 3.19, 3.20]). More generally, the set K of all functions in $B(X)$ which are increasing with respect to some pre-order on X satisfies the hypotheses of the Approximation Theorem and so the set $K_0$ of all functions in K which vanish on $X_0$ is an existence subset of $B(X)$ (cf. [5; Appendix, §5, Theorem 6]). In conclusion we remark that in the Approximation Theorem the hypothesis that K contain the constant functions cannot be omitted since not even every closed vector sublattice of $B(X)$ is an existence subset of $B(X)$ (see [1; 3.22-3.25]).

## References


[1] Blatter, J., Grothendieck spaces in approximation theory. Mem. Amer. Math. Soc. 120 (1972).

[2] Blatter, J. and G. L. Seever, Interposition of semi-

continuous functions by continuous functions. Proceedings of the Colloquium of Analysis held at the Universidade Federal do Rio de Janeiro, August 15-24, 1972.

[3] Katětov, M., On real-valued functions in topological spaces. Fund. Math. 38 (1951), 85-91.

[4] Katětov, M., Correction to "On real-valued functions in topological spaces," Fund. Math. 40 (1953), 203-205.

[5] Nachbin, L., Topology and Order. van Nostrand Mathematical Studies #4, van Nostrand, Princeton (1965).

[6] Tong, H., Some characterizations of normal and perfectly normal spaces. Bull. Amer. Math. Soc. 54 (1948), 65.

[7] Tong, H., Some characterizations of normal and perfectly normal spaces. Duke Math. J. 19 (1952), 289-292.

Institut für Angewandte Mathematik
Universität Bonn
5300 Bonn
West Germany

Department of Mathematics
The University of Texas at Austin
Austin, Texas 78712

# A SURVEY OF RECENT RESULTS AND PROBLEMS IN THE STUDY OF CONVOLUTION PRODUCTS OF SEQUENCES

R. Bojanic and Y. H. Lee

1. Let $p_n \geq 0$, and let $(a_n)$ be defined by

$$a_0 = 1, \; a_n = \sum_{k=1}^{n} p_k a_{n-k}, \qquad n=1,2,\ldots$$

The asymptotic properties of the sequence $(a_n)$, which has various interpretations in the theory of probability and renewal theory (see [1], [2]), have been studied by many authors. Under additional hypotheses that $\sum_{k=1}^{\infty} p_k = 1$ and that the greatest common divisor of the indices k such that $p_k > 0$ is one, P. Erdős, W. Feller and H. Pollard [3] proved that

$$\lim_{n\to\infty} a_n = 1/\sum_{k=1}^{\infty} kp_k,$$

where the right hand side is interpreted as zero if the sum in the denominator diverges. If $\lim_{n\to\infty} a_n = 0$, a precise asymptotic formula for $a_n$ was given later by A. Garsia and J. Lamperti [2] under additional hypotheses upon $p_n$. However, the main problem turned out to be the problem of existence of $\lim_{n\to\infty} (a_{n+1}/a_n)$. This problem was studied extensively by N. G de Bruijn and P. Erdős [5], [6], and, independently, by A. M. Garsia, S. Orey and E. Rodemich [7]. Assuming that $p_n > 0$ and $\limsup_{n\to\infty} \sqrt[n]{p_n} < \infty$, de Bruijn and Erdős proved that $\lim_{n\to\infty} (a_{n+1}/a_n)$ exists if and only if

$$\lim_{A\to\infty} \left\{ \limsup_{n\to\infty} \left| \sum_{k=A}^{n+1} p_k \frac{a_{n-k+1}}{a_n} - \gamma^{-1} \sum_{k=A}^{n} p_k \frac{a_{n-k}}{a_n} \right| \right\} = 0,$$

where $\gamma = \sup \{x : \sum_{k=1}^{\infty} p_k x^k \leq 1\}$.

De Bruijn and Erdös gave also several sufficient conditions for the existence of $\lim_{n\to\infty} (a_{n+1}/a_n)$. The simplest result of this type is probably the one which states that $\lim_{n\to\infty} (p_{n+1}/p_n) = \gamma^{-1}$ implies that $\lim_{n\to\infty} (a_{n+1}/a_n) = \gamma^{-1}$. They also stated the conjecture that in some cases $p_n = o(a_n)$ $(n\to\infty)$ is a necessary and sufficient condition for the existence of $\lim_{n\to\infty}(a_{n+1}/a_n)$ (see [4] and [5, p. 378]). It seems that this problem has not yet been completely solved.

On the other hand, assuming that $\sum_{k=1}^{\infty} p_k = 1$ and that the greatest common divisor of the indices k such that $p_k > 0$ is one, Garsia, Orey and Rodemich proved that $\lim_{n\to\infty}(a_{n+1}/a_n)$ exists if and only if

$$\lim_{A\to\infty} \left\{ \sup_{n\geq A} \sum_{k=A}^{n} p_k \frac{a_{n-k}}{a_n} \right\} = 0.$$

If $p_n > 0$, the simplest sufficient condition for the existence of $\lim_{n\to\infty} (a_{n+1}/a_n)$ in this case is $\limsup_{n\to\infty}(p_{n+1}/p_n) \leq 1$.

2. When one considers these problems from a slightly more general point of view, it is easy to see that there are two basic types of theorems in the study of convolution products of sequences.

A direct theorem states that $\lim_{n\to\infty} (a_{n+1}/a_n)$ implies the existence of

$$\lim_{n\to\infty} \sum_{k=0}^{n} p_k \frac{a_{n-k}}{a_n},$$

while a converse theorem consists in showing that

$$(1)\quad \lim_{n\to\infty} \sum_{k=0}^{n} p_k \frac{a_{n-k}}{a_n} = C \quad (0 < C < \infty)$$

in certain cases implies the existence of $\lim_{n\to\infty} (a_{n+1}/a_n)$.

A systematic study of convolution products from this point of view has been completed recently by Y. H. Lee [8]. A direct result obtained by Lee can be stated as follows:

Suppose that $(p_n)$ is a sequence of real numbers and that the series $\Sigma_{k=1}^{\infty} p_k x^k$ has a positive radius of convergence R. If $(a_n)$ is a sequence of positive numbers such that $\lim_{n\to\infty}(a_{n+1}/a_n) = \lambda^{-1}$ and if $0 < \lambda < R$, then

$$\lim_{n\to\infty} \sum_{k=0}^{n} p_k \frac{a_{n-k}}{a_n} = \sum_{k=0}^{\infty} p_k \lambda^k .$$

The first two converse results proved by Lee are essentially extensions of the results mentioned in the preceding section. The third one is a discrete analog of a result obtained earlier by D. Drasin (see [9, Th. 6]). These results can be summarized as follows:

Suppose that $(p_n)$ and $(a_n)$ are sequences of positive numbers satisfying the asymptotic relation (1). A necessary and sufficient condition for the existence of $\lim_{n\to\infty} (a_{n+1}/a_n)$ is

$$(2) \quad \lim_{n\to\infty} \left\{ \sum_{k=A}^{n+1} p_k \frac{a_{n-k+1}}{a_n} - \sigma \sum_{k=A}^{n} p_k \frac{a_{n-k}}{a_n} \right\} = 0$$

for every fixed $A=1,2,\ldots$ Here $\sigma = 1/R$ if $\Sigma_{k=1}^{\infty} p_k R^k \leq C$ and $\sigma = 1/\xi$ if $\Sigma_{k=1}^{\infty} p_k R^k > C$, where $\xi \in (0,R)$ is such that $\Sigma_{k=1}^{\infty} p_k \xi^k = C$.

The necessary and sufficient condition (2) can be simplified if one knows that $C \leq \Sigma_{k=1}^{\infty} p_k R^k$. In that case, a necessary and sufficient condition for the existence of $\lim_{n\to\infty}(a_{n+1}/a_n)$ is

$$\lim_{A\to\infty} \left\{ \limsup_{n\to\infty} \sum_{k=A}^{n} p_k \frac{a_{n-k}}{a_n} \right\} = 0.$$

This condition is satisfied, in particular, if $\limsup_{n\to\infty}(p_{n+1}/p_n) \leq \sigma$.

Finally, if one knows that $C < \Sigma_{k=1}^{\infty} p_k R^k$, then a necessary and sufficient condition for the existence of $\lim_{n\to\infty} (a_{n+1}/a_n)$ is $\limsup_{n\to\infty} \sqrt[n]{a_n} > 1/R$.

The method used for the proof of these results is

essentially an extension of the method of de Bruijn and Erdős. More general convolution products, such as

$$\sum_{k=-\infty}^{\infty} p_k a_{n-k}, \qquad n=1,2,\ldots$$

can be studied in a similar way.

3. More precise versions of the results mentioned in Sections 1 and 2, in which the asymptotic relations in both hypotheses and conclusions appear with remainder terms have not been studied so far

In a preliminary study of possible results of this type, R. Bojanic and Y. H. Lee [10] have considered direct theorems with remainder terms. A typical result of this type can be stated as follows:

If

$$\frac{a_n}{a_{n+1}} = \lambda + O(\delta_n) \qquad (n \to \infty)$$

where $0 < \lambda < R$ and $(\delta_n)$ is a sequence of positive numbers such that $\delta_n \to 0$ and $\delta_{n+1}/\delta_n \to 1$ $(n \to \infty)$, then

$$\sum_{k=0}^{n} p_k \frac{a_{n-k}}{a_n} = \sum_{k=0}^{\infty} p_k \lambda^k + O(\delta_n) \qquad (n \to \infty).$$

If $\lambda = R$, the same result is true under somewhat stronger hypotheses about the sequence $(\delta_n)$ and the nature of convergence of the series $\Sigma_{k=0}^{\infty} p_k x^k$ on the boundary of its circle of convergence. In this case it is necessary to assume that $\Sigma_{k=1}^{\infty} k^{\alpha} |p_k| R^k < \infty$ for every $\alpha \geq 1$ and that

$$n\delta_n = O(1), \quad n\left\{\frac{\delta_n}{\delta_{n-1}} - 1\right\} = O(1) \qquad (n \to \infty).$$

Results of this type indicate what one can expect if one studies the converse problem. In the converse problem we assume that $(p_n)$ and $(a_n)$ are sequences of positive numbers and that, instead of (1), we have

$$(3) \qquad \sum_{k=0}^{n} p_k \frac{a_{n-k}}{a_n} = C + O(\delta_n) \qquad (n \to \infty),$$

where $(\delta_n)$ is a sequence of positive numbers converging to zero, and we conclude that

$$\frac{a_n}{a_{n+1}} = \lambda + O(\delta_n) \qquad (n \to \infty),$$

where $\lambda$ is a solution of the equation $\Sigma_{k=0}^{\infty} p_k x^k = C$.

In the special case when $p_k = r^{-k} (0<r<\infty)$, $k=0,1,2,\ldots$, results of this type are easily available. We have the following simple result in that case:

Let $(a_n)$ be a sequence of positive numbers such that

$$\sum_{k=0}^{n} r^{-k} \frac{a_{n-k}}{a_n} = C + O(\delta_n) \qquad (n \to \infty)$$

where $1<C<\infty$ and $(\delta_n)$ is a sequence of positive numbers such that $\delta_n \to 0$ and $\delta_n/\delta_{n+1} = O(1)$ $(n\to\infty)$. Then

$$\frac{a_{n-1}}{a_n} = r(1 - \frac{1}{C}) + O(\delta_n) \qquad (n \to \infty).$$

To see this, let $A_n = \Sigma_{k=0}^{n} r^k a_k$. We have then

$$\frac{A_n}{r^n a_n} = C + O(\delta_n) \quad \text{and} \quad \frac{r^n a_n}{A_n} = \frac{1}{C} + O(\delta_n).$$

Since $r^n a_n = A_n - A_{n-1}$, we also have

$$\frac{A_{n-1}}{A_n} = 1 - \frac{1}{C} + O(\delta_n).$$

Hence

$$\frac{a_{n-1}}{a_n} = r \frac{r^{n-1} a_{n-1}}{A_{n-1}} \frac{A_{n-1}}{A_n} \frac{A_n}{r^n a_n}$$

$$= r(\frac{1}{C} + O(\delta_{n-1})).(1 - \frac{1}{C} + O(\delta_n)).(C + O(\delta_n))$$

$$= r(1 - \frac{1}{C}) + O(\delta_n).$$

No results of this type for more general sequences $(p_n)$, analog to the converse results of Section 2, seem to be known.

References


[1] Feller, W., Fluctuation theory of recurrent events. Trans. Am. Math. Soc. 67 (1949), 98-119.

[2] Garsia, A. and J. Lamperti, A discrete renewal theorem with infinite mean. Comm. Math. Helv. 37 (1963), 221-234.

[3] Erdős, P., W. Feller and H. Pollard, A property of power series with positive coefficients. Bull. Am. Math. Soc. 55 (1949), 201-204.

[4] de Bruijn, N. G. and P. Erdős, On a recursion formula and on some Tauberian theorems. Journal of Research of the Nat. Bureau of Standards, 50 (1953), 161-164.

[5] de Bruijn, N. G. and P. Erdős, Some linear and some quadratic recursion formulas I. Indag. Math. 13 (1951), 374-382.

[6] de Bruijn, N. G. and P. Erdős, Some linear and some quadratic recursion formulas II. Indag. Math. 14 (1952), 152-163.

[7] Garsia, A., S. Orey and E. Rodemich, Asymptotic behavior of successive coefficients of some power series. Ill. J. Math. 6 (1962), 620-629.

[8] Lee, Y. H., Asymptotic properties of convolution products of sequences. Ph.D Thesis, The Ohio State University, Columbus, Ohio, 1972.

[9] Drasin, D., Tauberian theorems and slowly varying functions. Trans Am. Math Soc. 133 (1968), 333-356.

[10] Bojanic, R. and Y. H. Lee, An estimate for the rate of convergence of convolution products of sequences. To appear.


Department of Mathematics
The Ohio State University
Columbus, Ohio 43210

# THE QUASI-INTERPOLANT AS A TOOL IN ELEMENTARY POLYNOMIAL SPLINE THEORY

Carl de Boor[1]

This talk is intended to demonstrate with the help of some examples that the quasi-interpolant of [2] is very convenient when it comes to proving even very elementary old and new facts about polynomial splines. The key is a formula which gives each B-spline expansion coefficient for a given spline in terms of the value of its derivatives at a point.

## 1. Definitions

Let $k \in \mathbb{N}$, let $\underline{t} := (t_i)_{-\infty}^{\infty}$ be real, nondecreasing $t_i < t_{i+k}$, all $i$, and set

$$a := \inf_i t_i,$$

and

$$b := \inf_i t_i.$$

For $i \in \mathbb{Z}$, <u>the</u> $i^{th}$ <u>B-spline of order</u> $k$ <u>with</u> (or, for the) <u>knot sequence</u> $\underline{t}$ is given by the rule

$$N_{ik}(t) := g_k(t_i,\ldots,t_{i+k};\, t)(t_{i+k} - t_i)$$

$$g_k(s;t) := (s - t)_+^{k-1}$$

taking, for each fixed $t$, the $k^{th}$ divided difference of $g(s) := g_k(s;t)$ at $t_i,\ldots t_{i+k}$ in the usual manner even when some or all of the $t_j$'s coincide. I leave unresolved any possible ambiguity when $t = t_j$ for some $j$, and concern myself only with left and right limits at such a point; i.e., I replace each $t = t_j$ by the "two points" $t_j^-$ and $t_j^+$.

As is well known,

$$N_{ik} > 0 \text{ on } (t_i, t_{i+k}), \text{ and } N_{ik} = 0 \text{ off } [t_i^+, t_{i+k}^-]$$

so that (since $t_i < t_{i+k}$, by assumption) $N_{ik}$ is not identically zero, while on the other hand, no more than k of the $N_{jk}$'s are nonzero at any particular point. Consequently, for an arbitrary $\underline{a} \in R^Z$, the rule

$$f(t) := \Sigma_i \, a_i N_{ik}(t)$$

defines a function on (a,b) if we take the sum to be <u>pointwise.</u> I call every such function a <u>polynomial spline of order</u> k <u>with knot sequence</u> t, and denote their collection by

$$S_{k,\underline{t}} .$$

The "quasi-interpolator" Q of interest here is given by the rule

$$Qf := \Sigma_i (\lambda_i f) N_{ik}$$

where

$$\lambda_i f := \lambda_{\tau_i, \psi_{ik}} f := \sum_{j<k} (-)^{k-1-j} \, \psi_{ik}^{(k-1-j)}(\tau_i) \, f^{(j)}(\tau_i)$$

$$\psi_{ik}(t) := (t_{i+1} - t) \ldots (t_{i+k-1} - t)/(k-1)!$$

and $\tau_i$ is an <u>arbitrary</u> point in $(t_i, t_{i+k})$. One verifies directly that [2]

$$\lambda_i \, N_{jk} = \delta_{ij}, \quad \text{all } i,j.$$

Consequently,

(i) Q is a linear projector with range $S_{k,\underline{t}}$;

(ii) every $f \in S_{k,\underline{t}}$ has a unique representation as a B-spline series;

(iii) if $f = \Sigma_i a_i N_{ik}$, then

$$a_i = \lambda_{\tau_i, \psi_{ik}} f \text{ for arbitrary } \tau_i \in (t_i, t_{i+k}).$$

## 2. Existence and Uniqueness of the B-spline Expansion

The rather curious freedom in the choice of $\tau_i$ above leads to the following short proof of

Theorem (Curry et Schoenberg [3]). $S_{k,\underline{t}}$ consists of exactly those f on (a,b) for which

(i) for all i, $f|_i \in P_k$ (:= polynomials of degree < k); and

(ii) if $t_s < t_{s+1} = \dots = t_{s+r} < t_{s+r+1}$, then $\text{jump}_{t_{s+1}} f^{(k-j)} = 0$ for all j > r.

In particular, any such f has exactly one B-spline expansion (in terms of the B-splines of order k with knots $\underline{t}$).

Here and below, we denote by $f|_i$ the restriction of f to $(t_i, t_{i+1})$. For the proof, I show that Qf = f for all such f:

(a) For all such f, and all i,

$$g(\tau) := \lambda_{\tau,\psi_{ik}} f = \sum_{j<k} (-)^{k-1-j} \psi_{ik}^{(k-1-j)}(\tau) f^{(j)}(\tau)$$

is constant on $\tau \in (t_i, t_{i+k})$ = support $N_{ik}$, since

(α) for $\psi \in P_k$ and smooth f,

$$(\lambda_{\tau,\psi} - \lambda_{\sigma,\psi})f = \int_\sigma^\tau \psi\, df^{(k-1)} \quad (= 0 \text{ if } f|[\sigma,\tau] \in P_k)$$

hence, as $f|_{(t_j,t_{j+1})} \in P_k$, g <u>is constant on each</u> $(t_j, t_{j+1})$;

and

(β) if $t_i \le t_s < t_{s+1} = \dots = t_{s+r} < t_{s+r+1} \le t_{i+k}$, then $t_{s+1}$ is an r-fold zero of $\psi_{ik}$, hence

$$\psi_{ik}^{(k-1-j)}(t_{s+1}) = 0, \quad \text{for } j=k-1, k-2, \dots, k-r,$$

while, by assumption on f,

$$\text{jump}_{t_{s+1}} f^{(j)} = 0, \quad \text{for } j=k-r-1, \dots, 0;$$

hence g <u>is continuous across each</u> $t_{s+1}$ <u>with</u> $t_i < t_{s+1} < t_{i+k}$.

(b) For all such f, and all j with $t_j < t_{j+1}$,

$$(Qf)|_j = f|_j .$$

For, $(Qf)|_j = \Sigma^j_{i=j+1-k} (\lambda_{\tau_i,\psi_{ik}} f)(N_{ik})|_j$. But I can assume by (a) without loss that $\tau_i \in (t_j, t_{j+1})$, $i=j+1-k,\ldots,j$; hence

$$(Qf)|_j = \sum_{i=j+1-k}^{j} \lambda_{\tau_i,\psi_{ik}} (f|_j)(N_{ik})|_j ,$$

while

$$\delta_{ir} = \lambda_{\tau_i,\psi_{ik}} N_{rk} = \lambda_{\tau_i,\psi_{ik}} (N_{rk}|_j), \qquad r=j+1-k,\ldots,j$$

shows the k-sequence $N_{ik}|_j$, $i=j+1-k,\ldots,j$, in $P_k$ to be independent, hence a basis for $P_k$. Consequently,

$$\sum_{i=j+1-k}^{j} (\lambda_{\tau_i,\psi_{ik}} h)(N_{ik})|_j = h, \text{ for all } h \in P_k .$$

## 3. Uniqueness of Odd-degree Spline Interpolation

In discussing the smooth extension of a real valued function defined on some closed subset of R to all of R, Golomb et Schoenberg [4] prove that, for $\underline{t}$ strictly increasing, every $f \in S_{2k,\underline{t}}$ which vanishes at the points of $\underline{t}$ and has square-integrable $k^{th}$ derivative must vanish identically. Their proof is not simple. In particular, the straightforward argument

> $\forall_i\ f(t_i)=0$, hence, $\forall_i\ 0=f(t_i,\ldots,t_{i+k}) = \int N_{ik}(t)f^{(k)}(t)dt/c_{ik}$ with $c_{ik} := (k-1)!(t_{i+k}-t_i)$; i.e., $f^{(k)}$ is <u>orthogonal</u> to every $N_{ik}$, while at the same time being in $S_{k,\underline{t}}$ which is spanned by the $N_{ik}$'s; hence $f^{(k)} = 0$, and so $f = 0$.

was not open to them since it requires $(N_{ik})$ to be a Schauder basis for $S_{k,\underline{t}} \cap L_2$, a fact they did not know.

<u>Theorem</u>. Let $1 \le p \le \infty$, and $N_{ikp} := (k/(t_{i+k}-t_i))^{1/p} N_{ik}$. Then

$$\Sigma_i b_i N_{ikp} \in L_p(a,b) \text{ iff } \|\underline{b}\|_p < \infty .$$

Precisely, there exists $D_{kp} > 0$ (independent of $\underline{t}$) so that

$$D_{kp}^{-1}\|\underline{b}\|_p \leq \|\Sigma_i b_i N_{ikp}\|_p \leq \|\underline{b}\|_p, \text{ for all } \underline{b} \in R^Z.$$

The second inequality is straightforward. As to the first, let $f := \Sigma_i a_i N_{ik} = \Sigma_i b_i N_{ikp}$, so that $a_i((t_{i+k}-t_i)/k)^{1/p} = b_i$, all i. Then, from Sec. 1, $|a_i| \leq \Sigma_{j<k} |\psi_{ik}^{(k-1-j)}(\tau_i)||f^{(j)}(\tau_i)|$.

Take I to be a largest interval among $(t_i,t_{i+1}),\ldots,(t_{i+k-1},t_{i+k})$, and choose $\tau_i \in I$. Then $|\psi_{ik}^{(k-1-j)}(\tau_i)| \leq A_{jk}|I|^j$ for some constants $A_{jk}$, while $|f^{(j)}(\tau_i)| \leq B_{jkp}|I|^{-j-1/p} \cdot (\int_I |f(t)|^p dt)^{1/p}$ since $f_{|I} \in P_k$. Hence,

$$|b_i|^p = |a_i|^p(t_{i+k}-t_i)/k \leq |a_i|^p|I| \leq (\Sigma_j A_{jk}B_{jkp})^p \int_I |f|^p$$
$$\leq C_{kp} \int_{t_i}^{t_{i+k}} |f|^p$$

which, after summing over i, gives the required inequality with $D_{kp} = (kC_{kp})^{1/p}$.

For a uniform knot sequence $\underline{t}$, this theorem has already been proved by Schoenberg in [5] using a special case of the above formula for the B-spline coefficients.

Corollary. For $1 \leq p < \infty$, $(N_{ikp})_{-\infty}^{\infty}$ is a Schauder basis for $S_{k,\underline{t}} \cap L_p(a,b)$.

Bolstered by this Corollary, the earlier argument establishes uniqueness of odd-degree spline interpolation even in the limiting case of repeated or osculatory interpolation at multiple knots.

## 4. Bounds for Least-squares Approximation by Splines

An attempt to bound the error in odd-degree spline interpolation to a smooth function in the uniform norm leads to the problem of bounding least-squares approximation by splines, considered as a map on $L_\infty$, independently of the knot sequence (cf. [1]), a question of interest in itself.

Let $n \in \mathbb{N}$, $S := \operatorname{span}\{N_{1k},\ldots,N_{nk}\}$, and denote by $Lf$ the Least-squares approximation to an $f \in L_\infty[t_1,t_{n+k}]$ by elements of S. Then, L is a linear projector, characterized by the fact that

(*) $Lf \in S$, and, for all $\lambda \in \Lambda$, $\lambda Lf = \lambda f$

with the "interpolation conditions"

$$\Lambda := \{\lambda \in L_\infty^* \mid \text{for some } \phi \in S \text{ and all } f,\ \lambda f = \int \phi f\}.$$

One verifies that (*) implies

$$\|L\| = \sup_{x \in S} \inf_{\lambda \in \Lambda} \|\lambda\|\,\|x\| / |\lambda x|.$$

But, in order to compute, one needs to coordinatize. Letting $(\lambda_i)$ and $(\phi_i)$ be bases for $\Lambda$ and S, respectively, we get that

$$\|L\| = \sup_{\underline{a}} \inf_{\underline{b}} \|\Sigma_i b_i \lambda_i\|\,\|\Sigma_j a_j \phi_j\| / |\Sigma_{ij} b_i \lambda_i \phi_j a_j|.$$

Take $\phi_i := N_{ik}$, $\lambda_i := k\int \cdot\, N_{ik}/(t_{i+k}-t_i)$, $i=1,\ldots n$. From the earlier theorem,

$$D_{k1}^{-1} D_{k\infty}^{-1} \|\underline{b}\|_1 \|\underline{a}\|_\infty \le \|\Sigma_i b_i \lambda_i\|\,\|\Sigma_j a_j \phi_j\| \le \|\underline{b}\|_1 \|\underline{a}\|_\infty$$

while

$$\sup_{\underline{a}} \inf_{\underline{b}} \|\underline{b}\|_1 \|\underline{a}\|_\infty / |\Sigma_{ij} b_i \lambda_i \phi_j a_j| = \|(\lambda_i \phi_j)^{-1}\|_\infty$$

with $\|A\|_p$ denoting the norm for the matrix A induced by the p-norm on vectors. This proves

<u>Proposition.</u> For some positive $C_k$ (independent of $\underline{t}$ and n),

$$C_k \|(\lambda_i \phi_j)^{-1}\|_\infty \le \|L\| \le \|(\lambda_i \phi_j)^{-1}\|_\infty$$

(considering L as a map on $L_\infty[t_1,t_{n+k}]$), with

(**) $\lambda_i \phi_j = k\int N_{ik} N_{jk}/(t_{i+k}-t_i)$, $\quad i,j=1,\ldots,n$.

It has been known for some time that L could be bounded if

only the Gramian $(\lambda_i \phi_j)$ could be bounded below (in the max-norm). This proposition adds that such bounding below of the Gramian is also necessary for bounding L. For this reason, I offer the modest sum of m-1972 ten dollar bills to the first person who communicates to me a proof or a counterexample (but not both) of his or her own making for the following conjecture (known to be true when k=2 or k=3):

Conjecture. For given n and $\underline{t}$, let $(\lambda_i \phi_j)$ be the $n \times n$ matrix whose entries are given by (**). Then

$$\sup_{n,\underline{t}} \|(\lambda_i \phi_j)^{-1}\|_\infty < \infty$$

Here, m is the year A.D. of such communication.

## 5. Estimates for $\operatorname{dist}(f, S_{k,\underline{t}})$

Let Qf be the quasi-interpolant to f as defined in Section 1. For a sufficiently smooth f,

$$f(t) - (Qf)(t) = \int E(t,s)\,df^{(k-1)}(s)$$

with $E(t,\cdot)$ a nonnegative function of small support. This makes Qf a convenient approximation when it comes to estimating the distance of such f from splines with fixed and with variable knots. Lack of space precludes, unfortunately, any discussion of this important aspect of the quasi-interpolant here.

---

[1]Supported by the United States Army under Contract DA-31-124-ARO-D-462.

## References

[1] de Boor, C., On the convergence of odd-degree spline interpolation. J. Approx. Theory 1 (1968), 452-463.

[2] de Boor, C. and G. J Fix, Spline approximation by quasi-interpolants. J. Approx. Theory 7 (1973)

[3] Curry, H. B. and I. J. Schoenberg, On Polya frequency functions IV; the fundamental spline functions and their limits. J. Analyse Math. 17 (1966), 71-107.

[4] Golomb, M. and I. J. Schoenberg, On $H^m$-extension of functions and spline interpolation. This paper exists only as a reference in other papers, e.g., in [I. J. Schoenberg, On spline interpolation at all integer points of the real axis, Mathematica 10 (33) (1968) 151-170] where its proposed content is outlined, and in [M. Golomb and J. Jerome, Linear ordinary differential equations with boundary conditions on arbitrary point sets, Trans. Amer. Mathem. Soc. 153 (1971), 235-264] in which it is incorrectly specified as a MRC Technical Summary Report and where its proposed content is generalized.

[5] Schoenberg, I. J., Cardinal interpolation and spline functions: II. Interpolation of data of power growth. J. Approx. Theory 6 (1972), 404-420.

Mathematics Research Center
University of Wisconsin
Madison, Wisconsin 53706.

# GLOBAL ANALYSIS AND CHEBYSHEV APPROXIMATION BY EXPONENTIALS

Dietrich Braess

The purpose of this paper is to present a solution of the following problem: How many local best approximations exist when approximating with exponentials in the Chebyshev norm? The method of attack is fairly general and may be applied to nonlinear families of functions which satisfy a local but not a global Haar condition. In general, there are more than one, but finitely many local best approximations. Therefore, there is no hope of success when applying results of convex programming. We must use methods from global analysis [3].

Let X be a compact interval and let C(X) be endowed with the uniform norm. We consider the exponentials of order $\leq N$:

$$(1)\quad V_N = \{F; F = \sum_{\nu=1}^{\ell} p_\nu(x)e^{t_\nu x},\ \sum_{\nu=1}^{\ell}(1+\partial p_\nu) = K \leq N,\ t_\nu \in R\}.$$

Here $p_\nu$ denotes a polynomial with degree $\partial p_\nu$ and the $t_\nu$'s are the characteristic numbers of F, which are free parameters, too. The family $V_N$ arises as the closure of the set of functions of the form

$$(2)\quad \sum_{\nu=1}^{N} \alpha_\nu e^{t_\nu x},\quad \alpha_\nu, t_\nu \in R$$

in C(X). It is known that $V_N$ does not have the varisolvency property when N is 2 or greater [1]. Moreover, the best approximation is not always unique.

<u>Example for Non-Uniqueness.</u> There are exactly 2 best approximations to $f(x) = \cos \frac{\pi}{2} x$ on $X = [-1,+1]$ in $V_2$.

We will estimate the number of best approximations by deriving a bound for the number of local best approximations. The latter bound is also of interest for the numerical construction of solutions.

Before we enter into the discussion of the global analysis we have to study the local structure of the family and introduce the tangent cones.

Definition 1. Let $V \subset C(X)$. Then the tangent cone $T_F V$ at F to V consists of the elements $h \in C(X)$ with the following property: there is a continuous mapping of $[0,1]$, $t \to F_t \in V_N$ such that $F_0 = F$ and $\|F_t - F - th\| = o(t)$ as $t \to 0$.

If F has the maximal order, i.e., $F \in V_N \setminus V_{N-1}$, then the tangent cone is a convex cone with the Haar property:

Definition 2. Let $v_1, v_2 \ldots v_n \in C(X)$ and $m \leq n$. The convex cone

$$\{u; u(x) = \sum_{i=1}^{n} \alpha_i v_i(x);\ \alpha_i \geq 0 \text{ for } i=m+1\ldots n\}$$

has the Haar property, if the functions $\{v_i, i \in J\}$ span a Haar subspace whenever $\{1,2,\ldots m\} \subset J \subset \{1,2,\ldots n\}$ holds.

Concerning Chebyshev approximation the classical results for the Haar subspaces may be extended to the cones with the Haar property:

Theorem 1. The best approximation in a cone with the Haar property is unique and satisfies a strong unicity condition.

Outline of proof: If $v = \Sigma_i\, \alpha_i u_i$ is a best approximation in the cone, then v is also optimal in the linear space spanned by $\{v_i, i \leq m \text{ or } \alpha_i > 0\}$. From this, unicity is immediate. Moreover, combining this argument with the techniques for the proof of the strong unicity for Haar subspaces, we obtain the second statement.

It was shown by Wulbert [5] that local best approximations in a manifold without boundary may be characterized by the approximation in the tangent space, provided that strong unicity holds. It is somewhat natural that we have to replace tangent spaces by tangent cones when we extend Wulbert's results to manifolds with boundaries and corners.

<u>Theorem 2</u>. Let V be a $C^1$-manifold (with corners) and let $T_F V$ be a cone with the Haar property. Then F is a local best approximation to f in V, iff 0 is a best approximation to f-F in $T_F V$.

Since the set of non-degenerate exponentials $V_M \setminus V_{N-1}$ satisfies these conditions, Theorem 2 can be applied immediately. On the other hand, $V_N$ is not a manifold.

Now we are in a position to ask: Which theory may yield a bound for the number of local best approximations? The answer is critical point theory, in particular Morse theory [4].

<u>Definition 3</u>. Let V be a $C^1$-manifold and $\varphi$: $V \to R$ a $C^1$-function. Then $u \in V$ is a critical point, if the differential $d_u \varphi$: $T_u V \to R$ vanishes.

When considering Chebyshev approximation, however, the function which is to be minimized, is generally not differentiable. This becomes obvious from an elementary example.

<u>Example</u>. Let $X = [-1,+1]$, $f(x) = x$ and let $V = \{F_a(x) \equiv a; a \in R\}$ be the set of constant functions. Then

$$\varphi(a) = \|f-F_a\| = \sup\{|x-a|;\ -1 \le x \le +1\} = 1 + |a|$$

is not a differentiable function.

For this reason, we cannot apply the results of the classical Morse theory directly. Nevertheless, though we have to abandon the concept of "steepest descent" we can use a "descent method." To this end critical points are redefined [2,3].

<u>Definition 4</u>. Let V be a $C^1$-manifold with corners. F is a critical point to f in V, iff 0 is a best approximation to the error curve f-F in $T_F V$.

If $F^o$ is not a critical point, then there is a path in V starting at $F^o$ such that the distance to f decreases along it. Indeed, let $\|f-F^o\| > \|f-F^o-h\|$ and choose $F_t$ to $h \in T_F V$ according to Definition 1. Then we get

$$\|f-F_t\| \leq \|f-F^o-th\| + \|F_t-F^o-th\| \leq \|f-F^o\| - \text{const.}\ t + o(t)$$
$$< \|f-F\|,$$

if t is sufficiently small. Since V is assumed to be a $C^1$-manifold, there is a continuous mapping in a neighborhood $U_o$ of $F^o$: $U_o \ni F \to h(F) \in T_F V$. Consequently, we have $\|f-F\| > \|f-F-h(F)\|$, if $U_o$ is sufficiently small. This yields a flow, i.e., an extension of the path given above:

$$(t,F) \to F_t \in V \qquad\qquad (t,F) \in [0,\varepsilon_o] \times U_o$$

such that $\|f-F_t\| < \|f-F\|$, if $t > 0$.

Now we introduce the usual denotation for level sets:

$$V^a = \{F \in V;\ \|f-F\| \leq a\}.$$

With this, the preceding discussion establishes the

Lemma 1. Let $F^o$ be a non-critical point in V and $\|f-F^o\| = c$. Then there is a neighborhood U of $F^o$, an $\varepsilon > 0$ and a flow

$$\Phi: [0,1] \times V \to V$$

such that $\|f-\Phi_t(F)\| \leq \|f-F\|$, $F \in V$ and $\Phi_1(U) \in V^{c-\varepsilon}$.

Proof: Referring to the flow in $U_o$ given above, choose $\chi \in C(V)$ with the following properties

$$\chi(F^o) = \varepsilon_o,$$

$$0 \leq \chi(F) \leq \varepsilon_o, F \in U_o,$$

$$\chi(F) = 0, F \in V \setminus U_o.$$

Then

$$\Phi(t,F) = \begin{cases} F_{t\cdot\chi(F)} & \text{if } F \in U_o, \\ F & \text{if } F \notin U_o, \end{cases}$$

has the properties required.

From this we obtain the

Defomation Theorem. Let $V$ be a $C^1$-manifold. If the set $\{F \epsilon V; a \leq \|f-F\| \leq b\}$ is compact and contains no critical point, then $V^a$ is a strong deformation retract of $V^b$.

The proof proceeds by covering the set $\{F \epsilon V;\ a \leq \|f-F\| < b\}$ by a finite number of open sets described in Lemma 1. By glueing the corresponding flows together in an appropriate way and using compactness once more we get for each $F \epsilon V^b$ a path of descent which ends in $V^a$.

Now, in the terminology of Morse theory [4] the Theorem 2 may be written as follows. If $V$ is a $C^1$-manifold and if all tangent cones satisfy the Haar property, then each critical point has an index of zero. For this reason we get a sharper result.

Uniqueness Theorem. Let $V$ be a $C^1$-manifold and let all tangent cones satisfy the Haar property. If $V^a$ is connected and compact, then $V^a$ contains only one critical point (local best approximation).

The proof is similar to that of the Deformation Theorem. We remark that we would also obtain this statement when we formally apply the Morse inequalities to the approximation problem.

Let us return to the family of exponentials. Since $V_N \setminus V_{N-1}$ (but not $V_N$ itself) satisfies the assumptions of the uniqueness theorem, we obtain:

Corollary 1. Given $a \epsilon \mathbf{R}$. Each component of $(V_N)^a$, which is disjoint from $V_{N-1}$, contains only one local best approximation.

Therefore, the enumeration of local best approximations is reduced to the problem of counting the components of level sets disjoint from $V_{N-1}$. This problem is solved by the following:

Construction and Theorem. Let $\hat{F}$ be a local best approximation to f in $V_{N-1}$ which is not a local best approximation in $V_N$. The characteristic numbers of F divide the real axis into at most N intervals. Then $F+0\cdot e^{t_0 x}$ is a formal exponential in $V_N$, and in every neighborhood of this element there is an element $F_0 \in V_N$ with $\|f-F_0\| < \|f-\hat{F}\|$. The component of $\{F\in V_N; \|f-F\| < \|f-F\|\}$ containing $F_0$ depends only on F and on the interval in which $t_0$ is chosen.

From this construction we know that to each local best approximation in $V_{N-1}$, there correspond at most N local best approximations in $V_N$. It does not matter that some of the components are counted twice. Thus, we have the final result:

Theorem 3. There are at most N! local best approximations in $V_N \setminus V_{N-1}$.

## References

[1] Braess, D., Chebyshev approximation by $\gamma$-polynomials, I and II. J. Approx. Theory. To appear.

[2] ______, On the number of best approximations in certain non-linear families of functions.

[3] ______, Kritische Punkte bei der nichtlinearen Tschebyscheff-Approximation. To appear.

[4] Milnor, J., Morse theory. Princeton University Press, Princeton 1963.

[5] Wulbert, D., Uniqueness and differential characterization of approximations from manifolds of functions. Amer. J. Math. 93 (1971), 350-366.

Institut für Numerische und
Instrumentelle Mathematik
Universität Münster
4400 Münster, Roxeler Str. 64
West Germany

Note added in proof. Notice was given that the tangent cone should be denoted as $C_F V$ instead of $T_F V$ to agree with differential geometry notation.

# A VARIATIONAL INEQUALITY AND SOME APPLICATIONS IN APPROXIMATION THEORY

B. Brosowski and K.-H. Hoffmann

Most of the well known theorems in optimization theory like those of Kuhn and Tucker are valid under certain assumptions of differentiability for the target function and the functions describing the side conditions. One cannot immediately use the results of optimization theory in approximation theory, because in general the target function is non-differentiable. We consider a special problem in optimization theory with a non-differentiable target function and prove a variational inequality. Consistency of this inequality is always a necessary condition for an unconstrained minimum of the target function. We use this result in getting global and local conditions in optimization as well as in approximation theory in a simple way.

Let [a,b] be a closed interval of the real line and T a compact Hausdorff space. We consider a real valued function $\Phi$: $[a,b] \times T \rightarrow R$, continuously partially differentiable in the first variable x, and we define

$$\Phi'(x,t) := \frac{\partial}{\partial x}\Phi(x,t).$$

Then we deal with the following optimization problem: Find a real number $x_0$ in [a,b], which realizes a local minimum of the function $\varphi(x) := \max_{t\in T} \Phi(x,t)$. The number $x_0$ is called a minimal point of the function $\varphi$ and $\varphi(x_0)$ the minimal value.

A necessary condition for $x_0$ to be a local minimal point of $\varphi$ is given by the following

Variational-Lemma. Let $x_o$ in [a,b] be a local minimum of the function $\varphi$, then the following inequality is consistent for all x in [a,b]:

$$\max_{t\in\Sigma_o} \Phi'(x_o.t)(x-x_o) \geq 0,$$

where $\Sigma_o := \{t\in T \mid \Phi(x_o,t) = \varphi(x_o)\}$.

Outline of proof: The function $\varphi$ attains its minimum on [a,b] because it is lower semicontinuous. Without loss of generality we assume the minimal value $\varphi(x_o)$ to be positive:

$$\varphi(x_o) =: E > 0.$$

Suppose there is a point $\overline{x}$ in [a,b] and a negative real number $\alpha$ satisfying the inequality

$$\max_{t\in\Sigma_o} \Phi'(x_o,t)(\overline{x}-x_o) \leq \alpha < 0.$$

The function $\Phi'$ was assumed to be continuous. Therefore we have

$$\bigwedge_{\varepsilon>0} \ \bigwedge_{t_o\in T} \ \bigvee_{\overline{\lambda}(t_o)>0} \ \bigvee_{U(t_o)\in\mathcal{U}(t_o)} \text{ open in } T$$

$$\{x_o+\lambda(\overline{x}-x_o) \in R \mid 0<\lambda\leq\overline{\lambda}(t_o)\} \times U(t_o)$$

$$\subset \Phi'^{-1}(\{r\in R \mid |\Phi'(x_o,t_o) - r| < \varepsilon\}).$$

Since T is compact, we conclude:

$$\bigvee_{\substack{t_1,t_2,\ldots,t_m\in T \\ m \text{ finite}}} \ \bigcup_{i=1}^{m} U(t_i) = T.$$

We define: $\overline{\lambda} := \min_{1\leq i\leq m} \overline{\lambda}_i$ .

Then for all $\lambda \in (0,\overline{\lambda}]$ and all $t \in T$ we have the inequality

$$|\Phi'(x_o,t) - \Phi'(x_o+\lambda(\overline{x}-x_o),t)| < \varepsilon.$$

Now let $U := \{t \in T \mid \Phi'(x_o,t)(\bar{x}-x_o) < \alpha/2\}$. U is an open neighborhood of $\Sigma_o$ in T. For all t in U one gets the estimate:

$$\bigwedge_{0<\lambda\leq\bar{\lambda}} \Phi(x_o+\lambda(\bar{x}-x_o),t) - \Phi(x_o,t) = \lambda(\bar{x}-x_o)\Phi'(x_o,t)$$
$$+ \lambda(\bar{x}-x_o)[\Phi'(x_o+\tau(t)\lambda(\bar{x}-x_o),t) - \Phi'(x_o,t)]$$
$$< \lambda\,\alpha/2 \;+ \lambda|\bar{x}-x_o|\varepsilon,$$

with $0 < \tau(t) < 1$.

We choose $\varepsilon < -\alpha/(4|\bar{x}-x_o|)$ and conclude:

$$\bigwedge_{t\in U} \bigwedge_{0<\lambda\leq\bar{\lambda}} \Phi(x_o+\lambda(\bar{x}-x_o),t) < \Phi(x_o,t) + \lambda\alpha/4.$$

Now let $t\in T\setminus U$. The set $T\setminus U$ is a compact set. Hence

$$\max_{t\in T\setminus U} \Phi(x_o,t) =: A < \varphi(x_o) = E$$

and

$$\bigwedge_{t\in T\setminus U} \Phi(x_o,t) < (A-E)/2 .$$

Because $\Phi$ is a continuous function, for each $t_o \in T\setminus U$ there exist open neighborhoods $W(t_o)$ in $T\setminus U$ and $V(x_o)$ in $[a,b]$ such that for all $(x,t) \in V(x_o) \times W(t_o)$ the inequality

$$\Phi(x,t) < (A+E)/2$$

is valid. The set $T\setminus U$ being compact, we can conclude in a similar way as we did at the beginning of the proof:

$$\bigwedge_{0<\lambda\leq\bar{\lambda}_1} \bigwedge_{t\in S\setminus U} \Phi(x_o+\lambda(\bar{x}-x_o),t) < (A+E)/2 < E.$$

In this formula we have defined

$$\bar{\lambda}_1 := \min(\min_{1\leq i\leq m'} \lambda(t_i) , \bar{\lambda}),$$

where $\bigcup_{i=1}^{m'} W(t_i) = T\setminus U$.

Therefore for all $t\in T$ we have the estimate

$$\Phi(x_o+\lambda(\bar{x}-x_o),t) < E.$$

This contradicts the assumption that $x_o$ is a minimal point of $\varphi$.

## Applications

### 1. Approximations by elements of a sun

Let X be a normed linear space and V a set in X which is a sun. That means

$$\bigwedge_{x\in X} (x_o \in P_V(x) \Rightarrow \bigwedge_{\mu\geq 1} x_o \in P_V(x+\mu(x-x_o))).$$

Here $P_V(x)$ is the set of all elements of best approximation to x out of V.

A well known result is stated as follows (see B. Brosowski [1])

$$\bigwedge_{x\in X} (x_o \in P_V(x) \Longleftrightarrow \bigwedge_{z\in V} \min_{x^*\in E_{x-x_o}} \mathrm{Re}\langle x^*, z-x_o\rangle \leq 0),$$

where $E_{x-x_o}$ is the set of extreme points of the weakly compact convex set $\Sigma_{x-x_o} := \{x^*\in X^* \mid \|x^*\| \leq 1, \langle x^*, x-x_o\rangle = \|x-x_o\|\}$.

It is known that it is sufficient to show the equivalence

$$\bigwedge_{x\in X} (x_o \in P_V(x) \Longleftrightarrow \bigwedge_{z\in V} \min_{x^*\in\Sigma_{x-x_o}} \mathrm{Re}\langle x^*, z-x_o\rangle \leq 0).$$

The implication from the right to the left is a simple conclusion. To show the necessity of the so-called Kolmogoroff condition one needs complicated considerations in a long proof. But we can use our variational inequality to get this result in a simple way.

We define:

$$T := K[\theta^*, 1] := \{x^*\in X^* \mid \|x^*\| \leq 1\},$$

$$[a,b] := [0,1],$$

$$\bigwedge_{z\in V} \Phi_z(\lambda, x^*) := \mathrm{Re}\langle x^*, x-(\lambda x_o+(1-\lambda)z)\rangle.$$

The function $\Phi_z$ satisfies all the assumptions of the variational lemma for each $z \in V$. Let $x_o \in V$ be a best approximation for an element $x \in X$. Then for each $z \in V$ the element $x_o$ is also a best approximation for $x$, out of the set $[x_o,z] := \{\rho x_o+(1-\rho)z \in X \mid \rho\in[0,1]\}$. This is easy to see: Assume there is a $z \in V$ and a $\tilde{x} \in [x_o,z]$ satisfying the inequality $\|x-\tilde{x}\| < \|x-x_o\|$. It follows:

$$\bigvee_{\mu>0} \quad z = x_o + \mu(\tilde{x}-x_o)$$

and

$$\|x_o+\mu(x-x_o)-z\| = \mu\|x-\tilde{x}\| < \mu\|x-x_o\| = \|x_o+\mu(x-x_o)-x_o\|.$$

This contradicts the fact that V is a sun.

Now we make use of our lemma. The assumption that $x_o \in V$ is a best approximation for $x \in X$ implies that $\lambda := 1$ is a minimal point of the function

$$\bigwedge_{z\in V} \quad \varphi_z(\lambda) := \max_{x^*\in K[\theta^*,1]} \operatorname{Re}\langle x^*, x-(\lambda x_o+(1-\lambda)z)\rangle .$$

Using the variational lemma we get

$$\bigwedge_{z\in V} \quad \bigwedge_{\lambda\in[0,1]} \quad \max_{x^*\in\Sigma_{x-x_o}} \operatorname{Re}\langle x^*, z-x_o\rangle(\lambda-1) \geq 0$$

and

$$\bigwedge_{z\in V} \quad \min_{x^*\in\Sigma_{x-x_o}} \operatorname{Re}\langle x^*, z-x_o\rangle \leq 0.$$

This completes the proof.

## 2. Nonlinear optimization with side conditions

Let $p: R^n \to R$ be a convex continuously differentiable function and $q: M \times R^n \to R$ a continuous mapping, where M is a compact Hausdorff space. We assume the mapping q to be convex and partially differentiable for each $\mu \in M$ and denote the derivative by grad $q(\cdot,x_o) : M \times R^n \to R$, which is assumed to be continuous. Then we consider the following convex optimization problem:

Find an element $x_0$ which minimizes the function p under the restrictions $q(\mu,x) \leq 0$ for all $\mu \in M$. As a corollary of our variational lemma we prove:

The element $x_0$ is a constrained minimum of the target function p only if the inequality

$$\bigwedge_{x \in R^n} \min_{\mu \in M_0} (\text{grad } p(x_0)(x_0-x), \text{grad } q(\mu,x_0)(x_0-x)) \leq 0$$

is valid, where we have defined

$$M_0 := \{\mu \in M \mid q(\mu,x_0) = 0\}.$$

In order to apply the variational lemma in proving the statement mentioned above, we define the function

$$\bigwedge_{x \in R^n} \Phi_x(\lambda,\mu) := \begin{cases} p(\lambda x_0+(1-\lambda)x) & \text{for } \mu = 0 \\ q(\mu,\lambda x_0+(1-\lambda)x)+p(x_0) & \text{for } \mu \in M. \end{cases}$$

This function satisfies all the conditions required in our variational lemma and we have

$$\bigwedge_{x \in R^n} \Phi_x'(\lambda,\mu) = \begin{cases} \text{grad } p(\lambda x_0+(1-\lambda)x)(x_0-x) & \text{for } \mu = 0 \\ \text{grad } q(\mu,\lambda x_0+(1-\lambda)x)(x_0-x) & \text{for } \mu \in M. \end{cases}$$

The function $\Phi_x$ was defined in such a way that $\lambda = 1$ is a minimal point of the function

$$\varphi_x(\lambda) := \max_{\mu \in M \cup \{0\}} \Phi_x(\lambda,\mu)$$

for all $x \in R^n$ whenever $x_0$ is a solution of the constrained optimization problem. Using the variational lemma we get the inequality

$$\bigwedge_{x \in R^n} \bigwedge_{\lambda \in [0,1]} \max_{\mu \in M_0} (\text{grad } p(x_0)(x_0-x), \text{grad } q(\mu,x_0)(x_0-x)) \cdot$$

$$\cdot (\lambda-1) \geq 0.$$

Then it follows:

$$\bigwedge_{x \in R^n} \min_{\mu \in M_0} (\text{grad } p(x_0)(x_0-x), \text{grad } q(\mu,x_o)(x_o-x)) \leq 0.$$

This completes the proof.

The inequality has the general form of a Kuhn-Tucker condition in optimization theory. One gets the well-known Kuhn-Tucker theorem by applying a theorem about linear inequalities. Assuming that the Slater condition is satisfied there exist nonnegative real numbers $\rho_\mu (\mu \in V_o)$ such that

$$- \text{grad } p(x_o) = \sum_{\mu \in M_c} \rho_\mu \text{ grad } q(\mu, x_o).$$

In our original paper [2] we proved the variational inequality in a generalized form. Therefore we were able to handle approximation problems in some metric linear spaces, too. In addition we gave many other applications. We remark that the variational inequality permits to treat local as well as global conditions and problems with as well as without side conditions.

## References

[1] Brosowski, B., Nichtlineare Approximation in normierten Vektorräumen. In Abstract Spaces and Approximation, ISNM 10, Birkhäuser, Basel 1970, pp. 140-159.

[2] Brosowski, B. and K.-H. Hoffmann, Eine Variationsungleichung und Anwendungen. To appear.

Lehrstühle für Numerische
und Angewandte Mathematik
Universität Göttingen
3400 Göttingen
West Germany

Mathematisches Institut
Universität München
8000 München 2
West Germany

# EXTREMAL POSITIVE SPLINES WITH APPLICATIONS

Hermann Burchard[1]

1. We consider an $n^{th}$ order Chebyshevian differential operator $L = w_0^{-1}Dw_1^{-1}\ldots Dw_n^{-1}$ and its formal adjoint $L^*$. The $w_i$ are positive functions in $C^n(R)$, [3]. For locally compact Hausdorff T let $M(T) = C_0(T)^*$, the Radon measures on T. In the following a subscript '0' denotes compact support, while superscript '+' indicates the positive cone. If I is an open interval and $u \in M(I)$ it can be shown that $Lu \in M(I)$ in the weak sense if and only if u has an $n^{th}$ distribution derivative in $M(I)$. The set of such u is $M^n(I)$.

Now let I be any interval and $\Omega = \text{int}(I)$. We define the Chebyshevian convexity cone $K(L,I) = \{u \in C(I) : L(u|\Omega) \in M(\Omega)^+\}$. The dual cone is $K^*(L,I) = \{\lambda \in M_0(I) : \int u\lambda \geq 0$ for all $u \in K(L,I)\}$, [3], [4].

<u>Lemma</u>. Let $\lambda \in M_0(I)$. Then $\lambda \in K^*(L,I)$ if and only if there is $\phi \in M_0(R)$, supp $\phi \subset I$, $\phi \geq 0$ and $\lambda = L^*\phi$; for such $\lambda$ and $\phi$, and for $u \in K(L,I)$ one has $\int u\lambda = \int \phi Lu$. $L^*$ is one-to-one on $M_0^n(R)$ and the inverse map is given by $\lambda \to \phi$, $\phi(t) = \int G(s,t)\lambda(ds)$, where $G(s,t)$ is a fundamental solution for L.

We assume $n \geq 2$ throughout, so that $K(L,I) \subset C(I)$.

<u>Definition</u>. Let $C$ be a convex cone in a linear space and let $0 \neq \lambda \in C$. We say $\lambda$ is <u>decomposable</u>, if there is a <u>decomposition</u> $\lambda = \lambda_1 + \lambda_2$ with linearly independent $\lambda_1, \lambda_2 \in C$. Otherwise $\lambda$ is <u>extremal in $C$</u>. If $C \subset M_0(T)$, T locally compact Hausdorff, then $\lambda$ is <u>P-decomposable</u> ($\nu$-decomposable for some $\nu \in M(T)$) in $C$ if there is a decomposition $\lambda = \lambda_1 + \lambda_2$ in $C$

such that $\int|\lambda| = \int|\lambda_1| + \int|\lambda_2|$ ($\lambda_i \ll \nu$); otherwise $\lambda$ is <u>P-extremal</u> (<u>$\nu$-extremal</u>) in $C$.

<u>Theorem 1</u>. (i) Let T be a compact Hausdorff space, $C$ a cone in $M(T)$, and $S^*$ the unit ball of the B-space $M(T) = C(T)^*$. Let $\lambda \in C$, $\lambda \neq 0$. Then $\lambda/\|\lambda\|$ is an extreme point of $C \cap S^*$ if and only if $\lambda$ is P-extremal in $C$.

(ii) Let $0 \neq \lambda \in C \subset M(T)$. Then $\lambda = \lambda_1 + \lambda_2$ is a P-decomposition in $C$ if and only if there are two nonconstant functions $h_i \in L^1(|\lambda|)$, $0 \leq h_i$, $h_1 + h_2 = 1$ a.e. $|\lambda|$ and $\lambda_i = h_i\lambda \in C$, i=1,2. In particular, if $\lambda$ is $|\lambda|$-extremal in $C$ then $\lambda$ is P-extremal in $C$.

L. deBranges in [2] proved the "only if" part of (i), in the form provided by part (ii) for the case of a $w^*$-closed linear subspace $C$. We term <u>Chebyshevian</u> <u>L-spline</u> with knots $s_1,\ldots,s_p$ any element $u \in M^n(R)$ with supp $Lu \subset \{s_1,\ldots,s_p\}$, cf. [3]. The space of Chebyshevian $L^*$-splines of compact support with knots $s_1,\ldots,s_p$ is denoted by $S_0(L^*,s_1,\ldots,s_p)$.

<u>Theorem 2</u>. (i) Let $s_1 < s_2 < \ldots < s_p$ and $t_1 \leq t_2 \leq \ldots \leq t_{p-n-1}$ be any reals selected subject to $p \geq n + 1$ and

(1) $s_{i+1} < t_i < s_{i+n}$, i=1,...,p-n-1;

(2) each distinct $t_i$ occurs with even multiplicity not larger than n - 1.

Then there is an $L^*$-spline $\phi$ satisfying

(3) $\phi \in S_0(L^*,s_1,\ldots,s_p)^+$, $\phi \neq 0$.

(4) On $(s_1,s_p)$, $\phi$ has p-n-1 zeros occurring at the $t_i$, counting multiplicities up to the $n^{th}$ order as usual.

(5) $$\phi(t) = G^*\begin{pmatrix} s_1,\ldots,s_n,s_{n+1},\ldots,s_{p-1},s_p \\ r_1,\ldots,r_n,t_1,\ldots,t_{p-n-1},t \end{pmatrix}, \quad t \neq t_i.$$

Here $r_1 < \ldots < r_n < s_1$. For the $G^*$-notation, cf. [3].

(ii) Let $\lambda \in M_0(I)$, $\lambda \neq 0$ and $S^-(\lambda) < \infty$. Then $\lambda$ is P-extremal in $K^*(L,I)$ if and only if $\lambda = cL^*\phi$ for some $c > 0$ and $\phi$ as in (i). In particular $\lambda$ has finite support, $\int u\lambda = \Sigma_{i=1}^p \cdot \alpha_i u(s_i)$, where $\alpha_i/c$ are the coefficients obtained when expanding (5) along the last column. The $\alpha_i$ alternate in sign.

2. Next we characterize those data sets $(x_i,y_i)_{i=1}^k$ which permit a generalized convex interpolation. Let $x_1 < x_2 < \ldots < x_k$ in the following.

<u>Theorem 3</u>. $\psi \in S_0(L^*,x_1,\ldots,x_k)$ is extremal in $S_0(L^*,x_1,\ldots,x_k)^+$ if and only if $\psi = c\phi$, $c > 0$, with $\phi$ as in (5), $s_1,\ldots,s_p$ being any consecutive points among the $x_i$. $S_0(L^*,x_1,\ldots,x_k)^+$ is spanned by its extremal elements.

In the author's thesis [1] it is shown that, if a generalized convex interpolation exists, all such interpolations lie between two extremal ones, which are Chebyshevian L-splines with at most k/2 knots, on $(x_1,x_k)$.

<u>Theorem 4</u>. A set of data points $(x_i,u_0(x_i))_{i=1}^k$ admits of $u \in K(L,I)$, $I \supset \{x_i\}_{i=1}^k$, such that $u(x_i) = u_0(x_i)$, $i=1,\ldots,k$, if and only if $\int u_0 L^* \phi \geq 0$ for all $\phi$ in (5) with $s_1,\ldots,s_p$ any consecutive ones among the $x_i$.

This result considerably improves an earlier one by T. Popoviciu. Next we characterize best uniform approximation by generalized convex functions. In [1] it is shown that such approximations exist on a compact interval.

<u>Theorem 5</u>. Let $u_0 \notin K(L,I)$, I a compact interval. Then there is a function $\phi$ as in Theorem 2, (5) such that if $u \in K(L,I)$, then u is a best approximation for $u_0$ from $K(L,I)$ in the uniform norm if and only if

(6) $u(s_i) = u_0(s_i) + (-1)^{p-i}\|u-u_0\|_\infty$, $i=1,\ldots,p$;

(7) on $[s_1,s_p]$, u is the unique Chebyshevian L-spline with simple knots at the distinct $t_i$ (zeros of $\phi$) satisfying (6).

Remark. The alternation and uniqueness conditions in the last theorem, established for approximation by functions satisfying $Lu \geq 0$, are seen to strongly resemble the Chebyshevian alternation and uniqueness theorem, valid for approximation by functions satisfying $Lu = 0$.

---

[1]Work supported in part by the Mathematics Research Center, the University of Wisconsin, under contract No. DA-31-124-ARO-D-462 and at Indiana University by AF-AFOSR grant No. 71-2205A.

## References

[1] Burchard, H. G., Interpolation and approximation by generalized convex functions. Thesis, Purdue University, 1968.

[2] deBranges, L., The Stone-Weierstrass theorem. Proc. Amer. Math Soc. 10 (1959), 822-824.

[3] Karlin, S and W. J. Studden, Tchebycheff Systems With Applications in Analysis and Statistics. New York, Interscience Publishers, 1966 .

[4] Ziegler, Z., On the characterization of measures of the cone dual to a generalized convexity cone. Pac. J. Math. 24 (1968), 603-626.

Department of Mathematics and Statistics
Oklahoma State University
Stillwater, Oklahoma 74074

# ON SIMULTANEOUS $L_1$ APPROXIMATION

M. P. Carroll

Let $I$ be a compact interval of the real line. Let $n$ be a fixed nonnegative integer and $P_n$ the set of polynomials of degree $n$ or less. Let $G$ be a set of real-valued functions defined on $I$ such that for some real number, $M$, $\sup_{x\in I} |g(x)| \leq M$ for all $g \in G$. Since E. Remes [5] in 1934 much work has been done on problems similar to minimizing the expression $\sup_{g\in G} \cdot \sup_{x\in I} [|g(x) - p(x)|]$, $p \in P_n$. Letting $F$ be a set of real-valued Lebesgue measurable functions such that $\int_I |f|d\mu \leq N$, $f \in F$, where $N$ is a real number and $\mu$ is Lebesgue measure, a parallel problem in the $L_1$ norm to the one above is that of minimizing $\sup_{f\in F} [\int_I |f - p|d\mu]$, $p \in P_n$.

For Lebesgue integrable functions $f_1,\ldots,f_m$ defined on $I$, another parallel problem to the above one is the problem of minimizing $\|(\|f_1 - p\|_{L_1}, \|f_2 - p\|_{L_1},\ldots,\|f_m - p\|_{L_1})\|_{\ell_1}$, $p \in P_n$, i.e., minimizing $\Sigma_{i=1}^m \int_I |f_i - p|d\mu$, $p \in P_n$. It is the purpose of this note to present known characterization and uniqueness results for these two $L_1$ problems.

First we will consider the problem of minimizing $\Sigma_{i=1}^m \int_I |f_i-p|d\mu$, $p \in P_n$. The characterization theorem is proven just like the classical $L_1$ theorem [6] and is indeed very similar to the classical theorem.

<u>Theorem 1</u>. Let $f_i \in L_1(I)$, $1\leq i\leq m$. An element $q \in P_n$ minimizes $\Sigma_{i=1}^m \int_I |f_i - p|d\mu$, $p \in P_n$, if and only if

$$\sum_{i=1}^m \int_{Z(f_i-q)} |p|d\mu \geq \left|\sum_{i=1}^m \int_I p\ \mathrm{sgn}(f_i - q)d\mu\right|,\ p \in P_n,$$

where $Z(f_i - q) = \{x | x \in I \text{ and } (f_i - q)(x) = 0\}$.

From this theorem the two uniqueness theorems for this problem follow, one for an even number of functions and one for an odd number.

Theorem 2 [1]. Let $f_1 \leq f_2 \leq \cdots \leq f_{2m}$ be continuous real-valued functions on I. Let $q \in P_n$ minimize $\Sigma_{i=1}^{2m} \int_I |f_i - q| d\mu$. If there exists $\bar{x} \in I$ such that $[f_m(\bar{x}) - q(\bar{x})][f_{m+1}(\bar{x}) - q(\bar{x})] > 0$, then q is unique.

If one has functions which are not ordered, then let $h_i(x) = \max_{\pi \in S_i} \min [f_{\pi(1)}(x), \ldots, f_{\pi(i)}(x)]$, $1 \leq i \leq 2m$, where $S_i$ denotes the set of all one to one mappings of the set of integers $[1, \ldots, i]$ into the set of integers $[1, \ldots, 2m]$. The $h_i$'s are ordered and if $q \in P_n$ minimizes $\Sigma_{i=1}^{2m} \int_I |f_i - p| d\mu$, $p \in P_n$, it minimizes $\Sigma_{i=1}^{2m} \int_I |h_i - p| d\mu$, $p \in P_n$, [1]. Thus, we have a uniqueness result for the $f_i$'s through the $h_i$'s.

Theorem 3 [1]. Let $f_1, \ldots, f_{2m+1}$ be continuous real valued functions on I. Let q be a polynomial in $P_n$ which minimizes $\Sigma_{i=1}^{2m+1} \int_I |f_i - q| d\mu$. Then q is unique.

For the problem of minimizing $\sup_{f \in F} [\int_I |f - p| d\mu]$, the characterization theorem is due to Laurent and Tuan [4].

Theorem 4. For Q a n-dimensional subspace of $L_1(I)$ and F a compact subset of $L_1(I)$, $q \in Q$, will minimize $\sup_{f \in F} [\int_I |f-p| d\mu]$, $p \in Q$, if and only if there exist m measurable functions, $1 \leq m \leq n+1$, $\phi_1, \phi_2, \ldots, \phi_m$, which take the values $\pm 1$ almost everywhere on I, m functions, $f_1, \ldots, f_m$ in F (not necessarily distinct), and scalars $\lambda_1, \lambda_2, \ldots, \lambda_m > 0$ such that

$$(1) \quad \int_I \phi_i (q - f_i) d\mu = \int_I |f_i - q| d\mu = \sup_{f \in F} [\int_I |f-q| d\mu], \; i=1,2,\ldots,m.$$

$$(2) \quad \sum_{i=1}^{m} \lambda_i \int_I \phi_i \, p \, d\mu = 0, \; p \in Q.$$

Using this result one can arrive at the following uniqueness theorem as seen in [2].

Theorem 5. Let F be a compact subset of $L_1(I)$ consisting of continuous real-valued functions on I. Let q be an element of $P_n$ which minimizes $\sup_{f \in F} [\int_I |f - q| d\mu]$. If there exists $\bar{x} \in I$ such that $\sup_{f \in F} f(\bar{x}) < q(\bar{x})$ or $\inf_{f \in F} f(\bar{x}) > q(\bar{x})$ then q is unique.

Further, all the previous results hold for linear combinations of a Haar system not just the polynomials of degree n or less [1], [2].

## References

[1] Carroll, M. P. and H. W. McLaughlin, $L_1$ Approximation of Vector Valued Functions. To appear in J. Approx. Theory.

[2] Carroll, M. P., Simultaneous $L_1$ Approximation of a Compact Set of Real-Valued Functions. Numer. Math. 19 (1972), 110-115.

[3] Diaz, J. B. and H. W. McLaughlin, Simultaneous Chebychev Approximation of a Set of Bounded Complex-Valued Functions. J. Approx. Theory 2 (1969), 419-433.

[4] Laurent, P. J. and Pham-Dinh-Tuan, Global Approximation of a Compact Set by Elements of a Convex Set in a Normed Space. M. R. C. Technical Summary Report 1052, December 1970, Madison, The University of Wisconsin.

[5] Remes, E., Sur la détermination des polynomes d'approximation de degré donné. Comm. Soc. Math. Kharkoff, (4) 10 (1934), 41-63.

[6] Rice, J. R., The Approximation of Functions, Vol. 1. Addison-Wesley, Reading, Mass., 1964.

Department of Mathematics
Virginia Polytechnic Institute
and State University
Blacksburg, Virginia 24061

# A KOROVKIN THEOREM FOR FINITELY DEFINED OPERATORS

A. S. Cavaretta, Jr.

The classical Korovkin theorem gives conditions for determining the convergence of a sequence of positive operators $A_m$ to I, the identity. Here we consider replacing the identity by certain positive operators of prescribed type. The possibility of such a replacement was first raised by G. G. Lorentz, and he and H Berens have obtained complete results with the identity operator replaced by a lattice preserving operator (see [1], these Proceedings). We replace I by positive operators $A: C(Q) \to C(Q)$ of the form

$$(1)\quad (Ag)(q) = \sum_{i=1}^{n} \psi_i(q)g(\varphi_i(q)),\ g \in C(Q),\ q \in Q$$

where the $\psi_i$ are non-negative continuous functions and the $\varphi_i$ are continuous mappings of a compact metric space Q into itself. Following Šaškin, who considers such operators in a somewhat related connection [6], we call A a finitely defined operator of order n.

Certain finite dimensional subspaces $X \subset C(Q)$ play a crucial role: we want X to contain non-negative functions which have prescribed zeros. Precisely, given any integer k, $1 \leq k \leq n$, and any k distinct points of Q, $q_1, \ldots, q_k$, we assume that there exist $g \in X$ such that $g(q) \geq 0$ and $g(q) = 0$ exactly when $q = q_i$ for $i=1,\ldots,k$. If X satisfies this assumption, we will say that X is a Korovkin space for finitely defined operators of order n, or briefly that X is an $n^{th}$ order Korovkin space.

This assumption on X can be rephrased as a geometric assumption on $X^*$. Let us assume that X has dimension m, so that

$\dim X^* = m$ and $X^{**} = X$. $X^*$ contains all evaluation functionals $\varepsilon_q(g) = g(q)$, restricted to $g \in X$. We consider the map $\Phi: Q \to X^*$ which assigns to each point $q \in Q$ the (restricted) evaluation functional $\varepsilon_q$. The map $\Phi$ is clearly continuous, and so the set $M = \Phi(Q)$ is a compact subset of $X^*$. Moreover, from the above condition on X, it is easily seen that $\Phi$ is one to one and $0 \notin M$. Thus M is actually homeomorphic to Q.

Let T equal the closed convex cone spanned by M and 0 in $X^*$. From the theorem of Carathéodory, we easily see that a point $x^* \in X^*$ belongs to T if and only if

$$x^* = \sum_{j=1}^{m+1} \lambda_j x_j^*,\ x_j^* \in M,\ \lambda_j \geq 0,\ \sum_{j=1}^{m+1} \lambda_j = c \geq 0.$$

Thus

$$x^*(g) = \sum_{j=1}^{m+1} \lambda_j\ \varepsilon_{q_j}(g) = \sum_{j=1}^{m+1} \lambda_j g(q_j).$$

Now suppose k is any integer $1 \leq k \leq n$ and $\varepsilon_{q_1}, \ldots, \varepsilon_{q_k}$ are distinct points of M. On the assumption that X is an $n^{th}$ order Korovkin space, we find $g \in X$ such that $g(q) \geq 0$ and equality holds whenever $q = q_i$, $i=1,\ldots,k$. But then g, considered as an element of $X^{**}$, defines a hyperphane in $X^*$ which supports T and intersects T only in the convex span of the k rays $c\varepsilon_{q_i}$, $i=1, \ldots, k$. And, of course, the converse also holds since any hyperplane in $X^*$ is defined by an element $g_0 \in X^{**} = X$. Thus we see that X is an $n^{th}$ order Korovkin space if and only if

(i) the function $\Phi$ is 1-1 and $0 \notin M$

(ii) given any k distinct points $x_1^*, \ldots, x_k^*$ of M, $1 \leq k \leq n$, there exists a hyperplane supporting T and intersecting T precisely in the convex span of the k rays $cx_i^*$, $i=1,\ldots,k$, $c > 0$.

Examples of $n^{th}$ order Korovkin spaces are readily available. Suppose Q is an interval $[a,b]$ or the circle, and let

$g_0, \ldots, g_{2n}$ be a complete Chebyshev system defined on Q with linear span X. Thus dim X = 2n + 1. Then by a theorem of M. G. Krein (see [2]), given any $k \leq n$ distinct points $q_1, \ldots, q_k$ of Q, there exists $g \in X$ with $g(q) \geq 0$ and $g(q) = 0$ precisely when $q = q_i$ for $i=1,\ldots,k$. Other examples can be constructed along the lines suggested by G. G. Lorentz in [4]. In particular, for a Korovkin set of order 2 on the sphere $S_{p-1}$ in $R^p$ we can use the functions

$$1, \{x^{(i)}\}_{i=1}^{p}, \{x^{(i)}x^{(j)}\}_{i,j=1}^{p}$$

where $x^{(i)}$ denotes the $i^{th}$ coordinate function on $R^p$.

We now turn to our theorems concerning finitely defined operators of order n, as given by (1).

Proposition. Suppose A is given by (1) and X is an $n^{th}$ order Korovkin space. If B: C(Q) → C(Q) is any positive operator and Bg = Ag for g ∈ X, then B ≡ A.

Proof: Fix q ∈ Q, and put $q_i = \varphi_i(q)$ for $i=1,\ldots,n$. By our assumptions on X, there exists a non-negative g ∈ X vanishing precisely at the points $q_i$. Denoting by $\varepsilon_q$ the evaluation functional at q, we conclude that $\varepsilon_q Ag = 0$. So $\varepsilon_q A$ is a positive functional represented by a measure carried on the $q_i$; i.e., $\varepsilon_q A = \Sigma_{i=1}^{n} \lambda_i \varepsilon_{q_i}$ for some choice of $\lambda_i \geq 0$. To determine $\lambda_i$, pick a non-negative function $g_i \in X$ vanishing precisely on the $q_j$ with $j \neq i$. Then $\varepsilon_q Ag_i = \lambda_i g_i(q_i)$. But from (1), $\varepsilon_q Ag_i = \psi_i(q) g_i(q_i)$. So $\lambda_i = \psi_i(q)$.

Since B is positive and agrees with A on X, exactly the same reasoning applies to $\varepsilon_q B$. We conclude that $\varepsilon_q B = \varepsilon_q A$ for each q ∈ Q, and therefore B ≡ A.

To state our next theorem concerning convergence of positive linear functionals, the following concept is convenient:

Definition. Let P be the cone of non-negative functions in C(Q). For each g ∈ P put

$$H(g) = \{f \in C^*(Q): \; f \text{ is positive and } f(g) = 0\}$$

$g$ is called quasi-smooth of order $k$ if the dimension of the linear span of $H$ is $k$.

This definition generalizes the well known concept of a smooth point of a convex body (see for example [7], p. 23, or [3]). Clearly $g$ is quasi-smooth of order $k$ whenever $g$ has precisely $k$ zeros. If $f \in H(g)$ we say that $f$ passes through $g$.

<u>Theorem 1.</u> Suppose $X$ is an $n^{th}$ order Korovkin space and let $g_o \in P$ be a given quasi-smooth point of order $k \leq n$. Let $\{f_m\}_{m=1}^{\infty}$ be a sequence of positive functionals such that

$$\lim_{m\to\infty} f_m(g_o) = 0$$

and $\lim_{m\to\infty} f_m(g)$ exists for $g \in X$. Then $f(g) = \lim_{m\to\infty} f_m(g)$ exists for all $g \in C(Q)$.

<u>Proof:</u> From the positivity of the $f_m$ and the convergence on $X$, it is easily seen that there exists $N > 0$ such that $\|f_m\| \leq N$ for all $m$. Now take any subsequence of $\{f_m\}_{m=1}^{\infty}$, which we continue to denote by $\{f_m\}_{m=1}^{\infty}$. By the weak $^*$ compactness of the unit ball in $C^*(Q)$, we can extract a weakly convergent subsequence $\{f_{m_j}\}_{j=1}^{\infty}$. Thus we find $f_o \in C^*(Q)$ and $f_o(g) = \lim_{j\to\infty} f_{m_j}(g)$ for all $g \in C(Q)$.

By hypothesis $f_o(g_o) = 0$. Hence it follows that $f_o = \Sigma_{i=1}^{k} \lambda_i \epsilon_{q_i}$, where the $q_i$ are the zeros of $g_o$, and the $\lambda_i$ must be determined. From our hypothesis on $X$, we find $g_\ell \in X$ such that $g_\ell(q_i) = 0$ for all $i \neq \ell$ and $g_\ell(q_\ell) > 0$. Put

$$\alpha_\ell = \lim_{m\to\infty} f_m(g_\ell) \qquad \ell=1,\ldots,k.$$

Then

$$\alpha_\ell = \lim_{m\to\infty} f_m(g_\ell) = \lim_{j\to\infty} f_{m_j}(g_\ell) = f_o(g_\ell) = \lambda_\ell g_\ell(q_\ell), \; \ell=1,\ldots,k.$$

So $\lambda_\ell = \alpha_\ell / g_\ell(q_\ell)$, $\ell=1,\ldots,k$. Notice that the $\lambda_\ell$ depend only on the $\alpha_\ell$ and the $q_\ell$; the $\lambda_\ell$ are independent of the particular subsequence of $\{f_m\}_{m=1}^{\infty}$ which we choose freely at the onset. Thus the original sequence $\{f_m\}_{n=1}^{\infty}$ has $f_o$ as a unique accumulation point. So putting $f = f_o$, we have the desired result.

We now can easily prove

Theorem 2. Let A be given by (1), let X be a Korovkin set of order n, and suppose $\{A_m\}_{m=1}^{\infty}$ is a given sequence of positive operators. If

$$\lim_{m\to\infty} A_m g = Ag \quad \text{uniformly, } g \in X$$

then $\lim_{m\to\infty} A_m g = Ag$ uniformly, $g \in C(Q)$.

Proof: We use the following lemma: For a given sequence $\{g_m\}_{m=1}^{\infty}$ and g in $C(Q)$, $g_m$ converges uniformly to g if and only if given any point $q \in Q$ and a sequence $\{q_m\}_{m=1}^{\infty}$ converging to q, we have $\lim_{m\to\infty} g_m(q_m) = g(q)$.

Now let $q \in Q$ be fixed, and consider $\varepsilon_q A$. As in the proposition, we find $g_o \in X$ such that $\varepsilon_q A g_o = 0$ and $g_o$ is quasi-smooth of order $\leq n$. If $q_m \to q$, then from our hypotheses and the lemma we see that

$$\lim_{n\to\infty} \varepsilon_{q_m} A_m g_o = \varepsilon_q A g_o = 0$$

and $\lim_{m\to\infty} \varepsilon_{q_m} A_m g = \varepsilon_q A g$ for $g \in X$. So our Theorem 1 implies that

$$\lim_{m\to\infty} \varepsilon_{q_m} A_m g = \varepsilon_q A g$$

for all $g \in C(Q)$. Using the lemma again we conclude that $\lim_{m\to\infty} A_m g = Ag$ uniformly for all $g \in C(Q)$.

Remarks.

(1) In [6] Šaškin poses the problem of finding the form of an operator A on $C(Q)$ which is completely determined by its action on m linearly independent functions $g_1, \ldots, g_m$. Thus he assumes that if another operator B satisfies

$Bg_i = Ag_i$, $i=1,\ldots,m$, then $B \equiv A$. He concludes that A is finitely defined of order m. His proof appears to be inconclusive, and it would be of interest to settle the question since only for such A's can we expect Korovkin type theorems.

(2) Results very similar to ours have also been given independently in [5] by C. A. Michelli in the case where X is spanned by a Chebyshev system.

(3) In the case that Q is the circle or interval, so that M is a path in $X^* = R^m$, I conjecture that $m \geq 2n + 1$. This would imply that for operators of order n one does best by checking convergence on a Chebyshev system of order $2n + 1$.

## References

[1] Berens, H. and G. G. Lorentz, Korovkin theorems for positive operators in Banach function spaces. These Proceedings.

[2] Karlin, S. and W. J. Studden, Tchebycheff systems with applications in analysis and statistics. Interscience Publishers, New York, 1966.

[3] Krasnosel'skiĭ, M. A., V. S. Klimov, and E. C. Lifšič, Convergence of positive functionals and operators. Dokl. Akad. Nauk SSSR 162 (1965), 258-261.

[4] Lorentz, G. G., Korovkin sets. Lecture Notes, Regional Conference at the University of California, Riverside, June 15-19, 1972.

[5] Micchelli, C. A., Chebyshev systems and convergence of positive linear operators. IBM Report RC 3994.

[6] Šaškin, Yu. A., Finitely defined linear operators in spaces of continuous functions. Uspeki Mat. Nauk 20 (1965), no. 6 (126), 175-180.

[7] Singer, I., Best approximation in normed linear spaces by elements of linear subspaces. Springer-Verlag, New York, 1970.

# A KOROVKIN THEOREM

Department of Mathematics
California Institute of Technology
Pasadena, California 91109

## APPROXIMATION OF FUNCTIONS FROM THEIR MEANS

Charles K. Chui and Chin-Hung Ching

Let $H^p$ be the Hardy spaces with norms $\| \|_p$ and let V be the Banach space of functions f holomorphic in the open unit disc U such that $f' \in H^1$ with the norm $\|f\|_V = \|f'\|_1 + |f(0)|$. Clearly, if f is in V, then f is in A, the space of functions holomorphic in U and continuous on $\overline{U}$. For $f \in A$, we consider its means on $T = \partial U$

$$s_n(f) = \frac{1}{n} \sum_{k=1}^{n} f(e^{i2\pi k/n}) ,$$

$s_\infty(f) = \lim s_n(f)$ and $r_n(f) = s_n(f) - s_\infty(f)$. In [1], some sufficient conditions on the Taylor coefficients of $f \in A$ are given to guarantee that f is uniquely determined by its means $s_n(f)$. By applying a result of H. Davenport [5,6], we now give a new method to obtain a different sufficient condition as in the following

<u>Theorem 1</u>. Every function f in V is uniquely determined by its means $s_n(f)$, n=1,2,....

To prove this result, we let {t} be the saw-tooth function on the real line defined by $\{t\} = -\sum_{k=1}^{\infty} \sin 2\pi kt/\pi k$. Then it follows that

$$(1) \quad r_n(f) = \frac{1}{n} \int_0^1 \{nt\} df(e^{i2\pi t})$$

for all n. Our idea of proof is to follow the observation that the saw-tooth functions {nt} behave like sin $2\pi nt$ (cf. [4]) and the sequence $r_n(f)$ behaves like the cosine coefficients $c_n(f) = \int_0^1 f(e^{i2\pi t})\cos 2\pi nt dt$ (cf. [2]). In particular, if

$s_n(f) = 0$ for all n, we obtain the inequality

$$|nc_n(f)| \le \|f'\|_1 \; \|\sin 2\pi nt + \sum_{k=1}^{N} \mu(k) \frac{\pi\{nkt\}}{k}\|_\infty$$

for all positive integers N, where $\mu(k)$ is the Möbius function (cf. [8]). The right side converges to zero as N tends to infinity by [6] and hence, $c_n(f) = 0$ for all $n=1,2,\ldots$. But $c_0(f) = s_\infty(f) = 0$. Thus, by the analyticity of f, we conclude that $\int_T \bar{z}^{n+1} f(z)dz = \int_T (z^{n-1} + \bar{z}^{n+1})f(z)dz = 4\pi i c_n(f) = 0$ for $n=1,2,\ldots$, so that f is the zero function.

We remark that the function $g(z) = z - 1/z$ satisfies $s_n(g) = 0$ for all n. We now approximate f from its means as in the following

Theorem 2. Let $f \in V$. Then the series

$$(2) \qquad \sum_{n=1}^{\infty} r_n(f)p_n(z) + s_\infty(f) ,$$

where $p_n(z) = \Sigma_{k|n} \mu(\frac{n}{k})z^k$, converges uniformly to f on $\overline{U}$. Also, for any finite p and any $h > 0$, it converges in $H^p$ with the rate $\mathcal{O}((\log n)^{-h})$.

We remark that in general the series $\Sigma\, r_n p_n(z)$ does not converge for any $z \neq 0$ in U even though $r_n = \mathcal{O}(1/n)$ (cf. [3]). But if $r_n(f) = \mathcal{O}(1/n^p)$ for some $p > 1$, then the series (2) converges uniformly to f in $\overline{U}$ with the rate $\mathcal{O}(1/n^\varepsilon)$ for some $\varepsilon = \varepsilon(p) > 0$, (cf. [3]).

To prove Theorem 2, we let

$$R_N(z) = f(z) - \sum_{n=1}^{N} r_n(f)p_n(z) - s_\infty(f).$$

Then,

$$R_N(z) = \sum_{k=1}^{\infty} a_k z^k - \sum_{n=1}^{N} r_n(f) \sum_{k|n} \mu(\tfrac{n}{k})z^k$$

$$= \sum_{k=1}^{N} \{a_k - \sum_{j=1}^{[N/k]} r_{kj}(f)\mu(j)\}z^k + \sum_{k=N+1}^{\infty} a_k z^k .$$

Here, since $f' \in H^1$, we can apply the Hardy's inequality (cf. [9]) to conclude that $\Sigma_{k=N+1}^{\infty} a_k z^k$ converges to zero uniformly on $\overline{U}$. Also, from (1) and by applying a result in [6], we have

$$(3)\quad \left|a_k - \sum_{j=1}^{[N/k]} r_{kj}(f)\mu(j)\right|$$

$$= \frac{1}{k}\left|\int_0^1 \left\{\frac{\sin 2\pi kt}{\pi} + \sum_{j=1}^{[N/k]} \frac{\mu(j)}{j}\{kjt\}\right\} df(e^{i2\pi t})\right|$$

$$\leq \frac{C_h}{k}(\log N)^{-h}\|f'\|_1$$

for any $h > 1$. Hence, $R_N(z) \to 0$ uniformly on $\overline{U}$. Moreover, for any finite $p \geq 2$, we have, by applying the Hardy-Littlewood inequality and (3),

$$\|R_N\|_p^p \leq A_p\left\{\sum_{k=1}^{N} k^{p-2}\left|a_k - \sum_{j=1}^{[N/k]} r_{kj}(f)\mu(j)\right|^p + \sum_{k=N+1}^{\infty} k^{p-2}|a_k|^p\right\}$$

$$\leq B\left\{\sum_{k=1}^{N} k^{p-2}\frac{(\log N)^{-hp}}{k^p} + \sum_{k=N+1}^{\infty} k^{p-2}\frac{1}{k^p}\right\} = O((\log N)^{-hp}).$$

We now consider the normed linear spaces $R_\alpha$ of functions f in A with the norms

$$\|f\|_{R_\alpha} = \sup_n |n^\alpha r_n(f)| + |s_\infty(f)| < \infty.$$

It is clear that in the spaces $R_\alpha$, the partial sums of the series (2) give the best polynomial approximation: That is, if $f \in R_\alpha$ and $P_N$ is any polynomial of degree no greater than N, then

$$\|f - P_N\|_{R_\alpha} \geq \left\|f - \sum_{n=1}^{N} r_n(f)p_n - s_\infty(f)\right\|_{R_\alpha}.$$

We now compare the norms $\|\ \|_{R_\alpha}$ with the norm $\|\ \|_V$ of V. By (1), it is clear that for all $f \in V$,

$$(4)\quad \|f\|_{R_\alpha} \leq C_\alpha\|f\|_V$$

for $\alpha \leq 1$ (with $C_\alpha = 2\pi$), and it is not difficult to construct examples to show that (4) does not hold for $\alpha > 1$, (cf. [2,9]). On the other hand, for each $\alpha > 3/2$, we have

(5) $\|f\|_V \leq C_\alpha \|f\|_{R_\alpha}$

for all $f \in R_\alpha$. Indeed, from a result in [3], we see that since $\alpha > 3/2$, the series (2) converges uniformly to f on $\overline{U}$, and from this we can conclude that

$$\|f'\|_1^2 \leq \|f'\|_2^2 \leq \sum_{k=1}^{\infty} k^2 \Big(\sum_{j=1}^{\infty} |r_{kj}|\Big)^2$$

$$\leq \sum_{k=1}^{\infty} k^2 \Big\{\sum_{j=1}^{\infty} (\|f\|_{R_\alpha} - |s_\infty(f)|)/(kj)^\alpha\Big\}^2$$

$$= C_\alpha^2 (\|f\|_{R_\alpha} - |s_\infty(f)|)^2$$

and hence, we obtain (5) with $C_\alpha = \{\Sigma k^{-2(\alpha-1)}\}^{1/2}\{\Sigma k^{-\alpha}\}$. However, (5) does not hold for each $\alpha < 3/2$. To see this we can take $1 < \alpha < 3/2$ and $2(\alpha-1) < \beta < 1$ and let $f(z) = \Sigma_{n=1}^{\infty} \cdot e^{in^\beta} z^n/n^\alpha$. Then $\|f\|_V = \infty$ (cf. [7]) but $\|f\|_{R_\alpha} \leq \Sigma n^{-\alpha}$.

We close by asking the following questions:

(a) Does Theorem 1 hold for functions in A?

(b) Can the rate of convergence in Theorem 2 be improved to $\mathcal{O}(1/n)$?

(c) Does (5) hold for $\alpha = 3/2$?

(d) Let $f \in A$ and $a_n = f^{(n)}(0)/n!$. Is it true that $a_n = \mathcal{O}(1/n) \Longleftrightarrow r_n(f) = \mathcal{O}(1/n)$ and $a_n = o(1/n) \Longleftrightarrow r_n(f) = o(1/n)$?

(e) Let $0 < t_n < 1$ and $t_n \to 1$, and for a holomorphic function f in U, let $S_n(f) = \frac{1}{n} \Sigma_{k=1}^{n} f(t_n e^{i2\pi k/n})$. When is f uniquely determined by the means $S_n(f)$?

(f) If f is uniquely determined by the $S_n(f)$, how do we reconstruct f from the $S_n(f)$?

## References


[1] Ching, C. H. and C. K. Chui, Uniqueness theorems determined by function values at the roots of unity. J. Approx. Theory. To appear.

[2] Ching, C. H. and C. K. Chui, Asymptotic similarities of Fourier and Riemann coefficients. J. Approx. Theory. To appear.

[3] Ching, C. H. and C. K. Chui, Mean boundary value problems and Riemann series. J. Approx. Theory. To appear.

[4] Chui, C. K., Concerning rates of convergence of Riemann sums. J. Approx. Theory 4 (1971), 279-288.

[5] Davenport, H., On some infinite series involving arithmetical functions. Quarterly J. of Math. 8 (1937), 8-13.

[6] Davenport, H., On some infinite series involving arithmetical functions II. Quarterly J. of Math. 8 (1937), 313-320.

[7] Hardy, G. H., A theorem concerning Taylor's series. Quarterly J. of Math. 44 (1913), 147-160.

[8] Hardy, G. H. and E. M. Wright, An introduction to the theory of numbers. Oxford University Press, Oxford 1954.

[9] Zygmund, A., Trigonometric Series, 2nd ed., Cambridge University Press, New York, 1959.



Department of Mathematics
Texas A&M University
College Station, Texas 77843

Department of Mathematics
University of Melbourne
Melbourne, Victoria
Australia

# EXTENDED ERROR BOUNDS FOR SPLINE AND L-SPLINE INTERPOLATION

Stephen Demko and Richard S. Varga

## 1. Introduction

The basic object of this paper is to extend many of the error bounds for spline and L-spline interpolation, as given in Swartz and Varga [1] and Scherer [2]. To describe these extensions, the following standard notation is used. If $-\infty < a < b < +\infty$, and if N is a positive integer, then

(1) $\Delta\colon\ a = x_0 < x_1 < \dots < x_N = b$

is a partition of the interval $[a,b]$ with knots $x_i$, and we set $\overline{\Delta} \equiv \max_{0\le i\le N-1} \{x_{i+1}-x_i\}$, and $\underline{\Delta} \equiv \min_{0\le i\le N-1} \{x_{i+1}-x_i\}$. For $\sigma \ge 1$, $P_\sigma(a,b)$ then denotes all partitions $\Delta$ of $[a,b]$ for which $\overline{\Delta}/\underline{\Delta} \le \sigma$. Next, if $\pi_n$ is the set of all real polynomials of degree at most n and $W_p^k[a,b]$ is the usual Sobolev space with norm $\|\cdot\|_{W_p^k[a,b]}$, then for nonnegative integers n and m with $n\ge m\ge 0$,

(2) $Sp(n,m,\Delta) \equiv \{s\in W_\infty^m[a,b]\colon s(x)\in\pi_n \text{ for } x\in[x_i,x_{i+1}],\ 0\le i\le N-1\}$

is the space of polynomial splines on $[a,b]$ relative to the partition $\Delta$. Similarly, if $z = (z_1,z_2,\dots,z_{N-1})$ is an $(N-1)$-tuple of positive integers $z_i$ with $1\le z_i\le m$, then as in [1], with $\mu \equiv \max_{1\le i\le N-1} \{z_i\}$,

(3) $Sp(L,\Delta,z) = \{s\in W_\infty^{2m-\mu}[a,b]\colon L^*L\, s(x)=0 \text{ on } (a,b) - \{x_i\}_{i=1}^{N-1},$

$\text{and } D^k s(x_i-) = D^k s(x_i-) \text{ for all } 0\le k\le 2m-1-z_i,\ 0<i<N\}$

is the space of L-splines on $[a,b]$, with incidence vector z.

A typical result from [1, Theorem 7.4] is the following.

Theorem 1. Given $f\in C^k[a,b]$ with $0\le k<2m$ and given $\Delta\in P_1(a,b)$, let s be the unique interpolant of f in $Sp(2m-1, 2m-1,\Delta)$ such that

$$(4)\quad \begin{cases} (f-s)(x_i) = 0, & 1\le i\le N-1, \\ D^j(f-s)(a) = D^j(f-s)(b) = 0 & \text{for } 0\le j\le \min(k,m-1), \\ D^j s(a) = D^j s(b) = 0 & \text{if } \min(k,m-1)<j\le m-1. \end{cases}$$

Then,

$$(5)\quad K(\overline{\Delta})^{k-j}\omega_\infty(D^k f,\overline{\Delta}) \ge \begin{cases} \|D^j(f-s)\|_{L_\infty[a,b]}, & 0\le j\le k, \\ \|D^j s\|_{L_\infty[a,b]}, & \text{if } k<j\le 2m-1, \end{cases}$$

where $\omega_\infty$ denotes the usual modulus of continuity.

From the above result, one deduces (cf. [1, Corollary 7.5])

Corollary 2. With the hypotheses of Theorem 1, if $f\in W_p^{k+1}[a,b]$ with $0\le k<2m$ and $2\le p\le\infty$, then for $p\le q\le\infty$,

$$(6)\quad K(\overline{\Delta})^{k+1-j+\frac{1}{q}-\frac{1}{p}}\|D^{k+1}f\|_{L_p[a,b]} \ge \begin{cases} \|D^j(f-s)\|_{L_q[a,b]}, & 0\le j\le k, \\ \|D^j s\|_{L_q[a,b]}, & \text{if } k<j\le 2m-1. \end{cases}$$

The basic results of this paper can be described then as extensions of Theorem 1, and improvements of Corollary 2.

## 2. Basic Comparison Theorem

Given any $f\in L_p[a,b]$ with $1\le p\le\infty$, let

$$(7)\quad \omega_p(f,t) \equiv \sup_{|h|\le t}\left\{\int_a^b |f(x+h) - f(x)|^p dx\right\}^{1/p}, \quad 0<t\le(b-a),$$

denote the $p^{th}$ modulus of continuity of f, where we assume that f has been suitably extended to an $L_p$-function on $[2a-b,2b-a]$. As is well known,

$$(8)\quad \lim_{t\to 0}\omega_p(f,t) = 0 \text{ if } \begin{cases} f\in L_p[a,b], & 1\le p<\infty,\\ f\in C^0[a,b], & p=\infty.\end{cases}$$

Moreover, if $f\in W_\tau^1[a,b]$ with $1\le\tau\le p$, then

$$(9)\quad \omega_p(f,t) \le 4t^{1+\frac{1}{p}-\frac{1}{\tau}}\|Df\|_{L_\tau[a,b]} .$$

Next, for any $0<h\le 2(b-a)$, we define from f the function

$$(10)\quad f_h(x) = \frac{1}{h}\int_{x-h/2}^{x+h/2} f(t)dt, \qquad x\in[a,b],$$

the so-called Stekloff function of f. If $f\in W_p^k[a,b]$, then $f_h\in W_p^{k+1}[a,b]$, and moreover, it can be verified that

$$(11)\quad \begin{cases} D^j f_h(x) = (D^j f)_h(x)\ , \quad x\in[a,b], & 0\le j\le k,\\ \|D^j(f-f_h)\|_{L_p[a,b]} \le \omega_p(D^j f,\ h/2), & 0\le j\le k,\\ \|D^{k+1}f_h\|_{L_p[a,b]} \le \frac{1}{h}\,\omega_p(D^k f,h). \end{cases}$$

With the function $f_h$, the following comparison theorem, the analogue of Swartz and Varga [1, Lemma 3.2], can be proved by means of a Peano Kernel Theorem argument.

<u>Theorem 3</u>. Given $f\in W_p^k[a,b]$ with $0\le k<2m$ and given $\Delta\in P_\sigma(a,b)$, let g be the unique interpolant of f in $Sp(4m+1,2m+1,\Delta)$ such that

$$(12)\quad \begin{cases} D^j(f-g)(x_i) = 0, \quad 0\le j\le k-1 \text{ if } k>0, & 0\le i\le N,\\ D^k(f_h-g)(x_i) = 0, & 0\le i\le N,\\ D^j g(x_i) = 0, \quad k<j\le 2m, & 0\le i\le N. \end{cases}$$

Then, with $h = \overline{\Delta}$,

$$(13)\quad K(\overline{\Delta})^{k-j+\frac{1}{q}-\frac{1}{p}}\omega_p(D^k f,\overline{\Delta}) \ge \begin{cases} \|D^j(f-g)\|_{L_q[a,b]}, & 0\le j\le k-1 \text{ if } k>0, p\le q\le\infty,\\ \|D^k(f-g)\|_{L_p[a,b]}, & j=k,\ p=q,\\ \|D^j g\|_{L_q[a,b]}, & k<j\le 2m,\ p\le q\le\infty. \end{cases}$$

With the above comparison theorem, we give the following result and sketch its proof.

Theorem 4. Given $f \in W_p^k[a,b]$ with $0 \le k < 2m$ and with $2 \le p \le \infty$, and given $\Delta \in \mathcal{P}_1(a,b)$, let s be the unique interpolant of f in $Sp(2m-1, 2m-1, \Delta)$ such that

$$(14)\quad \begin{cases} (f-s)(x_i) = 0, \quad 1 \le i \le N-1, \quad \text{if } k>0, \\ (f_h - s)(x_i) = 0, \quad 1 \le i \le N-1, \quad \text{if } k=0, \\ D^j(f-s)(a) = D^j(f-s)(b) = 0 \text{ for } 0 \le j \le \min(k-1, m-1) \text{ if } k>0, \\ D^k(f_h - s)(a) = D^k(f_h - s)(b) = 0 \quad \text{if } k \le m-1, \\ D^j s(a) = D^j s(b) = 0 \quad \text{if } k<j \le m-1. \end{cases}$$

Then, with $h = \overline{\Delta}$,

$$(15)\quad K(\overline{\Delta})^{k-j+\frac{1}{q}-\frac{1}{p}} \omega_p(D^k f, \overline{\Delta}) \ge \begin{cases} \|D^j(f-s)\|_{L_q[a,b]}, \ 0 \le j \le k-1 \text{ if } k>0, p \le q \le \infty, \\ \|D^k(f-s)\|_{L_p[a,b]}, \ j=k,\ p=q, \\ \|D^j s\|_{L_q[a,b]}, \text{ if } k<j \le 2m-1,\ p \le q \le \infty. \end{cases}$$

Proof: To sketch the proof of this result, write $f-s = (f-g)+(g-s)$, where g is the interpolant of f in $SP(4m+1, 2m+1, \Delta)$, in the sense of (12). Applying (13) of Theorem 3 then suitably bounds the derivatives of $(f-g)$. Next, because of the definition of g in (13), s, as defined in (14), is also the unique interpolant of g in $Sp(2m-1, 2m-1, \Delta)$, in the sense of (4). Applying the first inequality of (6) of Corollary 2 then yields

$$\|D^j(g-s)\|_{L_q[a,b]} \le K(\overline{\Delta})^{k+1-j+\frac{1}{q}-\frac{1}{p}} \|D^{k+1} g\|_{L_p[a,b]} \text{ for } 0 \le j \le k, \quad p \le q \le \infty.$$

But, from the last inequality of (13) for $j = k+1$, $\|D^{k+1} g\|_{L_p[a,b]} \le K(\overline{\Delta})^{-1} \omega_p(D^k f, \overline{\Delta})$, whence $\|D^j(g-s)\|_{L_q[a,b]} \le K(\overline{\Delta})^{k-j+\frac{1}{q}-\frac{1}{p}} \omega_p(D^k f, \overline{\Delta})$, from which the first two inequalities of (15) follow. Similarly, upon writing $s=g-(g-s)$, the above technique, i.e., using the bounds of (13) and (6), yields the third inequality of (15). Q.E.D.

Corollary 5. With the hypotheses of Theorem 4, let $\{\Delta_i\}_{i=1}^{\infty} \in P_1(a,b)$ with $\lim_{i\to\infty} \overline{\Delta}_i = 0$, and let $s_i$ be the unique interpolant of f in $Sp(2m-1,2m-1,\Delta_i)$, in the sense of (14) with $h = \overline{\Delta}_i$. Then,

$$(16) \quad \lim_{i\to\infty} \|D^k(f-s_i)\|_{L_p[a,b]} = 0.$$

It is interesting to remark that, since the error bounds of (15) for the special case $p=q=\infty$ essentially reduce for smooth functions to the bounds of (5), then Theorem 4 can be viewed as an extension of Theorem 1. Similarly, the error bounds of (15) represent a sharpening of the error bounds of (6) (with k replaced by k-1) in two ways. First, the added factor $\omega_p(D^k f,\overline{\Delta})$ in (15) tends to zero as $\overline{\Delta} \to 0$ if $2\le p<\infty$ (cf. (8)). Second, one obtains the additional estimate for $\|D^k(f-s)\|_{L_p[a,b]}$ in (15) which does not appear in (6) (with k replaced by k-1).

## 3. Extensions

The basic comparison function g, as defined in Theorem 3, can similarly be systematically used to obtain improved error bounds for L-spline, Hermite L-spline, and polynomial spline interpolation under general boundary conditions, thereby extending the general error bounds of Swartz and Varga [1], both for uniform and nonuniform partitions of [a,b]. The same can also be done to extend the stability-type error bounds of [1]. Finally, new interpolation error bounds for even-ordered polynomial splines, determined from integral-type interpolation conditions, as considered, for example, in Scherer [2], also can be deduced from the basic comparison Theorem 3.

## References

[1] Swartz, B. K. and R. S. Varga, Error bounds for spline and L-spline interpolation. J. Approx. Theory 6 (1972), 6-49.

[2] Scherer, K., A comparison approach to direct theorems for polynomial spline approximation. To appear in J. Approx. Theory.

Department of Mathematics
Kent State University
Kent, Ohio 44240

# CARDINALITY OF EXTREME POINTS OF THE UNIT SPHERE WITH APPLICATIONS TO NONEXISTENCE THEOREMS IN APPROXIMATION THEORY

Frank Deutsch

In this note we give an elementary proof of a generalization of a theorem of Garkavi [1], and some applications of this theorem regarding the nonexistence of subspaces having certain approximation properties.

If M is a (linear) subspace of the normed linear space X, the set of best approximations from M to a given x in X is defined by

$$P_M(x) = \{y \in M : \|x-y\| = \inf_{m \in M} \|x-m\|\}.$$

M is called <u>proximinal</u> (resp. <u>Chebyshev</u>) if $P_M(x)$ contains at least (resp. exactly) one element for every x in X.

<u>Definition.</u> A subspace M of the normed linear space X is called <u>admissibly compact</u> if M is proximinal and there is a locally convex topology $\tau$ on X such that $P_M(x)$ is $\tau$-compact for every x in X.

The following are a few examples of admissibly compact subspaces:

(1) Any Chebyshev subspace, or more generally, any proximinal subspace M such that $P_M(x)$ is finite-dimensional for every x in X

(2) Any weak$^*$ closed subspace of a dual space

(3) Any closed subspace of a reflexive space.

For any set A, we denote the cardinality of A by card A,

and the extreme points of A by ext A. The unit sphere in a normed linear space X is the set

$$S(X) = \{x \in X: \|x\| = 1\}.$$

Lemma 1. Let M be a subspace of the normed linear space X.

(1) If $x + M \in S(X/M)$, then $x - P_M(x) = S(X) \cap (x+M)$

(2) If $x + M \in \text{ext } S(X/M)$, then $S(X) \cap (x+M)$ is an extremal subset of $S(X)$

(3) If M is admissibly compact and $x + M \in \text{ext } S(X/M)$, then there exists $x_0 \in P_M(x)$ such that $x - x_0 \in \text{ext } S(X)$.

Proof: The proofs of (1) and (2) are simple consequences of the definitions involved. To prove (3), let $x + M \in \text{ext } S(X/M)$. Since M is admissibly compact, the Krein-Milman theorem implies that the $\tau$-compact set $x - P_M(x)$ contains an extreme point $y = x - x_0$, $x_0 \in P_M(x)$. By (1), y is an extreme point of $S(X) \cap (x+M)$ which, by (2), implies that $y \in \text{ext } S(X)$.

Theorem 1. Let M be an admissibly compact subspace of X. Then

$$\text{card}[\text{ext } S(X/M)] \leq \text{card}[\text{ext } S(X)].$$

Proof: By (3) of Lemma 1, for each $x + M \in \text{ext } S(X/M)$ there exists $y \in x+M$ such that $y \in \text{ext } S(X)$. The mapping $x + M \to y$ thus defined is clearly a one-to-one map from $\text{ext } S(X/M)$ into $\text{ext } S(X)$.

Remarks. (1) Garkavi [1] had proved Theorem 1 in the case when M is Chebyshev and X/M is reflexive by a different method.

(2) Theorem 1 is false in general if M is not admissibly compact, even if M is proximinal and has finite codimension. To see this, one need only construct a proximinal hyperplane M in $c_0$

$$(\text{e.g.}, M = \{x = (\xi_1, \xi_2, \ldots) \in c_0 : \xi_1 = 0\})$$

and recall that ext $S(c_0) = \emptyset$ but $c_0/M$ is one-dimensional so ext $S(c_0/M) \neq \emptyset$.

(3) The map $x + M \rightarrow y$ in the proof of Theorem 1 takes linearly independent sets in $X/M$ into linearly independent sets in X. Thus <u>if</u> M <u>is an admissibly compact subspace of</u> X, <u>there are at least as many linearly independent elements in</u> ext $S(X)$ <u>as there are in</u> ext $S(X/M)$. In particular,

<u>Corollary 1</u>. Let M be an admissibly compact subspace of X having finite codimension n. Then $S(X)$ contains at least n linearly independent extreme points.

Corollary 1 is false in general if M does not have finite codimension. For if we let $M = \{x \text{ in } L_1[0,1] : x = 0 \text{ on } [0,1/2]\}$, then M is a Chebyshev subspace of $L_1[0,1]$. But the unit ball in $L_1[0,1]$ has no extreme points.

Theorem 1 and Corollary 1 immediately imply a number of results concerning the nonexistence of subspaces having certain approximation properties. We call a subspace M factor reflexive if $X/M$ is reflexive. In particular, every subspace of finite codimension is factor reflexive. If M is factor reflexive, ext $S(X/M) \neq \emptyset$ since the unit ball in a reflexive space is weakly compact so must have extreme points. If T is a locally compact Hausdorff space, $C_0(T)$ will denote the real-valued continuous functions on T vanishing at infinity and endowed with the supremum norm. If T is compact, we set $C(T) = C_0(T)$.

<u>Corollary 2</u>. (1) If $S(X)$ has no extreme points, then X contains no admissibly compact subspace which is factor reflexive. (This is the case, for example, when $X = L_1[a,b]$ or $X = C_0(T)$ for T locally compact but not compact, e.g., $X = c_0$.)

(2) If $S(X)$ has only a finite number of linearly independent extreme points, then X contains no admissibly compact subspace which is factor reflexive and has infinite codimension.

(3) The space $C(T)$, $T$ compact and connected, contains no admissibly compact subspace which is factor reflexive and has codimension greater than one.

Corollary 1 may be sharpened in the particular case when $M$ is a Chebyshev subspace of $X$ (having finite codimension $n$). Then the unit ball in $X$ must contain $n$ linearly independent exposed points. (See also Singer [2].) This follows from the well-known fact that the exposed points of the unit ball in a finite-dimensional space are dense in the set of extreme points; and the fact that if $x + M$ is an exposed point of $S(X/M)$, then $x - P_M(x)$ is an exposed point of $S(X)$.

Problem. Is the converse of Theorem 1 valid? That is, if $M$ is a proximinal subspace of $X$ such that

$$\text{card}[\text{ext } S(X/M)] \leq \text{card}[\text{ext } S(X)],$$

must $M$ be admissibly compact?

## References

[1] Garkavi, A. L., On best approximation by elements from infinite dimensional subspaces of a certain class. Mat. Sbornik, 62 (1963), 104-120.

[2] Singer, I., On best approximation in normed linear spaces by elements of subspaces of finite codimension. Revue Roumaine de Math. Pures et Appl., to appear.

Department of Mathematics
The Pennsylvania State University
University Park, Pennsylvania 16802

Note added in proof: Walter Pollul, in a letter communication dated April 5, 1973, has shown that the converse of Theorem 1 is not valid. His example is $X = c_0$ and

$$M = \{x = (x_1, x_2, \dots) \in c_0 : \ x_{2n} = 0 \ (n=1,2,\dots)\}.$$

Then $M$ is proximinal and since

$$\|x + M\| = \sup\{|x_{2n}| : n=1,2,\dots\},$$

the mapping $x + M \to (x_2, x_4, x_6, \ldots)$ is an isometry of $X/M$ with $c_0$. Thus

$$\text{card}\,[\text{ext}\, S(X/M)] = \text{card}\,[\text{ext}\, S(X)] = 0.$$

Let $x = (1,0,0,\ldots)$. Then

$$P_M(x) = \{m \in M : \|m\| \leq 1\}.$$

If $M$ were admissibly compact, then $P_M(x)$ would have an extreme point by the Krein-Milman theorem, i.e., ext $S(M) \neq \emptyset$. But since the mapping $x \in M \to (x_1, x_3, x_5, \ldots) \in c_0$ is an isometry, $S(M)$ has no extreme points.

## AN EXTENSION OF BERNSTEIN'S INEQUALITY

Ronald DeVore[1]

### 1. Introduction

Let $\Pi_n$ denote the class of trigonometric polynomials of degree $\leq n$. If $T \in \Pi_n$, then

$$(1.1) \qquad \|T'\|_{L^\infty} \leq n\|T\|_{L^\infty},$$

where $\|\cdot\|_{L^\infty}$ is the $L^\infty$-norm on $[-\pi,\pi]$. This, of course, is the classical inequality of S. Bernstein which plays an important role in many areas of analysis, especially in approximation theory. There are several generalizations of Bernstein's inequality. These usually are obtained by either replacing $\Pi_n$ by another class of functions, (e.g., algebraic polynomials or functions of exponential type) or by replacing $L^\infty$ by other spaces (e.g., $L^p$, $1\leq p<\infty$).

Here, we are interested in a generalization of Bernstein's inequality obtained in the second manner. We will consider trigonometric polynomials as being members of certain dual spaces. Namely, if $\omega$ is a modulus of continuity and $1\leq p<\infty$, let $L^p_\omega$ denote the space of all $2\pi$-periodic functions in $L^p[-\pi,\pi]$ for which

$$(1.2) \qquad \|f\|_{L^p_\omega} = \max\left(\|f\|_{L^p}, \sup_{t>0}(\omega(t))^{-1}\|f(x+t)-f(x)\|_{L^p}\right)$$

is finite. Here $\|\cdot\|_{L^p}$ is the $L^p$-norm on $[-\pi,\pi]$. $(L^p_\omega, \|\cdot\|_{L^p_\omega})$ is then a Banach space and we denote its dual space by $(L^p_\omega)^*$. For $p=\infty$, we use $L^\infty_\omega$ to denote the space of $2\pi$-periodic continuous functions with $\|\cdot\|_{L^\infty_\omega}$ defined by (1.2), with $p=\infty$.

Although, this is some abuse of notation for $p=\infty$, it will provide simpler statements of our results.

Each trigonometric polynomial T determines in a natural way a functional $\ell_T$ in $(L^p_\omega)^*$ by

$$\ell_T(f) = \int_{-\pi}^{\pi} f(t)T(t)dt.$$

Our main result is Theorem 1 in Section 2 which states that there is an absolute constant $c>0$ such that for $n\geq 1$, $1\leq p\leq\infty$ and $\omega$ a modulus of continuity, we have

$$(1.3) \qquad \|\ell_T\|_{(L^p_\omega)^*} \geq c\omega(n^{-1})\|T\|_{L^q}$$

for all $T\in\Pi_n$, where $q = \frac{p}{p-1}$ is the conjugate index. Thus, (1.3) gives a lower bound for the norm of the functional $\ell_T$.

To see the connections between (1.3) and Bernstein's inequality, let us take $p=1$ and $\omega(t) = t$. Then $L^1_\omega$ is the collection of all $2\pi$-periodic functions of bounded variation (see Butzer and Nessel [1; p. 181 and p. 367]) and

$$\|f\|_{L^1_\omega} = \max(\|f\|_{L^1}, \int_{-\pi}^{\pi} |df|).$$

Now, let $T\in\Pi_n$, then

$$\|T\|_{L^\infty} = \sup_{\|f'\|_{L^1}\leq 1} \left\{ \int_{-\pi}^{\pi} f'(t)T(t)dt \right\} \geq \sup_{\|f\|_{L^1_\omega}\leq 1} \left\{ \int_{-\pi}^{\pi} f(t)T'(t)dt \right\}$$

$$= \|\ell_{T'}\|_{(L^1_\omega)^*} \geq \frac{c}{n}\|T'\|_{L^\infty}$$

where the last inequality is (1.3). So, we see that (1.3) retrieves Bernstein's inequality in a non-sharp form, since we don't have $c=1$. In a similar way, we could show that Bernstein's inequality gives (1.3) for $p=1$, $\omega(t)=t$. We should remark that we do not concern ourselves as to what is the largest constant c for which (1.3) holds.

Inequality (1.3) gives a lower estimate for the norm of the convolution operator $L_T(f) = f*T$ considered as a mapping from $L^p_\omega$ to $L^\infty$. That is,

$$\|L_T\| \geq c\omega(\tfrac{1}{n})\|T\|_{L^q} .$$

Such lower estimates combined with the Banach-Steinhaus theorem give necessary conditions for a smoothness criteria, like $f\in L^p_\omega$, to guarantee uniform approximations by convolution operators. In Section 3, we illustrate this for the $n^{th}$ partial sums of the Fourier series. We also illustrate how to obtain necessary conditions for a smoothness criteria to guarantee absolute convergence of the Fourier series. While the results of Section 3 are well-known, our approach provides some unification for theorems of this type.

## 2. Proof of Inequality (1.3)

In what follows n will always denote a positive integer, T a polynomial of degree $\leq n$, $1\leq p\leq\infty$, and $q=p/(p-1)$, the conjugate index to p. For the proof of inequality (1.3), we will use a representation for trigonometric polynomials which can be found in Zygmund [4, Ch.X].

If k is an integer, let $x_k = (2k\pi)(3n)^{-1}$. Then, T can be represented by

$$(2.1)\qquad T(x) = n^{-1}\sum_{k=0}^{3n-1} T(x_k)V_n(x-x_k)$$

where

$$(2.2)\qquad V_n(x) = \frac{\sin(3nx/2)\,\sin(nx/2)}{3n\,\sin^2(x/2)}$$

Also, there is an absolute constant $c_1 > 1$ such that

$$(2.3)\qquad \begin{aligned} c_1^{-1}\left(n^{-1}\sum_{k=0}^{3n-1}|T(x_k)|^p\right)^{\frac{1}{p}} &\leq \|T\|_{L^p} \leq c_1\left(n^{-1}\sum_{k=0}^{3n-1}|T(x_k)|^p\right)^{\frac{1}{p}}, \quad 1\leq p<\infty, \\ c_1^{-1}\max_{0\leq k\leq 3n-1}|T(x_k)| &\leq \|T\|_{L^\infty} \leq c_1\max_{0\leq k\leq 3n-1}|T(x_k)|. \end{aligned}$$

The idea of the proof of (1.3) is to construct a function g in $L^p_\omega$, with $\|g\|_{L^p_\omega} \leq 1$, such that

$$\int_{-\pi}^{\pi} g(x)T(x)dx \geq c\omega(\tfrac{1}{n})\|T\|_{L^q} .$$

We now proceed to the construction of g and some of its properties. If $0<\delta<1/4$, let $h_\delta$ denote the $2\pi$-periodic "roof" function defined on $[-\pi,\pi]$ by

$$(2.4) \qquad h_\delta(x) = \begin{cases} 1 - \frac{3n}{2\delta\pi}|x|, & |x| \leq \frac{2\delta\pi}{3n}, \\ 0 \quad , & \frac{2\delta\pi}{3n} \leq |x| \leq \pi. \end{cases}$$

<u>Lemma 1</u>. Let $0<\delta<1/4$ and $T\in\Pi_n$. If $1<p\leq\infty$, define

$$g_\delta(x) = A\omega(n^{-1}) \sum_{k=0}^{3n-1} |T(x_k)|^{q-1} \operatorname{sgn}T(x_k)h_\delta(x-x_k),$$

where

$$A^{-1} = \left(n^{-1} \sum_{k=0}^{3n-1} |T(x_k)|^q\right)^{\frac{1}{p}} .$$

For p=1, let $k_0$ be such that $T(x_{k_0}) = \max_k |T(x_k)|$ and define

$$g_\delta(x) = n\omega(n^{-1})\operatorname{sgn}T(x_{k_0})h_\delta(x-x_{k_0}).$$

Then, $g_\delta$ is in $L^p_\omega$ and there is a constant $c_2(\delta)$, depending only on $\delta$ for which

$$(2.5) \qquad \|g_\delta\|_{L^p_\omega} \leq c_2(\delta).$$

<u>Proof:</u> We consider only the case $1<p<\infty$, the other cases are handled similarly. If $3nt \geq 2\delta\pi$, then it follows from (2.4) that

$$\|h_\delta(x+t)-h_\delta(x)\|^p_{L^p} = \int_{-\pi}^{\pi} |h_\delta(x+t)-h_\delta(x)|^p\, dx \leq$$

$$\int_{-t-\frac{2\delta\pi}{3n}}^{-t+\frac{2\delta\pi}{3n}} |h_\delta(x+t)-h_\delta(x)|^p dx + \int_{\frac{-2\delta\pi}{3r}}^{\frac{2\delta\pi}{3n}} |h_\delta(x+t)-h_\delta(x)|^p dx \leq \frac{2^p\cdot 8\delta\pi}{3n} .$$

Here, we have used the fact that $|h_\delta(x)| \leq 1$, for all x.

Now, the functions $h_\delta(x-x_k)$, $k=0,\ldots,3n-1$, have disjoint supports and so

$$\|g_\delta(x+t)-g_\delta(x)\|_{L^p}=A\omega(n^{-1})\left(\sum_{k=0}^{3n-1}|T(x_k)|^q\|h_\delta(x+t-x_k)-h_\delta(x-x_k)\|_{L^p}^p\right)^{\frac{1}{p}}$$

$$\leq 2A\left(\frac{8\delta\pi}{3}\right)^{\frac{1}{p}}\omega(n^{-1})\left(n^{-1}\sum_{k=0}^{3n-1}|T(x_k)|^q\right)^{\frac{1}{p}} = 2\left(\frac{8\delta\pi}{3}\right)^{\frac{1}{p}}\omega(n^{-1})$$

$$\leq 2\left(1+\frac{8\delta\pi}{3}\right)\omega(n^{-1}),$$

where we have used the definition of A.

Since $\omega$ is a modulus of continuity, for $3nt \geq 2\delta\pi$

$$\omega(n^{-1}) \leq \left(1+\frac{3}{2\delta\pi}\right)\omega(t).$$

Therefore,

$$(2.6)\qquad \|g_\delta(x+t)-g_\delta(x)\|_{L^p} \leq c(\delta)\omega(t), \qquad t \geq \frac{2\delta\pi}{3n}.$$

When $3nt \leq 2\delta\pi$, we have

$$\|h_\delta(x+t)-h_\delta(x)\|_{L^p}^p \leq \int_{-\frac{4\delta\pi}{3n}}^{\frac{4\delta\pi}{3n}} |h_\delta(x+t)-h_\delta(x)|^p dx \leq \left(\frac{3n}{2\delta\pi}t\right)^p\cdot\frac{8\delta\pi}{3n}.$$

Hence,

$$(2.7)\qquad \|g_\delta(x+t)-g_\delta(x)\|_{L^p} \leq A\left(\frac{8\delta\pi}{3}\right)^{\frac{1}{p}}\frac{3nt}{2\delta\pi}\omega(n^{-1})\left(n^{-1}\sum_{k=0}^{3n-1}|T(x_k)|^q\right)^{\frac{1}{p}}$$

$$\leq c(\delta)nt\omega(n^{-1}) \leq 2c(\delta)\omega(t), \qquad t \leq \frac{2\delta\pi}{3n}.$$

In the last inequality, we have used the fact that $t_1^{-1}\omega(t_1) \leq 2t_2^{-1}\omega(t_2)$ when $t_2 < t_1$, which holds for moduli of continuity. The inequalities (2.6), (2.7), and the similar estimate $\|g_\delta\| \leq (n^{-1})$. prove (2.5).

Lemma 2. There are absolute constants $c_3$, $c_4 > 0$, such that

$$(2.8)\qquad \int_{-\pi}^{\pi} h_\delta(t)V_n(t) \geq c_3\delta \qquad \text{for each } 0 < \delta < 1/4,$$

(2.9) $$\int_{-\pi}^{\pi} h_\delta(t)|V_n(t-x_j)|dt \le c_4\delta^2(j-1/2)^{-2}, \quad 0<|x_j|<\pi.$$

Proof: From (2.4), it follows that $h_\delta(t) \ge \frac{1}{2}$ for $3n|t|\le\delta\pi$. From (2.2), we find that on this same interval

$$V_n(t) \ge \frac{\frac{2}{\pi}\cdot\frac{3nt}{2}\cdot\frac{2}{\pi}\cdot\frac{nt}{2}}{3n\left(\frac{t}{2}\right)^2} = 4\pi^{-2}n$$

Thus,

$$\int_{-\pi}^{\pi} h_\delta(t)V_n(t)dt = \int_{-\frac{2\delta\pi}{3n}}^{\frac{2\delta\pi}{3n}} h_\delta(t)V_n(t)dt$$

$$\ge \int_{-\frac{\delta\pi}{3n}}^{\frac{\delta\pi}{3n}} h_\delta(t)V_n(t) \ge \frac{1}{2}\cdot 4\pi^{-2}n\cdot\frac{2\delta\pi}{3n} = \frac{4\delta\pi^{-1}}{3}.$$

Note that $V_n(t) \ge 0$ for $|t| \le \frac{2\delta\pi}{3n}$, since $\delta < 1/4$.

To prove (2.9), we return to (2.2) to find, when $|x_j| < \pi$,

$$|V_n(t-x_j)| \le (3n)^{-1}\sin^{-2}\left(\frac{t-x_j}{2}\right)\sin(\delta\pi), \quad |t| \le \frac{2\delta\pi}{3n}.$$

Therefore,

$$\frac{1}{\pi\delta}\int_{-\pi}^{\pi} h_\delta(t)|V_n(t-x_j)|dt \le \int_{-\frac{2\delta\pi}{3n}}^{\frac{2\delta\pi}{3n}} (3n)^{-1}\sin^{-2}\left(\frac{t-x_j}{2}\right)dt$$

$$\le \frac{4\delta\pi}{9n^2}\sin^{-2}\frac{1}{2}\left(|x_j|-\frac{2\delta\pi}{3n}\right) \le \frac{4\delta\pi^3}{9n^2}\left(|x_j|-\frac{2\delta\pi}{3n}\right)^{-2} \le c_4\left(j-\frac{1}{2}\right)^{-2}\delta,$$

where in the second to last inequality we used the fact that $\sin^2 t/2 \ge \pi^{-2}t^2$ for $|t| \le \pi$ and in the last inequality we used our assumption that $\delta < 1/4$.

Theorem 1. There is an absolute constant $c>0$, such that for each $1\le p\le\infty$, $n>0$ and $T\in\Pi_n$, we have

(2.10) $$\|\ell_T\|_{(L^p_\omega)^*} \ge c\omega(n^{-1})\|T\|_{L^q},$$

where $q = p/p-1$.

Proof: We will prove the theorem for the case $1<p<\infty$, the other cases are handled similarly. Also, we will suppose that n is odd. It is clear that if (2.10) holds for $n = 2m+1$ then it also holds for $n = 2m$ with c replaced by $c/3$. Before we begin the proof it is useful to observe that because of periodicity, the sums in (2.1), (2.3) and the definition of $g_\delta$ can be taken over any set of 3n consecutive integers.

Now, let $g_\delta$ be the function introduced in Lemma 1 with $0<\delta<1/4$, to be prescribed later. We have from (2.1) that

$$\int_{-\pi}^{\pi} g_\delta(x)\ T(x)dx = n^{-1} \sum_{k=0}^{3n-1} T(x_k) \int_{-\pi}^{\pi} g_\delta(x)\ V_n(x-x_k)dx$$

For each k, let $I_k'$ denote the set of the 3n consecutive integers with middle term k and $I_k = I_k' - \{k\}$. Then, from Lemma 2, we find

$$(A\omega(n^{-1}))^{-1} T(x_k) \int_{-\pi}^{\pi} g_\delta(x)\ V_n(x-x_k)dx \geq |T(x_k)|^q \cdot$$

$$\int_{-\pi}^{\pi} h_\delta(x-x_k)V_n(x-x_k)dx - \sum_{j\in I_k} |T(x_k)|\ |T(x_j)|^{q-1} \int_{S_j} h_\delta(x-x_j)\cdot$$

$$V_n(x-x_k)dx \geq c_3\delta|T(x_k)|^q - c_4\delta^2 \sum_{j\in I_k} (j-k-\tfrac{1}{2})^{-2}\ T(x_k)|\cdot$$

$$|T(x_j)|^{q-1},$$

where $S_j = \{x: |x-x_j| \leq \frac{2\delta\pi}{3n}\}$.

This last estimate shows that

$$(2.12)\quad \int_{-\pi}^{\pi} g_\delta(x)T(x)dx \geq c_3\delta A\omega(n^{-1})n^{-1} \sum_{k=0}^{3n-1} |T(x_k)|^q$$

$$- c_4\delta^2 A\omega(n^{-1})n^{-1} \sum_{k=0}^{3n-1} \sum_{j\in I_k} (j-k-\tfrac{1}{2})^{-2}|T(x_k)|\,|T(x_j)|^{q-1}.$$

The second sum on the right hand side of (2.12) can be

estimated by letting $k-j = \nu$, to find

$$(2.13)\quad \sum_{k=0}^{3n-1} \sum_{j\in I_k} (j-k-\tfrac{1}{2})^{-2} |T(x_k)|\,|T(x_j)|^{q-1}$$

$$\le \sum_{0<|\nu|<\frac{3n-1}{2}} (\nu-\tfrac{1}{2})^{-2} \sum_{k=0}^{3n-1} |T(x_k)|\,|T(x_{k-\nu})|^{q-1}$$

$$\le \sum_{0<|\nu|\le\frac{3n-1}{2}} (\nu-\tfrac{1}{2})^{-2} \left(\sum_{k=0}^{3n-1} |T(x_k)|^q\right)^{\frac{1}{q}} \left(\sum_{k=0}^{3n-1} |T(x_k)|^q\right)^{\frac{1}{p}}$$

$$\le \sum_{\nu=1}^{\infty} (\nu-\tfrac{1}{2})^{-2} \sum_{k=0}^{3n-1} |T(x_k)|^q .$$

Now, choose $0<\delta<1/4$ sufficiently small that

$$c_3\delta - c_4\delta^2 \sum_{\nu=1}^{\infty} (\nu-\tfrac{1}{2})^{-2} = c_5 > 0$$

and fix $\delta$. Then we use (2.13) in (2.12) to obtain

$$\int_{-\pi}^{\pi} g_\delta(x)T(x)dx \ge c_5 A\omega(n^{-1})n^{-1} \sum_{k=0}^{3n-1} |T(x_k)|^q$$

$$= c_5\omega(n^{-1})\left(n^{-1} \sum_{k=0}^{3n-1} |T(x_k)|^q\right)^{\frac{1}{q}} \ge c_5 c_1^{-1}\omega(n^{-1})\|T\|_{L^q}.$$

We have used (2.3) and the definition of A. Since $\|g_\delta\| \le c(\delta)$, with $c(\delta)$ now an absolute constant, the theorem is proved.

## 3. Applications

As we have indicated in the introduction, we wish to use the Banach-Steinhaus theorem together with Theorem 1 to examine convergence theorems for convolution operators. To accomplish this, we must work with spaces in which trigonometric polynomials are dense. Accordingly, we introduce the space $L^p_{\omega,o}$ which is the closure of trigonometric polynomials in $L^p_\omega$. If $t^{-1}\omega(t) \to \infty$ as $t \to 0$, then $L^p_{\omega,o}$ is the collection of all functions $f$ in $L^p_\omega$ for which

$$\|f(x+t)-f(x)\|_{L^p} = o(\omega(t)) \qquad (t \to 0).$$

If $\omega(t) = O(t)$, then $L^p_{\omega,o}$ is the collection of all absolutely continuous functions $f$ for which $f' \in L^p$.

For our applications, we will consider $S_n(f)$ the $n^{th}$ partial sum of the Fourier series of $f$,

$$S_n(f) = \sum_{k=-n}^{n} \hat{f}(k)e^{ikx} = \frac{1}{\pi}\int_{-\pi}^{\pi} f(t)D_n(t-x)dt = f*D_n$$

where $D_n$ is the Dirichlet kernel

$$D_n(t) = \frac{\sin(n+1)t/2}{2\ \sin t/2}$$

We first ask what are necessary and sufficient conditions on $p$ and $\omega$ to guarantee that $f \in L^p_{\omega,o}$ implies $S_n(f)$ converges uniformly. From the Banach-Steinhaus Theorem, we see that a necessary and sufficient condition is that the operators $S_n$ as mappings from $L^p_{\omega,o}$ to $L^\infty$ be uniformly bounded. The norm of the operator $S_n$ is the same as the norm of the functional

$$S_n(f,0) = \frac{1}{\pi}\int_{-\pi}^{\pi} f(t)D_n(t)dt$$

in $(L^p_{\omega,o})^*$. Since any function $f$ in $L^p_\omega$ with $\|f\|_{L^p_\omega} \leq 1$ is the limit in $L^p$ of functions from the unit ball of $L^p_{\omega,o}$, we see that

$$(3.1) \qquad \|S_n\|_{L^p_{\omega,o}} = \|S_n\|_{L^p_\omega} = \frac{1}{\pi}\|\ell_{D_n}\|_{(L^p_\omega)^*} .$$

Therefore, what we are seeking is a necessary and sufficient condition for

$$(3.2) \qquad \|\ell_{D_n}\|_{(L^p_\omega)^*} = O(1)$$

Now, if we apply Theorem 1, we see there is a constant $c_1 > 0$, with

$$(3.3) \qquad c_1\omega(n^{-1})\|D_n\|_{L^q} \leq \|\ell_{D_n}\|_{(L^p_\omega)^*} .$$

It is also true that there is a constant $c_2>0$, such that

$$(3.4) \qquad \|\ell_{D_n}\|_{(L^p_\omega)^*} \le c_2\omega(n^{-1})\|D_n\|_{L^q} + 3$$

This can be shown by using the de la Vallee Poussin operators $V_m$, $m=[n/2]$ (see G. Lorentz [2, p. 93]). The operator $V_m$ as an operator from $L^p$ to $L^\infty$ has norm $\le 3$ and there is a constant $c_2 > 0$ such that

$$\|f-V_m(f)\|_{L^p} \le c_2\omega(n^{-1})\|f\|_{L^p_\omega}$$

Also, $V_m(f) = S_n(V_m(f))$. Hence, if $\|f\|_{L^p_\omega} \le 1$, then

$$\|S_n(f)\|_{L^\infty} \le \|S_n(f-V_m(f))\|_{L^\infty} + \|V_m(f)\|_{L^\infty} \le c_2\omega(n^{-1})\|D_n\|_{L^q} + 3$$

which is (3.4). The inequalities (3.3) and (3.4) show that (3.2) holds if and only if

$$\omega(n^{-1})\|D_n\|_{L^q} = O(1) .$$

Standard calculations give that

$$(3.5) \qquad \|D_n\|_{L^q} \sim \begin{cases} \ln n & , \quad p = \infty, \\ n^{\frac{1}{p}} & , \quad 1 \le p < \infty . \end{cases}$$

Hence, a necessary and sufficient condition for $f \in L^p_{\omega,0}$ to imply $S_n(f)$ converges uniformly is that

$$(3.6) \qquad \omega(n^{-1})\ln n = O(1), \qquad p = \infty ,$$

$$(3.7) \qquad \omega(n^{-1})n^{\frac{1}{p}} = O(1) , \qquad 1 \le p < \infty .$$

As a second example, we seek necessary conditions on p and $\omega$, so that $f \in L^p_{\omega,0}$ implies that the Fourier series of f converges absolutely.

Let $(T_n)$ be any sequence of trigonometric polynomials, $T_n \in \Pi_n$ with

$$|\hat{T}_n(k)| \leq 1, \qquad k = 0, \pm 1, \ldots, \pm n \ ,$$

If each $f \in L^p_{\omega,0}$ has an absolutely convergent Fourier series then

$$\|f*T_n\|_{L^\infty} \leq \sum_{-n}^{n} |\hat{f}(k)||\hat{T}_n(k)| \leq \sum_{-n}^{n} |\hat{f}(k)| \leq \sum_{-\infty}^{\infty} |\hat{f}(k)| < +\infty,$$

$$n = 1,2,\ldots$$

Thus, the convolution operators $L_n(f) = f*T_n$ considered as mappings from $L^p_{\omega,0}$ to $L^\infty$ are pointwise and hence uniformly bounded. It then follows from Theorem 1 that there is an absolute constant $M>0$ such that

$$c_1\omega(n^{-1})\|T_n\|_{L^q} \leq \|\ell_{T_n}\|_{(L^p_\omega)^*} = \frac{1}{\pi}\|L_{T_n}\| \leq M.$$

Since $M$ does not depend on $T_n$, we have

$$(3.8) \qquad \Lambda_{n,p}\, \omega(n^{-1}) = O(1),$$

where

$$\Lambda_{n,p} = \sup_{\substack{T \in \Pi_n \\ |\hat{T}(k)| \leq 1}} \|T\|_{L^q} \ .$$

That is, (3.8) is a necessary condition on $p$ and $\omega$ for each $f \in L^p_{\omega,0}$ to have an absolutely convergent Fourier series. The asymptotic behavior of $\Lambda_{n,p}$ is given by

$$(3.9) \qquad \Lambda_{n,p} \sim \begin{cases} n^{\frac{1}{2}} \ , & 2 \leq p \leq \infty \ , \\ n^{\frac{1}{p}} \ , & 1 \leq p \leq 2 \ , \end{cases}$$

The case $2 \leq p \leq \infty$ in (3.9) follows from a result of D. J. Newman [3, Theorem 1]. The case $1 \leq p \leq 2$ follows from the well known inequality $\|f\|_{L^q} \leq \|\hat{f}\|_{\ell^p}$ (see Zygmund [4, p. 101, Vol. II]) and the previous estimate (3.5). Hence, (3.9) shows that a necessary condition for each $f \epsilon L^p_{\omega,o}$ to have an absolutely convergent Fourier series is that

$$(3.10)\quad \begin{aligned} n^{\frac{1}{p}}\,\omega(n^{-1}) &= O(1)\ , \qquad 1 \leq p \leq 2. \\ n^{\frac{1}{2}}\,\omega(n^{-1}) &= O(1)\ , \qquad 2 \leq p \leq \infty. \end{aligned}$$

Most classical theorems on absolute convergence of Fourier series are stated for the classes $L^p_\omega$. Although (3.10) does give necessary conditions for each $f \epsilon L^p_\omega$ to have an absolutely convergent Fourier series, slightly stronger necessary conditions are known. We consider one example, $p=\infty$, which is typical, and refer the reader to Zygmund [4, p. 240, Vol. I] for a more detailed discussion of absolute convergence theorems. When $p=\infty$, there is a function $f \epsilon L^\infty_{t^{1/2}}$, whose Fourier series does not converge absolutely. Our result would only show that for each $\omega$ for which $\lim_{t \to 0} t^{-1/2}\,\omega(t) = \infty$, there is a function $f \epsilon L^\infty_{\omega,o}$ whose Fourier series does not converge absolutely.

---

[1]The author gratefully acknowledges NSF support under grant GP19620.

## References

[1] Butzer, P. L. and R. J. Nessel, Fourier Analysis and Approximation, I. Academic Press, New York, 1970.

[2] Lorentz, G. G., Approximation of Functions. Holt, Rinehart and Winston, New York, 1966.

[3] Newman, D. J., An $L^1$ extremal problem for polynomials. Proc. A.M.S. 16 (1965), 1287-1290.

[4] Zygmund, A., Trigonometric Series, I, II. Cambridge Univ. Press, Cambridge, 1959, 383 pp. and 364 pp.

Department of Mathematics
Oakland University
Rochester, Michigan 48063

# VARISOLVENT CHEBYSHEV APPROXIMATION ON SUBSETS

Charles B. Dunham

Let F be an approximating function unisolvent of variable degree on a closed interval $[\alpha,\beta]$ and let F be of bounded degree. All functions considered will be continuous, and for such functions we define the norm on a closed non-empty subset Y of $[\alpha,\beta]$ to be

$$\|g\|_Y = \max \{ |g(x)| : x \in Y\}.$$

(When Y is omitted, it is understood that $Y = [\alpha,\beta]$). The Chebyshev problem on Y is for a given continuous function f to find an element $F(A^*,.)$, $A^* \in P$, for which $\|f-F(A,.)\|_Y$ is minimal. Such an element $F(A^*,.)$ is called a best Chebyshev approximation to f on Y.

A definition of varisolvence and theory for approximation on an interval is given in [4]. We assume that the difficulty of [2] does not occur.

There may exist continuous f with no best approximation on $[\alpha,\beta]$. Even if best approximations exist to all continuous f on $[\alpha,\beta]$, best approximations may not exist on finite subsets of $[\alpha,\beta]$. This is well known in the case of approximation by polynomial rational functions $R_m^n[\alpha,\beta]$. We can, however, show that if the best approximation on $[\alpha,\beta]$ is of maximum degree, then there exists a best approximation on all sufficiently dense subsets.

Define the density of a subset Y of $[\alpha,\beta]$ to be

$$\sup \{\inf \{|x-y| : y \in Y\} : \alpha \le x \le \beta\}.$$

We say $\{X_k\} \to [\alpha,\beta]$ if $X_k \subset [\alpha,\beta]$ and for $x \in [\alpha,\beta]$, there is

$x_k \in X_k$ such that $\{x_k\} \to x$.

Lemmas similar to the following were first stated in [1]. The three lemmas are proven (in more general form) in [3].

Lemma 1. Let $F(A,.)$ be the best approximation to $f$ on $[\alpha,\beta]$ and let $F$ have degree $n$ at $A$. Let $\{x_0,\ldots,x_n\}$ be an ordered set of points on which $f-F(A,.)$ alternates $n$ times. Let $\varepsilon > 0$ be given. Then there exists $\delta$, $0<\delta<\varepsilon$, such that if $|x_i' - x_i| < \delta$ and $|f(x_i') - F(B,x_i')| \leq \|f-F(A,.)\|$ for $0\leq i\leq n$, then

$$(1)\quad \operatorname{sgn}[f(x_0) - F(A,x_0)](-1)^i\,[F(B,x_i') - F(A,x_i')] \geq -\varepsilon \qquad i=0,\ldots,n.$$

Lemma 2. If $F$ is of degree $n$ (maximal) at $A$, then for given $\varepsilon > 0$ there exists $\eta(\varepsilon)$ such that $\|F(A.,) - F(B,.)\| < \eta(\varepsilon)$ if (1) holds, and $\eta(\varepsilon) \to 0$ as $\varepsilon \to 0$.

Lemma 3. Let $F$ be unisolvent of degree $m$ at $A_k$, $k=0,1,\ldots$ and let $\{F(A_k,.)\}$ converge pointwise to $F(A_0,.)$ on $m$ distinct points then $\{F(A_k,.)\}$ converges uniformly to $F(A_0,.)$.

Theorem. Let $F$ be unisolvent of variable degree. Let $f$ have a best approximation $F(A,.)$ and $F$ be of degree $n$ (maximal) at $A$. There exists $\varepsilon > 0$ such that the density of $Y$ being less than $\varepsilon$ implies that $f$ has a best approximation on $Y$. If $\{X_k\} \to [\alpha,\beta]$ and $F(A_k,.)$ is best on $X_k$ to $f$, $\|F(A,.) - F(A_k,.)\| \to 0$.

Proof: Let $x_0,\ldots,x_n$ be as in Lemma 1. By definition of solvency of degree $n$ at $A$ there exists $\gamma > 0$ such that if $|y_k - F(A,x_k)| < \gamma$, $k=1,\ldots,n$, then there exists a parameter $B$ satisfying

$$(2)\quad F(B,x_k) = y_k \qquad k=1,\ldots,n.$$

Using property Z and maximality of $n$, it is easily seen that $F$ is unisolvent of degree $n$ at any such $B$, and hence $B$ is completely determined by (2). Choose $\varepsilon$ such that $\eta(\varepsilon) < \gamma/2$, then by Lemma 1 and 2, if the density of $Y$ is less than $\varepsilon$ and

$$\max\{|f(x_i') - F(B,x_i')| : i=0,\ldots,n\} \leq \|f-F(A,.)\|$$

then

$$\|F(A,.) - F(B,.)\| < \gamma/2.$$

Now choose Y with density less than $\varepsilon$. Let $\{B_k\}$ be a sequence of parameters such that

$$\{\|f-F(B_k,.)\|_Y\} \to \inf\{\|f-F(C,.)\|_Y : C \in P\} := \rho(f,Y).$$

If $\rho(f,Y) = \|f-F(A,.)\|$ then $F(A,.)$ is best to f on Y. Otherwise $\rho(f,Y) < \|f-F(A,.)\|$. Choose $x_i' \in Y$ such that $|x_i - x_i'| < \varepsilon$, $i=0,\ldots,n$. We can assume that

$$\max\{|f(x_i') - F(B_k,x_i')| : i=0,\ldots,n\} \leq \|f-F(A,.)\|.$$

It follows that $\|F(A,.) - F(B_k,.)\| < \gamma/2$ and the n-tuples of values of $F(B_k,.)$ at the points $x_1,\ldots,x_n$ form a bounded sequence with subsequence converging to an accumulation point $(y_1,\ldots,y_n)$, which determines a parameter B at which F is unisolvent of degree n. Using Lemma 3 we can show that for all $x \in Y$, $|f(x) - F(B,x)| \leq \rho(f,Y)$ and so $F(B,.)$ is a best approximation to f on Y. The first part of the theorem is proven. Now let $\{X_k\} \to [\alpha,\beta]$. Then for all k sufficiently large, there is a best approximation $F(A_k,.)$ to f on $X_k$. From Lemmas 1 and 2 it follows immediately that $\|F(A,.) - F(A_k,.)\| \to 0$.

If the best approximation to f on $[\alpha,\beta]$ is not of maximum degree, no such conclusions can be drawn.

Example. Let $[\alpha,\beta] = [-1,1]$, the approximating family be $R_1^0[-1,1]$, and $f(x) = x$. As f-C alternates once, 0 is best to f. Let $X_k = \{-1+1/k, -1+2/k,\ldots,1-1/k,1\}$. There is a sequence of approximants $F(A^j,.)$ such that $F(A^j,1) = 1$, $F(A^j,x) \to 0$ for $x \in X_k \setminus \{1\}$, and

$$\max\{|f(x) - F(A^j,x)| : x \in X_k\} \to 1-1/k.$$

Since approximants are of constant sign, there is no approximant for which the limiting error norm is attained and a best approximation on $X_k$ does not exist.

We have exactly the same result when the approximating family is given by

$$F(A,x) = a_1 \exp (a_2 x) .$$

If the best approximation is not of maximum degree, convergence of best approximations on subsets may not be uniform. An example is given in [5]. This matter is treated in more detail in [3;6].

## References

[1] Dunham, C. B., Continuity of the Varisolvent Chebyshev Operator. Bull. Amer. Math. Soc. 74 (1968), 606-608.

[2] Dunham, C. B., Necessity of Alternation. Can. Math. Bull. 10 (1968), 743-744.

[3] Dunham, C. B., Alternating Chebyshev Approximation. Trans. Amer. Math. Soc., To appear.

[4] Rice, John, Tchebycheff Approximations by Functions Unisolvent of Variable Degree. Trans. Amer. Math. Soc. 99 (1961), 298-302.

[5] Dunham, C. B., Minimax nonlinear approximation by approximation on subsets. Comm. ACM 15 (1972), 351.

[6] Dunham, C. B., Approximation by alternating families on subsets. Computing. 9 (1972), 261-265.

Computer Science Department
University of Western Ontario
London, Canada

BUTTERWORTH AND CHEBYSHEV SPLINES[1]

Rui J. P. de Figueiredo[2]

Two $Lg$ splines are introduced, motivated by the considerations mentioned at the end of this note. They will be called Butterworth and Chebyshev splines (and abbreviated simply as Bu- and Ch-splines) after the names of the filters [1-2] on the structures of which their respective differential operators $L$ are based.

In what follows, $\phi = \{\varphi_1, \ldots, \varphi_k\}$ will denote a set of continuous linear functionals on the Sobolev space $W^{n,2}[a,b]$, where $-\infty < a < b < \infty$ and $k \geq n$. Given a $n^{th}$ order linear differential operator $L$ with constant coefficients and letting $\tilde{\phi} \equiv \{\varphi_1, \ldots, \varphi_n\} \subset \phi$, we will write $\{L, \tilde{\phi}\} \in S$ if the elements of $\tilde{\phi}$ are linearly independent on the null space of $L$. Then the value at t of the unique $Lg$ spline interpolating a real k-tuple $r = \{r_1, \ldots, r_k\}$ with respect to $\phi$ will be denoted by $S(L, \phi, r, t)$.

Definition 1. Let

$$(1) \quad L_\lambda(D) = \prod_{j=1}^{n} (D - s_j), \; D = d/dt, \; t \in [a,b],$$

$$(2) \quad s_j = \lambda^{\frac{1}{2n}} e^{i(2j-1+n)\pi/2n}, \quad j=1, \ldots, n,$$

$$(3) \quad \lambda = \text{nonnegative constant.}$$

Given a $\phi$ such that $(L_\lambda, \tilde{\phi}) \in S$, the Bu-spline is defined as $S(L_\lambda, \phi, r, \cdot)$, where r is an arbitrary real k-tuple.

Definition 2. If in the preceding definition, (2) is replaced by (4a) and (4b) below:

(4a) $s_j = -\sinh\alpha \sin\left\{\frac{(2j-1)\pi}{2n}\right\} + i \cosh\alpha \cos\left\{\frac{(2j-1)\pi}{2n}\right\}$

$j = 1,\ldots,n,$

(4b) $\alpha = \frac{1}{n}\sinh^{-1}\sqrt{\lambda},$

the resultant $S(L_\lambda,\phi,r,\cdot)$ will be called Ch-spline.

Note that (within a multiplicative constant)

(5) $L_\lambda(-D)L_\lambda(D) = L(-D)L(D) + \lambda,$

where $L(D) = D^n$ for the Bu-spline and $L(D) = (-i)^n T_n(iD)$ for the Ch-spline, where $T_n(iD)$ = Chebyshev polynomial of the first kind of degree n in iD.

If $\gamma$ is a positive constant, let

(6) $\psi = \{f \in W^{n,2}[a,b]\colon \varphi_i(f) = r_i,\ i=1,\ldots,k;\ \|f\|^2_{L^2(a,b)} = \gamma^2\}.$

The following is a modification of the conventional minimum norm property [3-5] of spline functions resulting from the introduction of the additional equality constraint $\|f\|^2_{L^2(a,b)} = \gamma^2$. Suppose the adjoint $L^+(D) = L(-D)$.

<u>Theorem.</u> Let $L(D)$ be $D^n$ or $(-i)^n T_n(iD)$, and assume that we are given $\phi$, r, and $\gamma$ such that $\{L,\tilde{\phi}\} \in S$ and $\psi$ is nonempty. If there is a $\lambda^* \geq 0$ such that $\{L_{\lambda^*},\tilde{\phi}\} \in S$ (and hence $S(L_{\lambda^*},\phi,r,\cdot)$ is well defined) and

(7) $\int_a^b S^2(L_{\lambda^*},\phi,r,t)dt \equiv g(\lambda^*) = \gamma^2,$

then the problem

(8) $\inf_{f\in\psi} \|Lf\|^2_{L^2(a,b)}$

has a unique solution $f^*$ expressed by $S(L_{\lambda^*},\phi,r,\cdot)$.

The reconstruction of $f^*$, from given $\phi$ and r, by means of a Bu- or Ch-spline has two frequency domain attributes. It minimizes a generalized Gabor bandwidth [6]. It also represents a min-max estimate of an element of the class of smooth time-limited-essentially-band-limited functions, based on $\phi$ and r. The full version [7] of this note contains an elaboration of these considerations as well as a Fortran routine for Hermite-Birkhoff interpolation with a Bu-spline of order 2n up to 8.

---

[1]Sponsored in part by the United States Army under Contract No. DA-31-124-ARO-D-462, the National Science Foundation under Grant GK-36375, and NASA under Contract No. NAS-9-12776.

[2]On leave from Rice University, Houston, Texas 77001.

## References

[1] Butterworth, S., On the theory of filter amplifiers. Wireless Engr., 7 (1930), 536-541.

[2] Van Valkenburg, M. E., Introduction to Modern Network Synthesis. J. Wiley, New York (1960), p. 376.

[3] Holladay, J. C., Smoothest curve approximation. Math. Tables Aids Comput. 11 (1957), 233-243.

[4] Schoenberg, I. J., On interpolation by spline functions and its minimal properties. On Approximation Theory, P. L. Butzer, ed., Birkhäuser Verlag, Basel (1964), pp. 109-129.

[5] de Boor, C. and R. E. Lynch, On splines and their minimal properties. J. Math. Mech. 15 (1966), 953-969.

[6] Gabor, D., Theory of communication. J. IEE (London) 93 (part III) (1946), 429-458.

[7] Figueiredo, R. J. P. de, Butterworth and Chebyshev splines. MRC Tech Summ. Rept. #1327, Mathematics Research Center, Madison, Wisconsin (1973).

Mathematics Research Center
University of Wisconsin
Madison, Wisconsin 53706

# A PROOF OF CAUCHY'S INTEGRAL THEOREM USING BERNSTEIN POLYNOMIALS

L. Flatto and O. Shisha

Abstract

We give a simple proof of Cauchy's integral theorem viewed upon, as usual, as the foundation of complex analysis.

Theorem. Let f be holomorphic in an open set D of the (finite) complex plane. Let $c(t)$, $c_1(t)$ be complex functions, continuous and of bounded variation in $[0,1]$, mapping it into D, and satisfying $c(0) = c(1)$, $c_1(0) = c_1(1)$. Suppose that c can be deformed in D (as a loop) to $c_1$, namely, suppose there exists a complex function $C(t,s)$, continuous in the square $0 \le t \le 1$, $0 \le s \le 1$, and mapping it into D such that

$$C(0,s) = C(1,s) \text{ for every } s \in [0,1],$$

$$C(t,0) = c(t), \text{ and } C(t,1) = c_1(t), \text{ for every } t \in [0,1].$$

Then

$$\int_c f(z)dz = \int_{c_1} f(z)dz.$$

In particular, if $c_1$ is constant in $[0,1]$ so that c can be deformed in D to a point, then $\int_c f(z)dz = 0$.

The theorem is proved in three steps. First, we assume that f' is continuous in D and that $C(t,s)$ can be chosen so that its second order partial derivatives exist and are continuous at every point of $[0,1] \times [0,1]$. Next, we drop the restriction on $C(t,s)$ and approximate it by its Bernstein polynomials $B_n(t,s)$, making use of the variation diminishing property of (one variable) Bernstein polynomials. Finally, the explicit condition of continuity of f' is removed.

Department of Mathematics
Belfer Graduate School of Science
Yeshiva University
New York, New York 10033

Mathematics Research Center
Naval Research Laboratory
Washington, D. C. 20390

# DEGREE OF APPROXIMATION BY POLYNOMIALS ON COMPACT PLANE SETS

Tord H. Ganelius

## 1. Introduction

There are many different theorems on the degree of approximation by polynomials to continuous functions on compact sets. In the earlier work the boundary was supposed to be an analytic curve (Sewell [8]). More recently much weaker conditions on the boundary have been considered. In a sequence of papers [2, 3] Dzyadik and his co-workers have given a rather complete treatment of the field. Like Kövari [4], who recently considered sets bounded by a Jordan curve of bounded rotation, these authors obtain the direct theorems by construction of approximating polynomials.

The main purpose of this note is to use the elementary functional analysis approach to derive an explicit formula for

$$E_n(f) = \inf_{p \in \Pi_n} \sup_{z \in K} |f(z) - P(z)|,$$

where $\Pi_n$ is the set of polynomials of degree at most n and K is a set with rectifiable boundary. It seems reasonable to give a formulation which is not restricted to sets bounded by Jordan curves but also applies e.g., to arcs and union of arcs.

We shall see that our formula implies the central direct theorems given by Dzyadik. Our main point is not to claim important new results but to give a simple method which avoids the explicit construction of approximating polynomials.

## 2. The Fundamental Formula

It is well known that the best approximation to a function f on a set K by polynomials in $\Pi_n$ is given by

$$(1)\quad E_n(f) = \sup_{\mu\in Q_n} \left|\int f\, d\mu\right|,$$

where $Q_n$ is the set of complex measures on K of total variation 1 and satisfying

$$\int z^k\, d\mu = 0,\ k=0,1,2,\ldots,n.$$

If f is holomorphic in int K we may as well assume that the measure is supported by $\partial K$. (For the application of this method in the more difficult case of Mergelyan's theorem, see e.g., Carleson [1].)

The next step is to extend f continuously to a bounded harmonic function outside K. Let F be this function, so that $F|K = f$ and F is harmonic outside $\partial K$, in fact holomorphic in int K. To this function F we are going to apply the general Cauchy formula (see e.g., Lehto-Virtanen [5, p. 162]).

Without loss of generality we may assume that F is 0 at infinity so that the formula applied to the whole plane reads

$$F(\zeta) = -\frac{1}{\pi}\iint_D \frac{\partial F/\partial \bar{z}}{z-\zeta}\, dxdy,\quad z = x + iy.$$

This formula is easily proved directly in the case we are going to consider, when $\zeta$ belongs to the rectifiable boundary of K. In the interior of K we have $\partial f/\partial \bar{z} = 0$ since F is holomorphic there. For the complement of K we can write the harmonic function F as the sum of holomorphic functions H in z and G in $\bar{z}$, so that $F(x+iy) = H(z) + G(\bar{z})$. Denoting the complement of K by CK, our representation formula takes the form

$$f(\zeta) = -\frac{1}{\pi}\iint_{CK} G'(\bar{z})\,\frac{dxdy}{z-\zeta}.$$

We now change the variables by introduction of the mapping

function giving a conformal map of the exterior of the unit circle in the w-plane on the exterior of K, with correspondence between the points at infinity. Thus with $w = u + iv$ we find

$$(2)\quad F(\zeta) = -\frac{1}{\pi}\iint_{|w|\geq 1} G'(\overline{\psi}(\overline{w}))\ \overline{\psi}'(\overline{w})\ \frac{\psi'(w)}{\psi(w)-\zeta}\, du dv$$

$$= -\frac{1}{\pi i}\int_1^\infty \rho\, d\rho \int_{|w|=\rho} (G\circ\overline{\psi})'(\rho^2/w)\ \frac{\psi'(w)}{\psi(w)-\zeta}\cdot\frac{dw}{w},$$

if we perform the integration around circles first. If $1<\lambda<\rho=|w|$ it follows in an elementary way that

$$(3)\quad \frac{\psi'(w)}{\psi(w)-\zeta} = \frac{1}{\pi}\int_{s=0}^{2\pi}\frac{dv_\zeta(s)}{w-\lambda e^{is}}$$

with $v_\zeta(s) = \arg(\psi(\lambda e^{is})-\zeta)$. It is not necessary for our proof but we observe that (cf. Pommerenke [7, Lemma 1])

$$(4)\quad \int_0^{2\pi} e^{iks}\, dv_\zeta(s) = P_k(\zeta),$$

where $P_k$ is a polynomial of degree k in $\zeta$, the Faber polynomial except for normalization. If $v_\zeta$ is of uniformly bounded variation for all $\lambda > 1$, we can take $\lambda = 1$. That is, e.g., true if the boundary has bounded rotation in the generalized (non-Jordan) sense of Paatero [6] or if it is sufficiently smooth between a finite number of corners (that the boundary of a domain of bounded rotation is rectifiable has been pointed out to me by S. E. Warschawski). We put $\lambda = 1$ in the sequel. Introducing (3) in our formula (2) for f we get by the residue theorem applied to $|w| \leq \rho$ that

$$f(\zeta) = -\frac{2}{\pi}\int_{\rho=1}^{\infty} \rho\, d\rho \int_{s=0}^{2\pi} (G\circ\overline{\psi})'(\rho^2 e^{-is})e^{-is} dv_\zeta(s).$$

If we now integrate with respect to $\mu$ we find

$$(5)\quad \int f(\zeta)d\mu(\zeta) = -\frac{2}{\pi}\int_1^\infty \rho K(\rho)d\rho,$$

where

$$K(\rho) = \iint (G\circ\overline{\psi})'(\rho^2 e^{-is})e^{-is}dv_\zeta(s)d\mu(\zeta).$$

By (1) we have a formula for $E_n(f)$. The applications depend on the fact that K can be looked upon as a holomorphic function of $\rho$ for $|\rho| > 1$, and since (4) and $\mu\in Q_n$ imply that $K(\rho) = O(|\rho|^{-2n})$, Schwarz's lemma can be invoked to improve any estimate for $K(\rho)$ following from regularity assumptions on G, i.e , on f, and $\psi$. (For assumptions on $\partial K$ implying regularity of $\psi$ see Warschawski [9] and also some of his more recent papers.)

## 3. Application

We conclude by showing how to derive the following theorem (cf. [3] and [4]).

Theorem. Let the boundary of K be a rectifiable, not necessarily Jordan, curve $\partial K$ such that for all $\zeta\in\partial K$ the variation of

$$v_\zeta(s) = \arg(\psi(e^{is})-\zeta)$$

is uniformly bounded, $\psi$ being the exterior mapping function. Let f be continuous on K and holomorphic in the interior. If $f\circ\psi$ satisfies on the unit circle a Hölder condition of order $\alpha$, $0<\alpha\leq 1$, then

$$E_n(f) \leq Cn^{-\alpha}.$$

Proof: The assumption of Hölder continuity implies that $G\circ\overline{\psi}(r) = O((r-1)^{\alpha-1})$ if $0<\alpha<1$. (The case $\alpha=1$ is handled after an integration by parts with respect to $\rho$.) We get a similar estimate for K and by Schwarz's lemma we see that

$$|K(\rho)| \leq \begin{cases} C(\rho-1)^{\alpha-1}, & 1<\rho\leq 1+n^{-1}, \\ C\, n^{1-\alpha}((1+n^{-1})/\rho)^n, & 1+n^{-1} \leq \rho. \end{cases}$$

But then formula (5) gives

$$\sup_{\mu\in Q_n} |\int f d\mu| \leq C \int_1^{1+n^{-1}} \rho(\rho-1)^{\alpha-1}d\rho + Cn^{1-\alpha} \int_{1+n^{-1}}^{\infty} ((1+n^{-1})/\rho)^n d\rho \leq Cn^{-\alpha}.$$

## References

[1] Carleson, L., Mergelyan's theorem on uniform polynomial approximation. Math. Scand. 15 (1964), 167-175.

[2] Dzyadik, V. K., Investigations in the theory of approximations of analytic functions conducted at the Institute of Mathematics of the Ukrainan Academy of Sciences (Russian). Ukransk. Mat. Zh. 21 (1969), 173-192.

[3] Dzyadik, V. K. and G. A. Alibekov, Uniform approximation of functions of a complex variable on closed sets with corners. Mat. Sbornik (N.S.) 75 (117) (1968), 502-557. English translation Math. USSR Sbornik 4 (1968), 463-517 (1969).

[4] Kövari, T., On the order of polynomial approximation for closed Jordan domains. J. Approx. Theory 5 (1972), 362-373.

[5] Lehto, O., and K. I. Virtanen, Quasikonforme Abbildungen. Berlin 1965.

[6] Paatero, V., Über die konforme Abbildung von Gebieten, deren Ränder von beschränkter Drehung sind. Ann. Acad. Sci. Fenn. Ser A 33 (1931), 1-77.

[7] Pommerenke, Ch., Konforme Abbildung und Fekete-Punkte. Math. Z. 89 (1965), 422-438.

[8] Sewell, W. E., Degree of approximation by polynomials in the complex domain. Ann. Math. Studies, Princeton 1942.

[9] Warschawski, S. E., Über das Randverhalten der Ableitung der Abbildungsfunktion bei konformer Abbildung. Math. Z. 35 (1932), 321-456.

Department of Mathematics
University of California, San Diego
La Jolla, California 92037
and
University of Göteborg, Sweden

# PERMISSIBLE BOUNDS ON THE COEFFICIENTS OF GENERALIZED POLYNOMIALS

M. v. Golitschek

There has been some recent interest on permissible bounds on the coefficients of approximating polynomials (cf. [4,5,7]). The central problem being investigated here is stated as follows: How can the coefficients $a_k$ of the polynomial $\Sigma a_k x^k$ be restricted such that these polynomials form a dense subset of $C[0,1]$ or $C[a,b]$, $0<a<b$? Whereas the notion of (generalized) Bernstein polynomials plays a critical role in the papers [4,5], I shall use different methods. The following theorem improves the corresponding results of the three papers mentioned above.

Theorem 1. Let $\{m_q\}$ be a sequence of positive integers for which

$$(1) \quad 0 < m_1 < m_2 < \dots \text{ and } \sum_{q=1}^{\infty} 1/m_q = \infty .$$

If $\{w_q\}$ is a sequence of nonnegative real numbers for which $\lim w_q = \infty$ as $q \to \infty$, then for each function $f \in C[0,1]$, $f(0)=0$, and each real number $\varepsilon > 0$ there exists a polynomial $P(x) = \Sigma_{q=1}^{s} c_q x^{m_q}$ such that

$$(2) \quad \|f-P\|_{C[0,1]} < \varepsilon , \quad |c_q| < (\varepsilon w_q)^{m_q} \text{ for } q=1,2,\dots,s.$$

The proof of Theorem 1 is based upon three lemmas.

Lemma 1. Any given real numbers $c_j$, $j=r,\dots,s$, satisfy the inequality

$$(3) \quad |c_j| \le \sqrt{2m_j+1}\, \{\phi(r,s)\}^{1+2m_j} e^{3m_j/2} \left\| \sum_{q=r}^{s} c_q x^{m_q} \right\|_{C[0,1]} ,$$

where $\phi(r,s)$ is defined by $\phi(r,s) := \exp\,(\Sigma_{q=r}^{s} 1/m_q)$.

Proof: Let $\Pi_j := \sqrt{2m_j+1} \prod_{q=r,q\neq j}^{s} \frac{m_j+m_q+1}{|m_j-m_q|}$.

Using the identity

$$(4)\quad \inf_a \|x^{m_j} - \sum_{q=r,q\neq j}^{s} a_q x^{m_q}\|_{L_2(0,1)} = (\Pi_j)^{-1}$$

we obtain

$$(5)\quad |c_j| \leq \Pi_j \|\sum_{q=r}^{s} c_q x^{m_q}\|_{L_2(0,1)} \leq \Pi_j \|\sum_{q=r}^{s} c_q x^{m_q}\|_{C[0,1]}.$$

Estimating the product $\Pi_j$ in (5) completes the proof.

Lemma 2. Let i, r, and s be positive integers such that $0<i<m_r\leq m_s$. Then there exists a polynomial $Q_{is}$, $Q_{is}(x) = \Sigma_{q=r}^{s} c_{qis} x^{m_q}$, for which the inequality

$$(6)\quad \|x^i - Q_{is}(x)\|_{C[0,1]} \leq \prod_{q=r}^{s} \frac{m_q - i}{m_q + i} \leq \{\phi(r,s)\}^{-2i}$$

is satisfied.

Lemma 3. There exists a real number $C_o$ such that for each function $f \in C[0,1]$, $f(0) = 0$, there is a sequence of polynomials $R_n(x) = \Sigma_{i=1}^{n} a_{in} x^i$ for which

$$(7)\quad \|f-R_n\|_{C[0,1]} \leq C_o\omega(f;1/n), \quad n=1,2,\ldots,$$

$$(8)\quad |a_{in}| \leq C_o n^i \omega(f;1/n)/i!, \quad i=1,\ldots,n,$$

where $\omega(f;\cdot)$ is the modulus of continuity of f.

The preceding Lemmas 2 and 3 played an important role in my earlier papers on theorems of the Jackson-Müntz type (cf. [2]) and were proved there.

Proof of Theorem 1: For each pair (n,r) of positive integers, $2\leq n<m_r$, we can find a positive integer s, $s \geq r$, such that

(9) $\{\phi(r,s)\}^2 \le n \le 2\{\phi(r,s)\}^2$ .

For the polynomials $Q_{is}$ and $R_r$ of Lemma 2 and 3 we define the polynomial $T_s := \Sigma_{i=1}^{n} a_{in} Q_{is}$. $T_s$ is a polynomial in $x^{m_q}$, q=r, ...,s. Applying (6)-(9) we obtain the inequality

(10) $\|f-T_s\|_{C[0,1]} \le C\omega(f;1/n)$, $C:= e^2 C_o$ .

With the help of (3), (9), and (10), we deduce upper bounds for the coefficients of $T_s$:

(11) $|c_j| \le \sqrt{(2m_j+1)n}\,(ne^{3/2})^{m_j}\,(1+C)\|f\|_{C[0,1]}$, j=r,...,s.

Let $\varepsilon > 0$. We choose $n \in N$ such that $C\omega(f;1/n) < \varepsilon$. Then we choose $r \in N$ such that $m_r > n$ and such that the right half of the inequality (11) is less than $(\varepsilon w_j)^{m_j}$ for each $j \in N$, $j \ge r$. This choice of r is possible since $\lim w_j = \infty$. Finally, we select an integer s which satisfies (9). Then the polynomial $T_s$ constructed above has the property (2).

Theorem 1 seems to be the best possible in a certain sense, since the following "converse" theorem is well-known (cf. [1, 3,5,6]).

<u>Theorem 2</u>. Let $\Sigma_{q=1}^{\infty} 1/m_q < \infty$ Let $P_n(x) = \Sigma_{k=1}^{n} a_{kn} x^k$, n=1,2, .., be a sequence of polynomials satisfying

$$|a_{kn}| \le A^k \quad \text{for } k=1,2,\ldots,n;\ k \notin \{m_q\};\ n=1,2,\ldots,$$

where $A \ge 1$ is some constant. If the sequence $\{P_n\}$, n=1,..., converges to $f \in C[0,1]$ uniformly on [0,1], then f is analytic on the subinterval [0,1/A).

In the second part we consider the corresponding problem for intervals [a,b], 0<a<b. The next theorem shows that in this case the results are quite different.

<u>Theorem 3</u>. Let the sequence $\{m_q\}$ of positive integers satisfy (1). If 0<a<b, then for each function $F \in C[a,b]$, F(a) = 0, and each positive real number $\eta$ there exists a sequence of

polynomials $T_s(x) = \Sigma_{q=1}^{s} c_{qs} x^{m_q}$, s=1,2,..., such that

$$(12) \quad \lim_{s\to\infty} \|F-T_s\|_{C[a,b]} = 0$$

and

$$(13) \quad |c_{qs}| \le \|F\|_{C[a,b]} \left(\frac{1+\eta}{a}\right)^{m_q}, \quad q=1,\dots,s; \; s=1,2,\dots$$

Proof: The general proof of Theorem 3 and further results on permissible bounds on the coefficients of approximating polynomials will be published elsewhere. Here we shall only prove the important case $m_q := q$, q=1,2,...: For $\eta > 0$ and $s \in N$ we define $k \in N$ by $(2e)^{1/k} \le 1+\eta < (2e)^{1/(k-1)}$ and $m \in N$ by $m \le s/k < m+1$. Let $f \in C[0,1]$ be given by

$$f(x) := \begin{cases} F(bx^{1/k}), & R \le x \le 1, \\ 0 \qquad , & 0 \le x \le R, \end{cases}$$

where $R := (a/b)^k$. Then the Bernstein polynomial $B_m$ of f has the following properties:

$$(14) \quad B_m(f;x) = \sum_{i=0}^{m} b_{im} x^i := \sum_{j=0}^{m} \binom{m}{j} f(j/m) x^j (1-x)^{m-j} \; ;$$

$$(15) \quad \|f-B_m(f;\cdot)\|_{C[0,1]} \le \frac{5}{4}\,\omega(f;m^{-1/2}) \; .$$

From (14) it follows that $b_{im} = 0$ for $0 \le i \le Rm$ and

$$(16) \quad b_{im} = \binom{m}{i} \sum_{j=0}^{i} (-1)^{i-j} \binom{i}{j} f(j/m), \quad Rm < i \le m.$$

Since $\binom{m}{i} \le (em/i)^i$, we obtain from (16) for $Rm < i \le m$

$$|b_{im}| \le \binom{m}{i} 2^i \|f\|_{C[0,1]} \le (2e/R)^i \; \|F\|_{C[a,b]} \quad .$$

The polynomial $T_s(x) := B_m(f;(x/b)^k)$ satisfies the required conditions (12) and (13) for $m_q := q$, q=1,2,....

Theorem 3 leads us to a surprising result:

Corollary. Let $1<a<b$. Let the sequence $\{m_q\}$ of positive

integers satisfy (1). Then the subclass $\Omega$ of polynomials, defined by

$$\Omega := \left\{ \sum_{q=1}^{s} c_q x^{m_q} \;\middle|\; |c_q| \leq 1,\ q=1,\ldots,s;\ s=1,2,\ldots \right\},$$

is dense in C[a,b].

## References


[1] Clarkson, J. A. and P. Erdős, Approximation by polynomials. Duke Math. J. 10 (1943), 5-11.

[2] v. Golitschek, M., Erweiterung der Approximationssätze von Jackson im Sinne von Ch. Müntz. J. Approximation Theory 3 (1970), 72-86.

[3] v. Golitschek, M., Die Größenordnung der Ableitungen und der Koeffizienten algebraischer Approximationspolynome, Habilitationsschrift, Würzburg 1972.

[4] Roulier, J. A., Permissible bounds on the coefficients of approximating polynomials. J. Approximation Theory 3 (1970), 117-122.

[5] Roulier, J. A., Restrictions on the coefficients of approximating polynomials. J. Approximation Theory 6 (1972), 276-282.

[6] Schwartz, L., "Etude des sommes d'exponentielles," Actualités scientifiques et industrielles 959, Paris 1959.

[7] Stafney, J. D., A permissible restriction on the coefficients in uniform polynomial approximation to C[0,1]. Duke Math. J. 34 (1967), 393-396.



Institut für Angewandte Mathematik
Universität Würzburg
8700 Würzburg
West Germany

# BOHR TYPE INEQUALITIES FOR FOURIER EXPANSIONS IN BANACH SPACES

E. Görlich

The well-known Bohr inequality states that

$$(1)\qquad \|f\|_{X_{2\pi}} \leq A\, n^{-r}\, \|f^{(r)}\|_{X_{2\pi}} \qquad (r,n\in N)$$

for each $f\in X_{2\pi}$ for which $f',\ldots,f^{(r-1)}$ is absolutely continuous and $f^{(r)}\in X_{2\pi}$ provided the Fourier coefficients $f^\wedge(k) = (1/2\pi)\int_{-\pi}^{\pi} f(x)e^{-ikx}dx\ (k\in Z)$ vanish for $|f|<n$. Here $X_{2\pi}$ is one of the spaces $L^p_{2\pi}$, $1\leq p<\infty$, and $C_{2\pi}$ of $2\pi$-periodic functions with their usual norms, $A$ a constant, $Z$ the set of all integers, and $N$ denotes the naturals Several variants of (1) have recently been proved by L. Landberg [6], see also his references to a number of earlier papers. Other related contributions were given by J. Favard (1936), B. Lewitan (1937), B.Sz.-Nagy and A. Strausz (1938), Th. Bang (1941), G. Freud and J. Szabados (1971), and V. V. Žuk (1971).

In this note we indicate several further variants of Bohr's inequality which can easily be obtained in the setting of Fourier expansions in Banach spaces as outlined in Ch. 4 of P. L. Butzer's paper [2] in these proceedings (cf. also the literature cited there). Using the terminology of the latter paper we denote by $X$ a Banach space, $[X]$ the Banach algebra of all bounded linear operators of $X$ into itself, $P$ the set of all non-negative integers, $\{P_k\}_{k\in P} \subset [X]$ a total system of mutually orthogonal projections, $s$ the set of all sequences of scalars, and $M=M(X;\{P_k\})$ the set of all multiplier sequences for $X$, i.e., of sequences $\gamma=\{\gamma_k\}_{k\in P} \in s$ such that for each $f\in X$

there exists $f^\gamma \in X$ satisfying $\gamma_k P_k f = P_k f^\gamma$ for all $k \in P$. Moreover, $T \in [X]$ is called a multiplier operator if there exists $\gamma \in M$ such that $Tf = f^\gamma$ for each $f \in X$. The set $[X]_M \subset [X]$ of all such operators is isometrically isomorphic to M if the latter is equipped with the norm

(2) $\|\gamma\|_M = \sup\{\|f^\gamma\| \; ; \; f \in X, \; \|f\| \leq 1\}$.

For a given $\phi \in s$ let $X^\phi$ denote the set of all $f \in X$ for which there exists $f^\phi \in X$ such that $\phi_k P_k f = P_k f^\phi$ for all $k \in P$, and let $B^\phi$ be the closed linear operator on $X^\phi$ into X defined by $B^\phi f = f^\phi$. Then the proof of a Bohr type inequality may clearly be reduced to the verification of the multiplier condition (3) of the following

Theorem. Let $\phi \in s$ with $\phi_k > 0$ for all $k \in N$ and let there exist a non-negative function $\Omega(x)$, a family of sequences $\{\beta(n)\}_{n \in N} \subset s$ and a constant $A > 0$ such that $\beta_k(n) = \phi_k$ for $k \geq n$ and

(3) $\|\Omega(n)/\beta(n)\|_M \leq A$

uniformly for all $n \in N$. Then there holds the Bohr type inequality

(4) $\|f\| \leq A \Omega^{-1}(n) \; \|B^\phi f\|$

for each $f \in X^\phi / \oplus_{k=0}^{n-1} P_k(X)$ and $n \in N$.

Indeed, let $U_n \in [X]_M$ be the operator corresponding to the sequence $\{\beta_k^{-1}(n)\}_{k \in P}$. Then for each $f \in X^\phi / \oplus_{k=0}^{n-1} P_k(X)$ (i.e., for each $f \in X^\phi$ for which $P_k f = 0$ for $0 \leq k \leq n-1$) one has $P_k(U_n B^\phi f) = \beta_k^{-1}(n) \phi_k P_k f = P_k f$ for each $k \in P$, and hence $U_n(B^\phi f) = f$. This implies (4) since $\|U_n\|_{[X]} = \|\beta^{-1}(n)\|_M \leq A \Omega^{-1}(n)$ by (3).

In order to discuss (3) we proceed as in [5,I], cf. [2, Ch. 4]. Suppose that the Cesàro means of some fixed order $\alpha \in P$ of the Fourier series $\Sigma_{k=0}^{\infty} P_k f$ of f, defined by

$$(C,\alpha)_n f = (A_n^\alpha)^{-1} \sum_{k=0}^{n} A_{n-k}^\alpha P_k f, \quad A_n^\alpha = \binom{n+\alpha}{n} \qquad (n \in P)$$

are uniformly bounded, i.e., there exists a constant $C_\alpha$ such that

(5) $\|(C,\alpha)_n f\| \leq C_\alpha \|f\| \qquad (f \in X,\ n \in P).$

Then a sufficient condition for a family of sequences $\gamma(n) \in s$ to belong to M, uniformly with respect to $n \in N$, is the existence of functions $g_n$ and a constant B such that

(6) $\gamma_k(n) = g_n(k),\ g_n \in BV_{\alpha+1},\ \int_0^\infty x^\alpha |dg^{(\alpha)}(x)| \leq B < \infty$

uniformly for n where $BV_{\alpha+1}$ is the class of all bounded continuous functions f on $[0,\infty)$ for which $f, f', \ldots, f^{(\alpha-1)}$ are locally absolutely continuous and $f^{(\alpha)}$ is locally of bounded variation such that $\int_0^\infty x^\alpha |df^{(\alpha)}(x)| < \infty$. In particular, in case $\alpha=1$ condition (6) is satisfied for each family of non-negative sequences $\gamma(n)$ for which $\gamma_k(n) = g_n(k)$ with $g_n''(x) \geq 0$ on $(0,\infty)$ for $n \in N$ (thus $g_n(x)$ is convex), $\lim_{x\to\infty} g_n(x) = 0$, and $g_n(0) = O(1)$. This already implies that condition (3) can be verified in case of $(\omega > 0)$

(7) (i) $\phi = \{k^\omega\}$, (ii) $\phi = \{\log(1+k^2)^{1/2}\}$.

Indeed, set $\gamma(n) = \Omega(n)/\beta(n)$, $\Omega(n) = \phi_n$, $\beta_k(n) = \phi_k$ for $k \geq n$, and define $\gamma_k(n) = 1+(n-k)(\gamma_n - \gamma_{n+1})$ for $0 \leq k < n$. Then it is immediate that there are functions $g_n(x)$ having the required properties, in particular satisfying $g_n(0) = O(1)$ $(n\to\infty)$. Hence one has

<u>Corollary 1</u>. Let the system $\{P_k\}$ satisfy (5) for $\alpha = 1$, and let $\phi$ be one of the sequences (7). Then for each $f \in X^\phi / \oplus_{k=0}^{n-1} P_k(X)$

$$\|f\| \leq A\, \phi_n^{-1} \|B^\phi f\| \qquad (n \in N).$$

As an example, in the space $X_{2\pi}$ define the system $\{P_k\}_{k\in P}$ by

$$(P_0 f)(x) = f^\wedge(0),\quad (P_k f)(x) = f^\wedge(k)e^{ikx} + f^\wedge(-k)e^{-ikx} \qquad (k \in N).$$

Moreover, let $D_\omega$ denote the $\omega^{th}$ Riesz derivative of $f \in X_{2\pi}^\phi$ for

for $\phi = \{k^{\omega}\}$ (cf. [3, Sec. 11.5] for the definition). Then Corollary 1 gives

(8) $$\|f\| \leq A\ n^{-\omega}\|D_{\omega}f\| \qquad (n\in N)$$

for each $\omega>0$ and f such that $D_{\omega}f\in X_{2\pi}$ and $f^{\wedge}(k) = 0$ for all $|k|<n$. For the case $\phi = \{\log\ (1+k^2)^{1/2}\}$ we use the "logarithmic" derivative $D_L$ satisfying $P_k(D_Lf) = [\log\ (1+k^2)^{1/2}]\ P_kf$ for each $k\in P$ and having the representation (see [4], [2, Ch. II]).

$$(D_Lf)(x) = \frac{1}{2} \lim_{\varepsilon\to 0+} \int_{\varepsilon}^{\infty} \frac{f(x+t) - 2f(x) + f(x-t)}{te^t}\, dt ,$$

the limit existing in $X_{2\pi}$ iff $f\in X^{\phi}$. Hence $f\in X^{\phi}/\oplus_{k=0}^{n-1} P_k(X)$ implies

$$\|f\| \leq A\ [\log\ (1+n^2)^{1/2}]^{-1}\ \|D_Lf\|, \qquad (n\in N).$$

It should however be noted that the present approach does not yield Bohr type inequalities for rapidly increasing sequences such as $\phi_k = e^{\beta|k|}, \beta>0$, except for the trivial case when (5) is valid for $\alpha = 0$, i.e., $X_{2\pi} = L^p_{2\pi}$, $1<p<\infty$, in the present instance. Nevertheless, such a Bohr type inequality certainly holds in view of Bernstein's classical approximation theorem for functions analytic in a strip. Moreover, (8) coincides with (1) only in case of even values of r. The case of odd values is not covered for arbitrary $X_{2\pi}$-spaces.

Further examples may be treated as in [5, II], e.g., using expansions according to Jacobi polynomials or Bessel, Laguerre, Hermite, Haar and Walsh functions, spherical harmonics, etc. Also many other sequences $\phi$ can be considered. In the case when (5) holds for some $\alpha \neq 1$ a suitable choice of the family $\{\beta_n\}$ can be obtained by constructing a sufficiently smooth extension of $\{\phi_k^{-1}\}_{k=n}^{\infty}$ to the whole of P, see [7, p. 52].

Finally we note that Corollary 1 immediately implies a Jackson type inequality for the order of best approximation $E_n(f)$ of a function $f\in X^{\phi}$ where

$$E_n(f) = \inf\{\|f - p_n\|;\ p_n \in \oplus_{k=0}^{n-1} P_k(X)\}.$$

Corollary 2. Under the assumptions of Corollary 1 one has

$$E_n(f) = O(\phi_n^{-1}) \qquad (n\to\infty)$$

for each $f \in X^{\phi}$, $\phi$ being given by (7).

This follows by a familiar argument using de La Vallée Poussin's delayed means, see e.g., [1]. Clearly the case when (5) is satisfied for an integer $\alpha > 1$ is admissible here, too.

## References


[1] Achieser, N. and B. Lewitan, Über eine Anwendung der Ungleichung von H. Bohr und J. Favard. C. R. (Doklady) Acad. Sci. URSS 14 (1937), 419-421.

[2] Butzer, P. L., A survey of work on approximation at Aachen, 1968-72. These Proceedings.

[3] Butzer, P. L. and R. J. Nessel, Fourier Analysis and Approximation, Vol. I. Birkhäuser, Basel, and Academic Press, New York, 1971.

[4] Görlich, E., Logarithmic and exponential variants of Bernstein's inequality and generalized derivatives. In: Linear Operators and Approximation, P. L. Butzer, J. P. Kahane and B. Sz.-Nagy, eds., ISNM 20, Birkhäuser, Basel, 1972, pp. 325-337.

[5] Görlich, E., R. J. Nessel and W. Trebels, Bernstein-type inequalities for families of multiplier operators in Banach spaces with Cesàro decompositions, I: General theory, II: Applications. Acta Sci. Math. (Szeged). To appear.

[6] Landberg, L., Some variants of Bohr's inequality. Report No. 1971-17, Chalmers Inst. of Technology and the University of Göteborg 1971.

[7] Trebels, W., Multipliers for (C,α)-bounded Fourier expansions in Banach spaces and approximation theory. Habilitationsschrift, Techn. Univ. Aachen, 1973, 103 pp.


Lehrstuhl A für Mathematik
RWTH Aachen
5100 Aachen, Templergraben 55
West Germany

# THE SPLINE INTERPOLATION OF SEQUENCES SATISFYING A LINEAR RECURRENCE RELATION

T. N. E. Greville, I. J. Schoenberg and A. Sharma

## Abstract

Let

(1) $y = (y_j)$,

where j ranges over all the integers, be a sequence of reals satisfying a recurrence relation

(2) $\sum_{\nu=0}^{k} a_\nu y_{\nu+j} = 0 \qquad (a_0 a_k \neq 0,\ a_\nu \text{ real}).$

By a cardinal spline function we mean one whose knots are at the integers. Given a natural number n, we ask under what conditions there is a unique cardinal spline $S_n(x)$ of degree n that interpolates the sequence (1), i.e.,

(3) $S_n(j) = y_j$ <u>for all</u> j,

and also satisfies the functional equation

(4) $\sum_{\nu=0}^{k} a_\nu S_n(\nu + x) = 0$ <u>for all real</u> x.

The answer depends on the zeros of two polynomials. The first is the Euler-Frobenius polynomial of degree n - 1,

$$\Pi_n(x) \equiv n! \sum_{\nu=0}^{n-1} Q_{n+1}(\nu + 1)x^\nu,$$

where $Q_{n+1}(x)$ denotes the cardinal B-spline of degree n with the origin taken at the leftmost knot, given by

$$Q_{n+1}(x) = \frac{1}{n!} \Delta^{n+1} x_+^n,$$

in which

$$x_+ = \max(0,x).$$

The zeros of $\Pi_n(x)$ are known to be distinct and negative.

The second polynomial is the characteristic polynomial of the recurrence relation (2), namely

$$P(x) = \sum_{\nu=0}^{k} a_\nu x^\nu.$$

It is shown that there is a unique cardinal spline of degree n satisfying (3) and (4) if and only if $\Pi_n(x)$ and P(x) have no common zero.

Let the coefficients $a_\nu$ be normalized so that P(1) = 1, and let E be the "displacement operator" of the calculus of finite differences defined by

$$Ef(x) = f(x + 1).$$

Then, there exists a unique polynomial $p_n(x)$ of degree n, such that

$$P(E)\, p_n(x) = x^n.$$

Let $\mathcal{S}_{n,P}$ denote the class of cardinal splines of degree n satisfying (4). Then there is a unique element $S_{n,P}(x)$ of $\mathcal{S}_{n,P}$ such that

$$S_{n,P}(x) = p_n(x) \qquad (x \in [0,k]).$$

Conversely, any element of $\mathcal{S}_{n,P}$ having the property that on [0,k] it reduces to a simple polynomial in x (in other words, the knots expected at 1, 2, ..., k-1 are absent), is a constant multiple of $S_{n,P}(x)$. Moreover, the k cardinal splines $S_{n,P}(x+\nu)$ $(\nu=0,1,\ldots,k-1)$ form a basis for $\mathcal{S}_{n,P}$.

National Center for Health Statistics
5600 Fishers Lane
Rockville, Maryland 20852

Mathematics Research Center
University of Wisconsin
Madison, Wisconsin 53706

Department of Mathematics
University of Alberta
Edmonton, Alberta
Canada

# SOME SPACES WHERE BEST UNIFORM APPROXIMATION ALWAYS FAILS

Alfred P. Hallstrom

Let $X$ be a compact Hausdorff space and $E$ a uniformly closed subspace of $C(X)$ such that $E$ separates the points of $X$ and $1 \in E$. Suppose moreover that any nonzero measure $\mu$ which annihilates $E$ has the Shilov boundary, $\partial_E$, of $E$ for its closed support. If $f \in C(X)$, $f \notin E$, some point $x \in \partial_E$ is not in the Choquet boundary of the uniformly closed subspace, $[E,f]$, spanned by $E$ and $f$, and $\partial_{[E,f]} = \partial_E$, then there is no best uniform approximation to $f$ from $E$. For suppose $g \in E$ is a best approximation to $f$, i.e. $\|g-f\| = \inf_{h \in E} \|h-f\| = d$. By the Hahn-Banach and Riesz Representation theorems there exists a measure $\mu$ such that $\mu \perp E$, $\mu(f) = d$ and $\|\mu\| = 1$. Then.

$$d = \int f \, du = \int f-g \, du \leq \int |f-g| d|\mu| = d$$

so $|f-g| = d$ a.e. $\mu$ and, in particular, $|f-g| = d$ on $\partial_E$. Now since $p$ is not in the Choquet boundary of $[E,f]$ there exists a representing measure $\nu$ for $[E,f]$ at $p$ with $\nu(\{p\}) = 0$. Then $\partial_E = \overline{\text{supp } \nu}$ else $\nu - \delta_p$ would be an annihilating measure for $E$ whose closed support does not contain $\partial_E$. We may assume $(f-g)(p) = d$. Then

$$d = \int (f-g) d\nu \leq \int |f-g| d\nu = \int d \, d\nu = d$$

so $f-g = d$ a.e. $\nu$, and hence $f-g = d$ on $\partial_E$. But then $\mu(f) = \mu(f-g) = \mu(d) = 0$, a contradiction.

It is not hard to find examples which satisfy the above conditions. Let $X$ be a compact plane set such that the area of the boundary of $X$, $m(\partial X)$, is zero. Designate by $R(X)$ the uniform closure of the rational functions with poles off X and by A(X) those functions in C(X) which are analytic in the interior $X^o$ of X. Let $\mu$ be a measure on X and define

$$\hat{\mu}(z) = \frac{1}{2\pi i} \int_X \frac{d\mu(\zeta)}{\zeta - z} .$$

Then it is well-known that $\mu = 0$ iff $\hat{\mu} = 0$ a.e. m. Now if $0 \neq \mu \perp R(X)$ then $\hat{\mu} = 0$ off X and is analytic off $\overline{\text{supp}\ \mu}$ so that if $X^o$ is connected and dense in X, then $\overline{\text{supp}\ \mu} = \partial X = \partial_{R(X)} = \partial_{A(X)}$.

Now given any compact plane set Y such that $m(\partial Y) = 0$ and $A(Y) \neq R(Y)$, it is possible to construct a compact plane set X by enlarging Y a little so that $m(\partial X) = 0$, $A(X) \neq R(X)$, $X^o$ is connected and dense in X and there exists a point $p \in \partial X$ which is not in the Choquet boundary of A(X). Then no function in A(X) which is not in R(X) has a best uniform approximation from R(X).

Suitable X's as well as details for most of the above assertions may be found in [3].

In [4] Hintzman showed that best uniform approximations exist for certain functions in $C(\Delta = \{|z| \leq 1\})$ from $R(\Delta)$. This does not contradict the above since each point in $\partial_{R(\Delta)}$ is in the Choquet boundary of $R(\Delta)$. It prompts one to ask, however, whether best uniform approximation can always fail from E to F, $E \subset F \subset C(X)$ and $F \neq C(X)$ if the Choquet boundary of $F = \partial_F = \partial_E$. We suspect there are examples for which this is so. In fact an example would be given by an example of Davie in which he exhibits a compact plane set X such that $R(X) \neq A(X)$ and each point of X is a peak point for R(X), see [2], if the following conjecture is true.

<u>Conjecture.</u> Best uniform approximation is never possible from R(X) to A(X).

## References

[1] Carleson, L. and S. Jacobs, Best uniform approximation by analytic functions. Ark. Mat. 10 (1972), 219-229.

[2] Davie, A., An example on rational approximation. Bull. London Math. Soc. 2 (1970), 83-86.

[3] Gamelin, T., Uniform Algebras. Prentice-Hall, Englewood Cliffs, New Jersey, 1969.

[4] Hintzman, W., Best uniform approximations via annihilating measures. Bull. Amer. Math. Soc. 76 (1970), 1062-1066.

Department of Mathematics
University of Washington
Seattle, Washington 98195

# ON THE CONSTRUCTION OF MULTIVARIATE SPLINE SYSTEMS

Werner Haussmann and Heinz Josef Münch

An efficient tool to deal with multivariate spline functions and their minimum properties within the framework of functional analysis is the concept of a spline system. This notion is due to Delvos and Schempp [3]. Here we shall discuss certain generalizations of this notion which have been developed partly in [10].

<u>Definition 1.</u>

(i) A quadruple $(X,P,U,H)$ will be called a <u>spline system</u> provided that the following conditions hold:

(S1) $X$ is a (real or complex) vector space, $H$ is a (real resp. complex) prehilbert space with scalar product $(h,\tilde{h}) \mapsto (h|\tilde{h})$.

(S2) $P: X \to X$ is a linear idempotent mapping.

(S3) $U: X \to H$ is a linear mapping.

(S4) The orthogonality relation $\mathrm{Im}\, UP \perp \mathrm{Im}\, UP'$ holds true.

Here 'Im' and later 'Ker' designate the image resp. kernel of the corresponding linear mapping, and $P'$ is the supplementary projection of $P$, i.e., $P' = \mathrm{id} - P$. $\perp$ means the orthogonality defined by the scalar product in $H$.

(ii) A spline system is called <u>unique</u> if we have

(S5) $\mathrm{Ker}\, P \cap \mathrm{Ker}\, U = (0)$.

(iii) A (not necessarily unique) spline system $(X,P,U,H)$ is designated as a <u>topological spline system</u>, if the following property holds:

(S6) X is a locally convex Hausdorff topological vector space, and P as well as U are continuous mappings.

Given (X,P,U,H), $x \in X$, then let $T_P(x) := \{t \in X : Pt = Px\}$. An $s \in X$ is said to be a spline element belonging to x if

$$Ps = Px \qquad (\text{i.e., } s \in T_P(x)),$$

and

$$\|Us\| \leq \|Ut\| \qquad \text{for all } t \in T_P(x).$$

$\|\cdot\|$ designates the canonical norm in H.

Theorem 2.

(i) Let $\mathcal{P} := (X,P,U,H)$ satisfy (S1)-(S3). Then $\mathcal{P}$ is a spline system if and only if Px is a spline element belonging to x for each $x \in X$.

(ii) Given a spline system $\mathcal{P} := (X,P,U,H)$, $x \in X$, $s_o$ a spline element belonging to x, then these minimum properties hold:

(M1) $\|Us_o\| \leq \|Ut\|$ for all $t \in x + \text{Ker } P$,

(M2) $\|U(x-s_o)\| \leq \|U(x-s)\|$ for all $s \in \text{Im } P \oplus (\text{Ker } P \cap \text{Ker } U)$.

(iii) Let $\mathcal{P} := (X,P,U,H)$ be a spline system. Px is the only spline element belonging to $x \in X$ if and only if $\mathcal{P}$ is unique, or equivalently, iff $\text{Ker } U \subset \text{Im } P$.

As to the proof of Theorem 2 we refer to [10]. (M1) and (M2) are generalizations of well known spline interpolation minimum properties (see e.g., de Boor and Lynch [1] or further papers cited in [10]). The uniqueness condition (S5) was considered in the case of Lg-splines by Jerome and Schumaker [12].

We are now going to investigate construction principles to get new spline systems from known ones with special emphasis to the construction of multivariate spline systems and to the extension of the minimum properties (M1) and (M2) to multivariate spline interpolation. These construction principles are

(α) completion of spline systems (cf. [11]),

($\beta$) tensor products of spline systems (cf. [10],[11]).

As to a construction principle using additivity properties we refer to [10].

Let $\mathcal{P} := (X,P,U,H)$ be a topological spline system, $\hat{X}$ and $\hat{H}$ the completions of X resp. H, and $\hat{P}$ and $\hat{U}$ the respective extensions of P and U. Then the following theorem holds true:

Theorem 3. Given the topological spline system $\mathcal{P} := (X,P,U,H)$, then its completion $\hat{\mathcal{P}} := (\hat{X},\hat{P},\hat{U},\hat{H})$ is also a topological spline system.

Proof: Since (S1)-(S3) are immediate we point out the validity of (S4). For any $x \in X$ we have $\hat{U}\hat{P}x = UPx = \widehat{UP}x$. With the aid of $\widehat{UP}(\hat{X}) \subset \overline{UP(X)} = \widehat{\text{Im UP}}$ we have

(1) $\text{Im } \hat{U}\hat{P} \subset \widehat{\text{Im UP}}$.

The mapping

$$\omega : \text{Im UP} \times \text{Im UP}' \ni (x,y) \mapsto (x|y) \in R \text{ (resp. C)}$$

vanishes identically by (S4). Since $\omega$ is uniformly continuous there is one and only one extension $\hat{\omega}$ to $\widehat{\text{Im UP} \times \text{Im UP}'} = \widehat{\text{Im UP}} \times \widehat{\text{Im UP}'}$, i.e., $\hat{\omega} = 0$. Now, (1) yields $\text{Im } \hat{U}\hat{P} \perp \text{Im } \hat{U}\hat{P}'$.

Remark. If a topological spline system $\mathcal{P}$ is unique then its completion $\hat{\mathcal{P}}$ does not need to be unique again (cf. [11]).

Now we consider tensor products of topological spline systems in order to treat multivariate spline problems. For notational convenience we restrict our investigations to the bivariate case. For higher dimensions one can use induction arguments.

In applications there occur three main topologies on the tensor product $X \otimes Y$ of given locally convex Hausdorff topological vector spaces X and Y (cf. Robertson and Robertson [13], Treves [17] and Dixmier [6]): The projective topology $\pi$, the topology $\varepsilon$ of equicontinuous convergence, and in the case when

X and Y are prehilbert spaces, the topology $\alpha$ which is induced by the canonical prehilbert structure on $X \otimes Y$ (see [9,10]). We shall write $X \otimes_\pi Y$, $X \otimes_\varepsilon Y$ etc., and $X \hat{\otimes}_\pi Y$, etc., for the topological tensor products resp. its completions

It is important to consider the completions $X \hat{\otimes}_\tau Y$ ($\tau = \alpha, \varepsilon, \pi$) since in concrete applications the multivariate spline ground spaces are completed tensor products. E.g., given two Hilbert spaces $L^2(I)$ and $L^2(J)$ (I and J compact real intervals), we have $L^2(I \times J) \cong L^2(I) \hat{\otimes}_\alpha L^2(J)$.

<u>Theorem 4</u>. Given two topological spline systems $\mathcal{P} := (X,P,U,H)$ and $\mathcal{Q} := (Y,Q,V,K)$ then the tensor product

$$\mathcal{P} \otimes_\tau \mathcal{Q} := (X \otimes_\tau Y,\ P \otimes Z,\ U \otimes V,\ H \otimes_\alpha K)$$

as well as its completion

$$\mathcal{P} \hat{\otimes}_\tau \mathcal{Q} := (X \hat{\otimes}_\tau Y,\ P \hat{\otimes} Q,\ U \hat{\otimes} V,\ H \hat{\otimes}_\alpha K)$$

are also topological spline systems provided that either $\tau=\pi$ or $\tau=\alpha$ in which latter case X and Y have to be prehilbert spaces.

<u>Proof</u>: By Theorem 3 and [10, Theorem 5] we only have to verify (S6) for $\mathcal{P} \otimes_\tau \mathcal{Q}$. $X \otimes_\pi Y$ resp. $X \otimes_\alpha Y$ and $H \otimes_\alpha K$ are locally convex Hausdorff topological vector spaces resp. prehilbert spaces. The tensor product $P \otimes Q$: $X \otimes_\pi Y \to X \otimes_\pi Y$ of two continuous linear mappings P and Q is continuous again. Since moreover the identity mapping $id_H \otimes id_K$: $H \otimes_\pi K \to H \otimes_\alpha K$ is continuous (the $\pi$-topology is finer than the $\alpha$-topology) it follows that $U \otimes V$ is continuous. Hence $\mathcal{P} \otimes_\pi \mathcal{Q}$ satisfies (S6).

If X and Y are prehilbert spaces then a similar argument shows that $\mathcal{P} \otimes_\alpha \mathcal{Q}$ satisfies (S6).

Theorem 4 guarantees the validity of (M1) and (M2) in the bi- and multivariate case. In general tensor products of unique spline systems fail to be unique again (cf. [10]). If the projections P and Q arise from interpolation conditions then the bivariate spline elements satisfy the corresponding bivariate interpolation conditions (see [10, Section 5]).

In some concrete applications one can use in Theorem 4 also $\tau=\varepsilon$ and not only $\tau=\pi$ or $\alpha$ to provide $X \otimes Y$ with an appropriate topology (cf. [11]).

Applications. The natural spline functions (Greville [8], de Boor and Lynch [1]), the L-splines (Schultz and Varga [16], Delvos and Schempp [4]) and the theory of Sard (Sard [14], Delvos [2], Delvos and Schempp [5]) may be considered in the framework of spline systems. In either case one obtains corresponding bi- and multivariate spline problems and their minimum properties by the method described above. Using an additivity construction principle (see [10, Section 2]) one can show that the spline blended surface interpolation that goes back to Gordon [7] may be dealt with within the framework of spline systems as done by Delvos [2] and in [10].

## References

[1] de Boor, C. R. and R. E. Lynch, On splines and their minimum properties. J. Math. Mech. 15 (1966), 953-969.

[2] Delvos, F. J., Über die Konstruktion von Spline Systemen. Doctoral dissertation, Ruhr-Universität, Bochum, 1972.

[3] Delvos, F. J. and W. Schempp, On spline systems. Monatsh. Math. 74 (1970), 399-409.

[4] Delvos, F. J. and W. Schempp, On spline systems: $L_m$-splines. Math. Z. 126 (1972), 154-170.

[5] Delvos, F. J. and W. Schempp, Sard's method and the theory of spline systems. To appear.

[6] Dixmier, J., Les algèbres d'operateurs dans l'espace Hilbertien. Gauthier-Villars, Paris, 1957.

[7] Gordon, W. J., Spline-blended surface interpolation through curve networks. J. Math. Mech. 18 (1969), 931-952.

[8] Greville, T. N. E., Spline functions, interpolation and numerical quadrature. In Mathematical Methods for Digital Computers, A. Ralston and H. S. Wilf, eds., Vol. 2, Wiley, New York, 1967, 156-168.

[9] Haussmann, W., Zur Theorie der Spline-Systeme. Habilitationsschrift, Ruhr-Universität, Bochum, 1970.

[10] Haussmann, W., On multivariate spline systems. To appear in J. Approx. Theory.

[11] Haussmann, W. and H. J. Münch, Topological spline systems. To appear.

[12] Jerome, J. W. and L. L. Schumaker, On Lg-splines. J. Approx. Theory 2 (1969), 29-49.

[13] Robertson, A. P. and W. Robertson, Topological vector spaces. University Press, Cambridge, 1964.

[14] Sard, A., Optimal approximation. J. Functional Anal. 1 (1967), 222-244, and 2 (1968), 368-369.

[15] Scheffold, E., Das Spline-Problem als ein Approximationsproblem. To appear.

[16] Schultz, M. H. and R. S. Varga, L-splines. Numer. Math. 10 (1967), 345-369.

[17] Treves, F., Topological vector spaces, distributions and kernels. Academic Press, New York-London, 1967.

Institut für Mathematik
Ruhr-Universität
463 Bochum, Germany

Rechenzentrum
Ruhr-Universität
463 Bochum, Germany

# ON THE EXISTENCE OF BEST ANALYTIC APPROXIMATIONS

William Hintzman

It is known that a large class of harmonic functions have unique best uniform analytic approximations to them [2]. In this paper we show that best analytic approximations to harmonic functions do not always exist.

Let D be the closed unit disk. Let E be the normed linear space of all g in C(D) which are harmonic in the interior of D where $\|g\| = \max_D |g(z)|$. Let A be the subspace of E consisting of all functions f which are analytic in the interior of D. For every g in E the Hahn-Banach theorem guarantees the existence of a linear functional L such that

$$\|g\|_A = \inf_{f\in A} \|g - f\| = L(g)$$

where $L(A) = \{0\}$ and $\|L\| = 1$. By the Riesz Representation theorem this L corresponds to a unique regular Borel measure on D with polar decomposition $\phi d\mu$. That is,

$$L(g) = \int_D g\phi d\mu, \ d\mu \geq 0, \ |\phi| = 1 \text{ a.e. } d\mu, \ \int_D d\mu = 1 .$$

Consequently if $f_0$ is a best uniform approximation to g from A, it must satisfy the equation

(1) $$g = \|g\|_A \overline{\phi} + f_0 \quad \text{a.e. } d\mu.$$

Equation (1) was used to construct best approximations [2]. A normal family argument was used to show that there is an $f_0$ in $H^\infty$ which satisfies (1). Also, $f_0$ is unique. For certain harmonic functions g, $\overline{\phi}$ is a continuous function on $\partial D$, the support of $d\mu$, and therefore $f_0$ is also continuous on $\partial D$ which proves $f_0$ is in A.

We shall now construct a function $\overline{\phi}$ which is in $L^\infty$ on $\partial D$ and a function $f_0$ in $H^\infty$ but not in A such that g in (1) is continuous.

Consider the function $f(z) = \Sigma_2^\infty -iz^n/(n\log n)$. On $\partial D$,

$$f(e^{i\theta}) = u(\theta) + iv(\theta) = \sum_2^\infty \frac{\sin n\theta}{n\log n} + i\sum_2^\infty \frac{-\cos n\theta}{n\log n}.$$

The behavior of $u(\theta)$ and $v(\theta)$ is well known [3]. $u(\theta)$ is continuous on $\partial D$ and $v(\theta)$ is continuous on $\partial D - \{1\}$, but as $\theta$ approaches 0, $|v(\theta)|$ approaches $\infty$.

Let

$$(2) \quad g(e^{i\theta}) = e^{-i\theta}e^{-iv} - e^{-(u+iv)} = e^{-iv}(e^{-i\theta} - e^{-u})$$

for $\theta \neq 0$ and $g(1) = 0$. Then $g(e^{i\theta})$ is continuous on $\partial D$ since $e^{-iv}$ is bounded for $\theta \neq 0$ and $(e^{-i\theta} - e^{-u})$ is continuous on $\partial D$ and approaches 0 as $\theta$ approaches 0. Equation (2) shows that for the harmonic extension of g to D, $\|g\|_A \leq 1$ since $e^{-(u+iv)}$ is in $H^\infty$.

On the other hand, the measure $\phi d\mu = (1/2\pi)e^{i\theta}e^{u+iv}d\theta$ on $\partial D$ annihilates A. If $\alpha = (1/2\pi)\int_{\partial D} e^u d\theta$ is the total variation of the measure $d\mu$, then

$$L(\cdot) = \frac{1}{2\pi\alpha}\int_{\partial D} (\cdot)e^{i\theta}e^{u+iv}d\theta$$

annihilates A, has norm 1, and $L(g) = 1$. Therefore $\|g\|_A \geq 1$ by [1]. Hence $\|g\|_A = 1$ and $f_0 = -e^{-(u+iv)}$ is the unique $H^\infty$ function which satisfies (1). Since $f_0$ is not in A there does not exist a best uniform approximation to g from A. Consequently,

<u>Theorem.</u> There exist harmonic functions for which best analytic approximation in the uniform norm do not exist.

## References

[1] Buck, R. C., Applications of duality in approximation theory. Proc. Sympos. Approximation of Functions (General Motors Res. Lab., 1964) Elsevier, Amsterdam, 1965, pp. 27-42. MR 33 #4554.

[2] Hintzman, W., Best uniform approximations via annihilating measures. Bull. Amer. Math Soc. 76 (1970), 1062-1066. MR 41 #7352.

[3] Zygmund, A., "Trigonometric Series," 2nd Ed., Cambridge University Press, New York, 1959, vol. 1, p. 253.

Department of Mathematics
California State University, San Diego
San Diego, California 92115

# EXISTENCE OF BEST APPROXIMATIONS BY EXPONENTIAL SUMS IN SEVERAL INDEPENDENT VARIABLES

David W. Kammler

## 1. Introduction

Let the complex valued function $y(t)$ and its partial derivatives of all orders be defined and continuous for all $t = (t_1,\dots,t_m) \in R^m$ and let $L[y]$ denote the corresponding linear space (with complex scalars) which is generated by the functions

$$[D_1^{i_1} \dots D_m^{i_m}]y(t), \qquad i_1,\dots,i_m = 0,1,\dots$$

where

$$D_i = \frac{\partial}{\partial t_i}, \qquad i = 1,\dots,m.$$

If $L[y]$ has finite dimension n we say that y is an exponential sum with order n. For such an exponential sum y we know that $D_i^n y$ can be written as a linear combination of $y, D_i y, \dots, D_i^{n-1} y$, $i=1,\dots,m$, so that y satisfies some system of partial differential equations of the form

$$(1) \quad [(D_i - \lambda_{i1}) \dots (D_i - \lambda_{in})]\, y(t) = 0, \qquad i=1,\dots,m.$$

Given a set $S \subseteq C$ we define $V_n(S)$ to be the set of those exponential sums y with order at most n which satisfy some system (1) for which

$$\lambda_{ij} \in S, \qquad i=1,\dots,m, \qquad j=1,\dots,n.$$

Our main result is the following m dimensional generalization of Theorem 2 from [1].

Theorem. Let $\mathcal{D}$ be a bounded nonvoid open subset of $R^m$, let

$1 \leq p \leq \infty$, let S be a closed subset of $C$, and let n be a positive integer. Then every $f \in L_p(\mathcal{D})$ has a best $\| \|_p$-approximation from $V_n(S)$.

## 2. Three Basic Definitions

Essentially the same proof which is used in [1] for the special case where m=1 and $\mathcal{D}$ is an interval can be used in the present situation provided that in addition to the above definition of $V_n(S)$ we formulate appropriate extensions of the parametrization $y_n(b,\lambda)$, of the seminorms $\| \|_{p,\alpha}$, and of the concepts of U,V,W-sequences which are introduced in [1].

In general, any solution of the system (1) is an exponential sum of order at most $n^m$, and we may specify a particular solution of (1) by assigning the $n^m$ initial conditions

$$(2)\quad [D_1^{i_1} \ldots D_m^{i_m}]y(0) = b_{i_1 \ldots i_m},\ 0 \leq i_1, \ldots, i_m \leq n-1.$$

We denote this unique solution by $y_n(b,\lambda)$ where $b,\lambda$ represent the parameters $b_{i_1 \ldots i_m}$ and $\lambda_{ij}$ appearing in (2) and (1), respectively. Thus each $y \in V_n(C)$ can be parametrized in the form $y = y_n(b,\lambda)$ for some choice of $b,\lambda$ although this parametrization is not necessarily unique.

For each $\alpha \geq 0$ we define

$$\mathcal{D}_\alpha = \{t \in \mathcal{D}: d(t, R^m \backslash \mathcal{D}) > \alpha\}$$

where d is the usual Euclidean distance function, and we define the seminorm $\| \|_{p,\alpha}$ on $L_p(\mathcal{D})$ such that

$$\|f\|_{p,\alpha} = \|f \cdot \chi_\alpha\|_p$$

where $\chi_\alpha$ is the characteristic function of $\mathcal{D}_\alpha$. Since $\mathcal{D}$ has nonvoid interior, we may select some $\alpha_0 > 0$ such that $\mathcal{D}_{\alpha_0}$ also has nonvoid interior.

Given an exponential sum $y \in V_n(C)$ we define

$$\Lambda_i[y] = \cap \{\lambda_{i1},\ldots,\lambda_{in}\}, \qquad i=1,\ldots,m,$$

with the intersections being taken over all possible choices of the exponential parameters $\lambda_{ij}$ for which (1) holds. A sequence of exponential sums, $\{y_\nu\}$, from $V_n(C)$ will be called a U-sequence, a V-sequence, or a W-sequence according as the corresponding spectral sets $\Lambda_i[y_\nu]$ satisfy the respective conditions

$$\max_i \liminf_\nu \{|\mathrm{Re}\lambda| : \lambda \in \Lambda_i[y_\nu]\} = +\infty,$$

$$\max_i \sup \{|\lambda| : \lambda \in \bigcup_{\nu=1}^{\infty} \Lambda_i[y_\nu]\} < +\infty,$$

or both of

$$\max_i \liminf_\nu \{|\mathrm{Im}\lambda| : \lambda \in \Lambda_i[y_\nu]\} = +\infty$$

$$\max_i \sup \{|\mathrm{Re}\lambda| : \lambda \in \bigcup_{\nu=1}^{\infty} \Lambda_i[y_\nu]\} < +\infty.$$

## 3. Fundamental Properties of U,V,W-Sequences

The following lemma presents four basic properties possessed by U,V,W-sequences.

Lemma. Let $\{u_\nu\}$, $\{v_\nu\}$, $\{w_\nu\}$ be U,V,W-sequences, respectively, from $V_n(C)$ and let $1\leq p\leq\infty$.

(i): If $\{u_\nu+v_\nu+w_\nu\}$ is a $\|\ \|_p$-bounded sequence from $V_n(C)$, then the component sequences $\{u_\nu\}$, $\{v_\nu\}$, $\{w_\nu\}$ are individually $\|\ \|_p$-bounded.

(ii): If $\{v_\nu\}$ is $\|\ \|_p$-bounded, then there is some subsequence of $\{v_\nu\}$ which uniformly converges on $\bar{\mathcal{D}}$ to some $v \in V_n(C)$.

(iii): If $\{u_\nu\}$ is $\|\ \|_p$-bounded, then $\{u_\nu\}$ converges uniformly to zero on every compact subset of $\mathcal{D}$.

(iv): If $\{u_\nu + w_\nu\}$ is a sequence from $V_n(C)$ and $f \in L_p(\mathcal{D})$, then

$$\underline{\lim}\ \|f + u_\nu + w_\nu\|_p \geq \|f\|_p.$$

The above lemma may be proved by using arguments analogous to those given in [1] for the special case where m=1. The desired existence theorem can then be established by applying in turn sections (i), (ii), and (iv) of the lemma to a minimizing sequence $y_\nu = u_\nu + v_\nu + w_\nu$ for $\|f-y\|_p$ from $V_n(S)$. The details of the argument will appear elsewhere.

## References

[1] Kammler, D. W., Existence of best approximations by sums of exponentials. J. Approx. Theory 6 (1973). To appear.

Mathematics Department
Southern Illinois University
Carbondale, Illionis 62901

# APPROXIMATION ON CURVES BY LINEAR COMBINATIONS OF EXPONENTIALS

J. Korevaar

## 1. Results and Problems Relative to Müntz-Szász Type Approximation on Curves

Let $p_1 < p_2 < \ldots$ be positive numbers tending to infinity; although not always necessary, we will think of them as positive integers. A classical result of Müntz [11] and Szász [12] implies that the exponentials $\exp(p_n s)$ span the space $C_o[-\infty,0]$ (of continuous functions vanishing at $-\infty$) if and only if the series $\Sigma\ 1/p_n$ diverges. Instead of $[-\infty,0]$, we will consider simple curves $\gamma$ in the complex plane which start at 0 and may extend to infinity in a westerly direction. Depending on $\gamma$ and the numbers $p_n$, the exponentials $\exp(p_n\zeta)$ may or may not span the space $C_o(\gamma)$ (which we interpret as $C(\gamma)$ when $\gamma$ is bounded). For example, if $\gamma$ is such that $e^{\gamma}$ is a Jordan arc, then by a theorem of Walsh [13], the exponentials $e^{n\zeta}$ span.

One has the feeling that, except perhaps in pathological cases, divergence of $\Sigma\ 1/p_n$ is a necessary condition for a spanning set:

<u>Theorem 1.</u> Let $\gamma$ be an arbitrary analytic arc, and let $\Sigma\ 1/p_n$ be convergent. Then the exponentials $\exp(p_n\zeta)$ fail to span $C(\gamma)$.

We are interested in conditions on $\gamma$ under which the divergence of $\Sigma\ 1/p_n$ is precisely the right condition for a spanning set. Thus we want no "verticality" in $\gamma$. Indeed, if $\gamma$ were a vertical segment, one would get into the realm of trigonometric approximation; in that case, some sort of positive density of the sequence $\{p_n\}$ would be required (cf. [7], [2]).

If $\gamma$ would contain two points that are a multiple of $2\pi i$ apart, the exponentials could not span because of their periodicity. Accordingly, our positive result is restricted to curves which "keep moving to the left":

Theorem 2. Let $\gamma$ be a curve starting at 0 and possibly extending to infinity. We require that there be a positive constant $\alpha < \pi/2$ such that the angles between the chords of $\gamma$ and the negative real axis do not exceed $\alpha$. Let $p_1 < p_2 < \ldots$ be positive integers such that the series $\Sigma\, 1/p_n$ diverges. Then the exponentials $\exp(p_n \zeta)$ span $C_o(\gamma)$.

Theorems 1 and 2 were obtained by the author during a year at Imperial College, London (1970-71). Theorem 1 has in the meantime also been announced by Malliavin and Siddiqi [9]. The requirement of analyticity in Theorem 1 can be weakened; it seems likely that continuous differentiability is more than sufficient. It would be important to have precise results on the relation between the nature of $\gamma$ and the degree of non-approximability of certain functions when $\Sigma\, 1/p_n$ converges. Such results would have a bearing on Macintyre's conjecture: A nonzero entire function given by a gap series

$$\Sigma\, a_n z^{p_n}, \text{ with } \Sigma \frac{1}{p_n} \text{ convergent}$$

cannot be bounded on any curve going to infinity [8]. Several mathematicians have been interested in this conjecture; the best known results to date are due to Gaier [5] and Kövari [6].

## 2. Corresponding Results and Problems in Complex Analysis

The approximation problems of Section 1 ask if Laplace transforms

$$f(z) = \mathcal{L} d\mu = \int_\gamma e^{z\zeta} d\mu(\zeta), \text{ with } d\mu \neq 0,\ \mu(\infty) = 0,$$

can have zeros at the points $p_n$. It is of considerable interest to ask more general questions about the distribution of the zeros of such functions.

Let us take $\gamma$ bounded for a moment. Unless the measure $d\mu$ is concentrated at a point, the entire function f(z) of exponential type must have "many" zeros: in particular, it cannot be of convergence type (if it were, its indicator diagram would reduce to a point, cf. [3]). However, are there angles (depending on $\gamma$) in which f(z) necessarily has "few" zeros $z_n$? "Few" zeros would mean convergence of the series $\Sigma\, 1/|z_n|$; in view of the following theorem, one could not expect "fewer" zeros in any angle.

Theorem 1*. Let $\gamma$ be an arbitrary analytic arc, and let $\{z_n\}$ be any infinite sequence of complex numbers (different from naught) such that $\Sigma\, 1/|z_n|$ converges. Then there is a measure $d\mu \neq 0$ on $\gamma$ whose Laplace transform f(z) vanishes at the points $z_n$.

In special cases, there are angles in which $L d\mu$ must have few zeros. For example, when $\gamma = [-A, 0]$, with $0<A\leq\infty$, the function $f(z) = L d\mu$ will be bounded in the half-plane Re $z>0$. Thus by a theorem of Carleman (cf. [3]), the zeros $z_n$ of f(z) with Re $z_n>0$ satisfy the condition $\Sigma\, \mathrm{Re}(1/z_n)<\infty$. It follows that for the zeros in $|\arg z| \leq \frac{1}{2}\pi - \varepsilon$ one has $\Sigma\, 1/|z_n|<\infty$. There is further evidence (cf. Sec. 3) for the following

Conjectured Theorem 2*. Let $\gamma$ be a bounded or unbounded curve as in Theorem 2, with associated constant $\alpha$. Let $d\mu$ be a non-zero measure on $\gamma$ (with $\mu(\infty)=0$) and f(z) its Laplace transform. Choosing $0 < \varepsilon < \frac{1}{2}\pi - \alpha$, we denote the zeros of f(z) in the angle $|\arg z| \leq \frac{1}{2}\pi - \alpha - \varepsilon$ by $z_1, z_2, \ldots$. Then for the $z_n \neq 0$,

$$\Sigma \frac{1}{|z_n|} < \infty .$$

## 3. Indication of Proofs

It would be very desirable to obtain complex-variable proofs of Theorems 1* and 2*. The approach sketched below is rather complicated (and incomplete). In order to represent functions, obtained by multiplying or dividing Laplace transforms by certain infinite products $\Pi(1 - z^2/z_n^2)$, as new Laplace transforms on $\gamma$, we "operate under the integral sign" and apply infinite order differential operators $\Pi(1 - D^2/z_n^2)$ to functions $\varphi$ on the curve, or we solve infinite order differential equations $\Pi(1 - D^2/z_n^2)\varphi = d\mu$ on $\gamma$. This technique requires the introduction of suitable non-quasianalytic and quasianalytic classes $C\{M_n\}$ on $\gamma$, and appropriate extensions of the Denjoy-Carleman theory [4] to curves. A function $\varphi(\zeta)$ on $\gamma$ will be called differentiable if $D\varphi = \lim \Delta\varphi/\Delta\zeta$ exists along $\gamma$. The class $C\{M_n\}$ on $\gamma$ will consist of those $C^\infty$ functions for which there are constants A and B such that $|D^n\varphi| \leq AB^nM_n$, $n=0,1,2,\ldots$

Proof of Theorem 1*: Let $\{z_n\}$ be any sequence as in that theorem. Forming

$$g(z) = \Pi(1 + z^2/|z_n|^2) = \Sigma\, z^{2k}/A_{2k},$$

the integral $\int_0^\infty \{\log g(r)\}\, dr/r^2$ will converge (it is equal to $2\pi\int_0^\infty (1/t)\, dn(t) = 2\pi\, \Sigma\, 1/|z_n|$). Thus by one form of the Denjoy-Carleman theorem, the least nonincreasing majorant of the series $\Sigma A_{2k}^{-1/2k}$ must converge. In other words, setting $\alpha_{2k} = \inf_{p\geq k} A_{2p}^{1/2p}$, one has $\alpha_{2k}\uparrow$ and $\Sigma\, 1/\alpha_{2k}$ convergent (cf. [10]). We now determine numbers $\beta_{2k}\uparrow$ such that $\beta_{2k}/\alpha_{2k} \to 0$, $\Sigma 1/\beta_{2k}$ converges, and $\beta_{2k+2} \leq 2\beta_{2k}$. Then we define

$$M_{2k}^{1/2k} = M_{2k+1}^{1/(2k+1)} = \beta_{2k}\ .$$

It follows that the series $\Sigma\, M_{2k}z^{2k}/A_{2k}$ and $\Sigma\, M_{2k+1}z^{2k}/A_{2k}$ have infinite radius of convergence, and that $M_n^{1/n}\uparrow$, $\Sigma\, M_n^{-1/n}$ converges, and $M_{n+1} \leq K^nM_n$.

Suppose now that $\gamma$ is a rectifiable arc which carries a nonzero function $\varphi$ of class $C\{M_n\}$ such that $D^{\nu}\varphi = 0$ at the endpoints of $\gamma$, $\nu=0,1,2,\ldots$. Then one can form the continuous nonzero function

$$\psi = \lim \psi_N, \quad \psi_N = \prod_{n=1}^{N} (1 - D^2/z_n^2)\varphi .$$

Indeed, the functions $\psi_N$ and $D\psi_N$ are majorized by $A \Sigma M_{2k}B^{2k}/A_{2k}$ and $AB\Sigma M_{2k+1}B^{2k}/A_{2k}$, hence the $\psi_N$ form a bounded equicontinuous family. Integration by parts shows that $L\psi_N$ is equal to $\Pi_1^N(1-z^2/z_n^2)L\varphi$. It follows that all convergent subsequences of $\{\psi_N\}$ have the same limit function $\psi$:

$$L\psi = \Pi(1 - z^2/z_n^2)L\varphi .$$

One concludes that $\psi \neq 0$ and that $L\psi$ vanishes at the points $z_n$.

For analytic and certain other arcs $\gamma$ it is possible to construct a function $\varphi$ with the desired properties by starting with a function $\Phi$ of related form in the non-quasianalytic class $C\{M_n\}$ on $[0,1]$ (cf. [4]).

Theorem 1 is an immediate consequence of Theorem 1*.

Partial proof of Theorem 2*: The following argument works in the case $\varepsilon = \frac{1}{4}\pi$ which requires that $\alpha \leq \frac{1}{4}\pi$. It is convenient (and no restriction) to assume that the (locally rectifiable) curve $\gamma$ extends to infinity. Suppose now that $f(z) = Ld\mu$, with $d\mu \neq 0$ and $\mu(\infty) = 0$, vanishes at points $z_n \neq 0$ in $|\arg z| \leq \frac{1}{4}\pi - \alpha$ such that $\Sigma\, 1/|z_n| = \infty$. It may be assumed that $r_n = |z_n| \uparrow$ and that $r_n/n \to \infty$. Taking $\varphi_0 = d\mu$, there exist unique bounded, integrable functions $\varphi_k$ on $\gamma$ such that

$$(1 - D^2/z_k^2)\varphi_k = \varphi_{k-1}, \quad \varphi_k(0) = D\varphi_k(0) = 0,$$

$k=1,2,\ldots$. One can write

$$\varphi_k(\zeta) = \int_{\gamma} \frac{1}{2} z_k \exp\{-z_k(\zeta-\eta)\varepsilon(\zeta-\eta)\}\varphi_{k-1}(\eta)d\eta = \sigma_k * d\mu,$$

where $\varepsilon(\xi) = 1$ for $\mathrm{Re}\,\xi > 0$, $\varepsilon(\xi) = -1$ for $\mathrm{Re}\,\xi < 0$, and $\sigma_k(\xi)$ is analytic and bounded in the angles $|\arg \xi\varepsilon(\xi)| < \alpha$. Integration by parts shows that $L d\mu = \Pi_1^k(1 - z^2/z_n^2) L\varphi_k$, hence $L\sigma_k = 1/\Pi_1^k(1 - z^2/z_n^2)$. The functions

$$\sigma_k(\xi) = \frac{1}{2\pi i} \int_\Gamma \frac{1}{\Pi_1^k(1-z^2/z_n^2)} e^{\xi z} dz$$

(where the path of integration $\Gamma$ may be chosen so as to make $\arg z + \arg \xi = \frac{1}{2}\pi$, mod $\pi$) converge to a function $\sigma(\xi)$ which is analytic and bounded for $|\arg \xi\varepsilon(\xi)| < \alpha$, and such that $L\sigma = 1/\Pi(1-z^2/z_n^2)$. It follows that the functions $\varphi_k = \sigma_k * d\mu$ converge to the limit $\varphi = \sigma * d\mu$ on $\gamma$; the function $\varphi$ will be infinitely differentiable and such that $D^\nu\varphi(0) = 0, \nu=0,1,2,\ldots.$ The formula $\varphi = \sigma * d\mu$ shows also that $\varphi$ is of class $C\{M_n\}$, where

$$M_{2k} = r_2^2 \cdots r_{k+1}^2 = M_{2k-1} r_{k+1} .$$

Since $\rho_n = M_n/M_{n-1} \uparrow$ and $\Sigma\, 1/\rho_n = \infty$, the class $C\{M_n\}$ is quasi-analytic on $\gamma$ (cf. [1]). It follows that $\varphi = 0$, hence $L\varphi = 0$. Thus $L d\mu = 0$ and consequently, $d\mu = 0$. This contradiction shows that $L d\mu$ cannot have so many zeros $z_n$ in $|\arg z| \leq \frac{1}{4}\pi - \alpha$ that $\Sigma 1/|z_n|$ diverges.

The above proof implies Theorem 2 for the case $\alpha \leq \frac{1}{4}\pi$. When $\alpha > \frac{1}{4}\pi$, the proof of Theorem 2 is more complicated; using the present methods, the case $\varepsilon < \frac{1}{4}\pi$ of Theorem 2* seems to require a separation condition on the $z_n$.

## References


[1] Bang, T., On quasi-analytiske funktioner. Thesis, Univ. of Copenhagen, 1946.

[2] Beurling, A. and P. Malliavin, On the closure of characters and the zeros of entire functions. Acta Math. 118 (1967) 79-93.

[3] Boas, R. P. Jr., Entire functions. Acad. Press, New York, 1954.

[4] Carleman, T., Les fonctions quasi-analytiques. Gauthier-

Villars, Paris, 1926.

[5] Gaier, D., Der allgemeine Lückenumkehrsatz für das Borel-Verfahren. Math. Z. 88 (1965) 410-417.

[6] Kövari, T., On the asymptotic paths of entire functions with gap power series. J. Analyse Math. 15 (1965) 281-286.

[7] Levinson, N., Gap and density theorems. Amer. Math. Soc. Colloq. Publ., vol. 26, New York, 1940.

[8] Macintyre, A. J., Asymptotic paths of integral functions with gap power series. Proc. London Math. Soc. 2 (1952) 286-296.

[9] Malliavin, P. and J. A. Siddiqi, Approximation polynomiale sur un arc analytique dans le plan complexe. C. R. Acad. Sci. Paris 273 (1971) 105-108.

[10] Mandelbrojt, S., Fonctions entières et transformées de Fourier. Applications. Math. Soc. Japan, 1967.

[11] Müntz, C. H., Über den Approximationssatz von Weierstrass. H. A. Schwarz Festschrift, Berlin, 1914, 303-312.

[12] Szász, O., Über die Approximation stetiger Funktionen durch lineare Aggregate von Potenzen. Math. Ann. 77 (1915) 482-496.

[13] Walsh, J. L., Interpolation and approximation by rational functions in the complex domain. Amer. Math. Soc. Colloq. Publ., vol. 20, New York, 1935.

Department of Mathematics
University of California, San Diego
La Jolla, California 92037

DISCRETIZATION OF APPROXIMATION PROBLEMS IN THE VIEW OF OPTIMIZATION

Werner Krabs

## 1. Introduction

In order to solve approximately the uniform approximation problem for continuous real-valued functions on a compact metric space the method of discretization is frequently used. It consists of replacing the underlying space by a finite subset thereof and solving the given problem with respect to this subset which, in many cases, can be handled more easily. This method, however, is only reasonable when the minimal distances and the solutions of the discretized problem get close to those of the "continuous" problem, if the subset is chosen sufficiently dense in the given metric space.

To what extent this is true in the case of linear approximation problems has been investigated by T. J. Rivlin and E. W. Cheney in [11] (see also [3] and [10]). H. Werner deals in [13] and [14] with ordinary and generalized rational approximation respectively, where he points out the necessity of the so-called "normality," and C. B. Dunham is concerned in [5] with rational approximation problems with nonnegative denominator functions. In [8] we have shown that normality is dispensable, if the denominator functions are required to be uniformly bounded from below by a positive constant. Furthermore, the discretization of exponential approximation problems has been investigated by B. H. Rosman in [12] and D. Braess in [2]. In a recent paper, [6], C. B. Dunham treats the case of varisolvent Chebyshev approximation on subsets. Here, the maximum degree plays the same role as normality in rational approximation.

Since the details in each of the papers quoted above are rather different, it seems to be useful to give a general model for all these convergence proofs by which the underlying ideas become more transparent. This, perhaps, can be done best by considering the approximation problem as a problem of minimizing a functional on a given set. The discretization then leads to the following situation.

Let E be a normed linear space. Further, let $\varphi : E \to R$ and $\varphi_m : E \to R$, for all positive integers m, be continuous functionals such that

(1.1) $\varphi(g) \geq \varphi_m(g)$ for all m and all $g \in E$.

Finally, let W be a nonempty subset of E and, for each m, let $W_m$ be a subset of E such that $W \subseteq W_m$ for all m. If we put

(1.2) $\rho = \inf_{w \in W} \varphi(w)$ and $\rho_m = \inf_{w \in W_m} \varphi_m(w)$,

then we have

$$\rho_m \leq \rho \quad \text{for all m.}$$

Hence, for each m, the set

(1.3) $W_m^* = \{w \in W_m : \varphi_m(w) \leq \rho + \frac{1}{m}\}$

is nonempty, and obviously

(1.4) $\rho_m = \inf_{w \in W_m^*} \varphi_m(w)$.

Let $\hat{w} \in W$ be given such that $\varphi(\hat{w}) = \rho$. Then we ask for conditions under which the following statements are true:

(a) $\lim_{m \to \infty} \rho_m = \rho$.

(b) For each sufficiently large m there exists $\hat{w}_m \in W_m^*$ such that

$$\varphi_m(\hat{w}_m) = \rho_m.$$

(c) Each such sequence has an accumulation point $w^* \in W$ and for each such $w^* \in W$ we have $\varphi(w^*) = \rho$.

If, in addition, we assume $\hat{w} \in W$ with $\varphi(\hat{w}) = \rho$ to be unique, then a simple consequence of these three statements is that for each sequence $\{\hat{w}_m\}$, $\hat{w}_m \in W_m^*$, with $\varphi_m(\hat{w}_m) = \rho_m$ for sufficiently large m we have $\hat{w} = \lim_{m\to\infty} \hat{w}_m$.

In a recent paper, [1], Anselone and Davis have also treated the discretization of approximation problems, however, from a different point of view and by using different concepts which they adopt in a generalized form from a discretization theory in the book [4] of Daniel. The theory of Daniel is primarily motivated by variational problems and the approximate solution of differential and integral equations.

In another recent paper, [7], Esser is also concerned with the discretization of extremum problems and gives an application to a problem of optimal control.

In this connection as well as in relation to general problems of optimization more papers could be mentioned. We confine ourselves to a hint to [9] where a list of further references can be found.

## 2. A Convergence Theorem

The purpose of this section is to prove the following

Theorem. Let $\hat{w} \in W$ be given such that $\varphi(\hat{w}) = \rho$.

Assumption 1. There exists a compact neighborhood $\hat{W}$ of $\hat{w}$ in E and a positive integer $m_o$ such that

(2.1) $W_m^* \subseteq \hat{W}$ for all $m \geq m_o$,

and we have

(2.2) $\lim_{m\to\infty} \max_{w \in \hat{W}} [\varphi(w) - \varphi_m(w)] = 0.$

Assumption 2. One of the following statements holds:

(α) The sets W and $W_m$ are closed for all $m \geq m_o$ and for each

$\delta > 0$ there is an integer $m(\delta) \geq m_0$ such that for all $m \geq m(\delta)$ we have:

For each $w \in W_m^*$ there is a $\tilde{w} \in W \cap \hat{W}$ such that $\|w - \tilde{w}\| \leq \delta$.

($\beta$) $\hat{W} \subseteq W$.

<u>Assertion.</u> Under these assumptions the three statements (a), (b), and (c) at the end of section 1 are true.

<u>Proof:</u> It can be easily shown that, under the assumptions 1 and 2, for each $m \geq m_0$ there exists a $\hat{w}_m \in W_m^*$ such that $\varphi_m(\hat{w}_m) = \rho_m$, i.e. the statement (b) of section 1 holds.

To prove the statement (a) we consider any such sequence $\{\hat{w}_m\}$, $\hat{w}_m \in W_m^*$. Let $\varepsilon > 0$ be given. Then, by Lemma 3.2 in [9], there is a positive integer $m_1(\varepsilon) \geq m_0$ and a real number $\delta(\varepsilon) > 0$ such that

$$\|w - \hat{w}_m\| \leq \delta(\varepsilon),\ w \in \hat{W} \Rightarrow |\varphi_m(w) - \varphi_m(\hat{w}_m)| \leq \frac{\varepsilon}{2}$$

$$\text{for all } m \geq m_1(\varepsilon).$$

Therefore, by assumption 2, there is an integer $m(\delta(\varepsilon)) \geq m_0$ and a sequence $\{\tilde{w}_m\}$, $\tilde{w}_m \in \hat{W} \cap W$ for $m \geq m(\delta(\varepsilon))$, such that

$$|\varphi_m(\tilde{w}_m) - \varphi_m(\hat{w}_m)| < \frac{\varepsilon}{2} \text{ for all } m \geq \max(m_1(\varepsilon), m(\delta(\varepsilon)).$$

Further, by (2.2) there is an integer $m_2(\varepsilon) \geq m_0$ such that

$$\varphi(\tilde{w}_m) - \varphi_m(\tilde{w}_m) < \frac{\varepsilon}{2} \text{ for all } m \geq m_2(\varepsilon).$$

Hence

$$0 \leq \rho - \rho_m \leq \varphi(\tilde{w}_m) - \varphi_m(\tilde{w}_m) + \varphi_m(\tilde{w}_m) - \varphi_m(\hat{w}_m) \leq \varepsilon$$

for all $m \geq \max(m_1(\varepsilon), m(\delta(\varepsilon)), m_2(\varepsilon))$ which completes the proof of (a).

Further, by (2.1) each sequence $\{\hat{w}_m\}$ with $m \geq m_0$, $\hat{w}_m \in W_m^*$, and $\varphi_m(\hat{w}_m) = \rho_m$ has an accumulation point $w^* \in \hat{W}$ and each such $w^*$ is also in W. If we put $w^* = \lim_{i\to\infty} \hat{w}_{m_i}$, then we obtain by observing $\varphi_{m_i}(\hat{w}_{m_i}) \leq \varphi_{m_i}(\hat{w})$ for all i and by using (1.1)

$$
\begin{aligned}
(2.3)\quad 0 &\le \varphi(w^*) - \varphi(\hat{w}) \\
&= \varphi(w^*) - \varphi_{m_i}(w^*) + \varphi_{m_i}(w^*) - \varphi_{m_i}(\hat{w}_{m_i}) \\
&\quad + \varphi_{m_i}(\hat{w}_{m_i}) - \varphi_{m_i}(\hat{w}) + \varphi_{m_i}(\hat{w}) - \varphi(\hat{w}) \\
&\le \varphi(w^*) - \varphi_{m_i}(w^*) + |\varphi_{m_i}(w^*) - \varphi_{m_i}(\hat{w}_{m_i})| .
\end{aligned}
$$

Again using (2.2) and Lemma 3.2 in [9] we deduce from (2.3) that $\varphi(w^*) = \varphi(\hat{w}) = \rho$ which completes the proof.

## 3. Application to Uniform Approximation

Let $T$ be a compact metric space and $E = C(T)$ the vector space of the real valued continuous functions on $T$ provided with the maximum norm

$$\|g\| = \max_{t\in T} |g(t)|,\ g \in E.$$

Further, let $W$ be a nonempty subset of $E$ and $f \in E$ be a given function. If we define $\varphi(g) = \|g - f\|$ for each $g \in E$, then $\varphi : E \to R$ is a continuous functional and the problem of minimizing $\varphi$ on $W$ is equivalent to finding a best approximant to $f$ in $W$. This is a problem of uniform approximation. In order to discretize this problem we consider a sequence $\{T_m\}$ of finite subsets $T_m$ of $T$ such that

$$(3.1)\quad \lim_{m\to\infty} \max_{t\in T} \min_{s\in T_m} d(t,s) = 0$$

holds where $d$ denotes the metric in $T$.

Further, we assume that to each $T_m$ there corresponds a subset $W_m$ of $E$ such that $W \subseteq W_m$ for all $m$.

Defining $\varphi_m(g) = \|g - f\|_m$, $g \in E$, where $\|g\|_m = \max_{t\in T_m} |g(t)|$ we obtain, for each $m$, a continuous functional $\varphi_m : E \to R$ such that (1.1) holds.

To minimize $\varphi_n$ on $W_n$ then is referred to as a discretization of the above problem of uniform approximation.

If for each $w \in C(T)$ we choose $\hat{t} \in T$ with $|w(\hat{t})-f(\hat{t})| = \|w-f\|$, then for each m we have

$$\varphi(w) - \varphi_m(w) \leq |w(\hat{t}) - w(s)| + |f(\hat{t}) - f(s)| \text{ for all } s \in T_m.$$

Therefore, if w varies in a compact subset $\hat{W}$ of C(T), then, by virtue of the uniform equicontinuity of W, the uniform continuity of f on T, and the assumption (3.1) we have that (2.2) is automatically satisfied. Hence, in the convergence theorem of the section 2 we need only check (2.1) and the assumption 2.

## References


[1] Anselone, P. M. and J. Davis, Perturbations of Best Approximation Problems. To appear in Numerische Mathematik.

[2] Braess, D., Chebyshev Approximation by Exponentials on Finite Subsets. To appear.

[3] Cheney, E. W., Introduction to Approximation Theory. McGraw-Hill, New York, 1966.

[4] Daniel, J. W., The Approximate Minimization of Functionals. Prentice Hall, Englewood Cliffs, N.J., 1971.

[5] Dunham, C. B., Rational Approximation on Subsets. J. Approx. Theory 1 (1968), 484-487.

[6] Dunham, C. B., Varisolvent Chebyshev Approximation on Subsets. To appear.

[7] Esser, H., Zur Diskretisierung von Extremalproblemen. To appear.

[8] Krabs, W., On Discretization in Generalized Rational Approximation. To appear in Abh. Sem. Univ. Hamburg.

[9] Krabs, W., Stabilität und Stetigkeit bei nichtlinearer Optimierung. To appear in Methods of Operations Research.

[10] Rice, J. R., The Approximation of Functions. Addison-Wesley, Reading, 1964.

[11] Rivlin, T. J. and E. W. Cheney, A Comparison of Uniform Approximation on an Interval and a Finite Subset thereof. J. SIAM Numer. Anal. 3 (1966), 311-320.

[12] Rosman, B. H., Exponential Chebyshev Approximation on Finite Subsets of [0,1]. Math. Comp. 25 (1971), 575-577.

[13] Werner, H., Die Bedeutung der Normalität bei rationaler Tschebysheff-Approximation. Computing 2 (1967), 34-52.

[14] Werner, H., Diskretisierung bei Tschebysheff-Approximation mit verallgemeinerten rationalen Funktionen. ISNM 9 (1968), 381-391.

Fachbereich Mathematik
Technische Hochschule, Darmstadt
6100 Darmstadt, Hochschulstr. 1
West Germany

# EXCHANGE ALGORITHM IN CONVEX ANALYSIS

P. J. Laurent

## Introduction

We give here a general algorithm for the minimization of a convex functional f defined on a linear space X over an n-dimensional subspace of X. This algorithm consists in replacing at each iteration the functional f by the supremum of n+1 affine functionals. When f is a norm, we obtain the generalized Rémès-algorithm. When X is finite dimensional and f is polyhedral, the algorithm becomes the dual simplex algorithm. The convergence is studied and an example is given.

## 1. Statement of the Problem

Let X and Y be two locally convex linear topological spaces in duality. Given a continuous convex functional f defined on X with values in $\mathbb{R}$ and a linear subspace V of X, we study the following minimization problem:

(P) $\alpha = \text{Inf } (f(x) \mid x \in V)$.

We suppose that $\alpha$ is a finite number. Let $f^*$ be the conjugate of f (defined on Y with values in $\mathbb{R} \cup \{+\infty\}$) and $V^\circ \subset Y$, the polar subspace of V. A classical dual problem of (P) is the following one (see [3], p. 412):

(Q) $\alpha = \max ( -f^* (y) \mid y \in V^\circ)$.

With our hypothesis, the dual problem has the same value $\alpha$ and has at least one solution. An element $\bar{x} \in V$ is a solution of (P) iff there exists $\bar{y} \in V^\circ$ such that $\bar{y} \in \partial f(\bar{x})$ (such an element $\bar{y}$ is then a solution of (Q)), where $\partial f(\bar{x})$ denotes the subdifferential of f at $\bar{x}$.

Throughout this study, we will suppose that V is an n-dimensional subspace. In that case, the dual problem (Q), as well as the characterization theorem, can be written more explicitly using the extreme points of the epigraph of $f^*$ (which we denote by epi $(f^*)$). This is based on the following result:

Lemma. Each extreme point of epi $(f^*) \cap (V° \times R)$ is a convex combination of k (with $1 \leq k \leq n+1$) extreme points of epi $(f^*)$; i.e., if $[y, f^*(y)]$ is an extreme point of epi $(f^*) \cap (V° \times R)$, then there exist $y_i$, $i=1,\dots,k$ $(1 \leq k \leq n+1)$ such that $[y_i, f^*(y_i)]$ is an extreme point of epi $(f^*)$ and there exist $\rho_i > 0$, $\Sigma_{i=1}^k \rho_i = 1$, such that $y = \Sigma_{i=1}^k \rho_i y_i$ and $f^*(y) = \Sigma_{i=1}^k \rho_i f^*(y_i)$.

If we denote by $E \subset Y$ a set containing all $y \in Y$ such that $[y, f^*(y)]$ is an extreme point of epi $(f^*)$, the dual problem becomes:

$$(Q) \quad \alpha = \max \left\{ -\sum_{i=1}^{k} \rho_i f^*(y_i) \,\middle|\, y_i \in E,\ \rho_i > 0,\ \sum_{i=1}^{k} \rho_i = 1,\ 1 \leq k \leq n+1,\ \sum_{i=1}^{k} \rho_i y_i \in V° \right\}.$$

We have also the following characterization theorem ([3], p. 439):

Theorem. An element $\bar{x} \in V$ is a solution of (P) iff there exist $y_1, \dots, y_k \in E$ $(1 \leq k \leq n+1)$, $\rho_1, \dots, \rho_k > 0$ such that:

1°/ $y_i \in \partial f(\bar{x})$, $i=1,\dots,k$,

2°/ $\displaystyle\sum_{i=1}^{k} \rho_i y_i \in V°$.

For the algorithm we will need two additional hypotheses:

$(H_1)$ We assume that the functional $f^*$ is bounded; as f is continuous, $f^*$ is $\sigma(Y,X)$-inf-compact and thus, $(H_1)$ implies that dom$(f^*)$ is $\sigma(Y,X)$-compact (with dom$(f^*)$ denoting the set $\{y \in Y \mid f^*(y) < \infty\}$).

($H_2$) All the sets $S_c = \{x \in V \mid f(x) \leq c\}$ that are non-empty have the same recession cone ([5], p. 70); we assume that <u>this recession cone</u> $(S_c)_\infty$ is <u>a linear subspace</u>.

The hypothesis ($H_2$) can be replaced by a stronger condition:

($H_2'$) There exists $c \in R$ such that $S_c$ is non-empty and bounded.

<u>Remark</u>. The condition ($H_2$) (or ($H_2'$)) implies that $0 \in \text{ri}(\text{dom}(f^*) + V^\circ)$ and consequently the existence of solutions for (P). It also implies that there exists $c > 0$ such that $\text{dom}(f^*) \cup (-\text{dom}(f^*)) \subset c\, \text{dom}(f^*) + V^\circ$. This inclusion has the following consequence:

$|\langle x,y\rangle| \leq c(f(x) + \omega)$, for all $x \in V$ and $y \in \text{dom}(f^*)$ with $\omega = \text{Sup}(|f^*(y)| \mid y \in \text{dom}(f^*))$; this majoration is fundamental for the convergence proof.

## 2. Algorithm

Suppose we have, at the $\nu^{\text{th}}$ iteration, n+1 elements $y_1^\nu, \ldots, y_{n+1}^\nu \in E$ and $\rho_1^\nu, \ldots, \rho_{n+1}^\nu \geq 0$, $\sum_{i=1}^{n+1} \rho_i^\nu = 1$, such that

$$y^\nu = \sum_{i=1}^{n+1} \rho_i^\nu y_i^\nu \in V^\circ.$$

We assume that the $y_i^\nu$ satisfy the Haar condition on V, i.e., if $x_j$, j=1,...,n, denotes a basis of V, all the determinants of rank n of the n(n+1) matrix $[\langle x_j, y_i^\nu\rangle]$ are nonzero.

This implies that the $\rho_i^\nu$ are strictly positive and uniquely determined. Put

$$f_i^\nu(x) = \langle x, y_i^\nu\rangle - f^*(y_i^\nu)$$

and

$$f^{\nu}(x) = \max\{f_i^{\nu}(x) \mid i=1,\ldots,n+1\}$$

and consider the discrete minimization problem:

$$(P^{\nu}) \quad \alpha^{\nu} = \operatorname{Inf}(f^{\nu}(x) \mid x \in V).$$

The only extreme points of epi $(f^{\nu *})$ are $[y_i^{\nu}, f^*(y_i^{\nu})]$, $i=1,\ldots,n+1$, so that the dual problem of $(P^{\nu})$ is reduced to:

$$(Q^{\nu}) \quad \alpha^{\nu} = - \sum_{i=1}^{n+1} \rho_i^{\nu} f^*(y_i^{\nu}).$$

Applying the characterization theorem to the problem $(P^{\nu})$, we see that an element $\overline{x}^{\nu} \in V$ is solution iff $y_i^{\nu} \in \partial f^{\nu}(\overline{x}^{\nu})$, $i=1,\ldots,n+1$, i.e., iff $f_i^{\nu}(\overline{x}^{\nu}) = f^{\nu}(\overline{x}^{\nu})$, $i=1,\ldots,n+1$.

If we set $\overline{x}^{\nu} = \Sigma_{j=1}^{n} a_j^{\nu} x_j$, as we have $f^{\nu}(\overline{x}^{\nu}) = \alpha^{\nu}$, the coefficients $a_j^{\nu}$ can be obtained by resolving the linear algebraic system:

$$\sum_{j=1}^{n} a_j^{\nu} \langle x_j, y_i^{\nu} \rangle = f^*(y_i^{\nu}) + \alpha^{\nu}, \quad i=1,\ldots,n.$$

(The last condition $(i=n+1)$ is then automatically satisfied.)

The solution $\overline{x}^{\nu}$ of $(P^{\nu})$ is unique.

We have of course the double inequality:

$$\alpha^{\nu} \le \alpha \le f(\overline{x}^{\nu}).$$

The iteration will consist in choosing a new element $\hat{y}^{\nu} \in \partial f(\overline{x}^{\nu}) \cap E$. As $\partial f(\overline{x}^{\nu})$ is $\sigma(Y,X)$-compact and non-empty, it has at least one extreme point $\hat{y}^{\nu}$ and it is easy to prove that $[\hat{y}^{\nu}, f^*(\hat{y}^{\nu})]$ is then an extreme point of epi $(f^*)$.

By the exchange theorem (see [3], p. 462), there exists an index $\hat{i}^{\nu}$ such that if,

$$y_i^{\nu+1} = \begin{cases} y_i^{\nu} & \text{if } i \neq \hat{i}^{\nu}, \\ \hat{y}^{\nu} & \text{if } i = \hat{i}^{\nu}, \end{cases}$$

then there exist coefficients $\rho_i^{\nu+1} \geq 0$ $(\Sigma_{i=1}^{n+1} \rho_i^{\nu+1} = 1)$ such that $y^{\nu+1} = \Sigma_{i=1}^{n+1} \rho_i^{\nu+1} y_i^{\nu+1} \in V°$.

We assume again that the $y_i^{\nu+1}$ satisfy the Haar condition on V. We have then $\rho_i^{\nu+1} > 0$ and these coefficients are unique. We consider the problem $(P^{\nu+1})$ associated with the $y_i^{\nu+1}$ as above. The value of $(P^{\nu+1})$ is

$$\alpha^{\nu+1} = -\sum_{i=1}^{n+1} \rho_i^{\nu+1} f^*(y_i^{\nu+1}) = \sum_{i=1}^{n+1} \rho^{\nu+1}(\langle \bar{x}^\nu, y_i^{\nu+1}\rangle - f^*(y_i^{\nu+1})).$$

As

$$\langle \bar{x}^\nu, y_i^{\nu+1}\rangle - f^*(y_i^{\nu+1}) = \begin{cases} \alpha^\nu & \text{if } i \neq \hat{i}^\nu \\ f(\bar{x}^\nu) & \text{if } i = \hat{i}^\nu \end{cases}$$

we have

$$\alpha^{\nu+1} = \alpha^\nu + \rho_{\hat{i}^\nu}^{\nu+1} (f(\bar{x}^\nu) - \alpha^\nu).$$

Thus, if $\bar{x}^\nu$ is not a solution of (P), then $\alpha^\nu < \alpha^{\nu+1} \leq \alpha$.

We will say that the algorithm is <u>iterative</u> if at each iteration the set $\{y_i^\nu, i=1,\ldots,n+1\}$ satisfies the Haar condition on V.

## 3. Convergence

Following the result obtained by C. Carasso ([1], [2]) for the Rémès-algorithm, we can prove the following convergence theorem:

<u>Theorem</u>. Under the assumptions $(H_1)$, $(H_2)$, if the algorithm is iterative, then it is convergent, i.e., $\lim_{\nu\to\infty} \alpha^\nu = \alpha$ and the sequence $f(\bar{x}^\nu)$ has a subsequence converging towards $\alpha$. (The proof of this result being rather long, we cannot give it here.)

We will give now another convergence theorem using the following assumption:

$(H_3)$ for all $y_1,\ldots,y_n \in \bar{E}$, non-zero and not proportional, one has $\det(\langle x_j, y_i\rangle) \neq 0$.

The condition $(H_3)$ is very strong: it is the direct generalization of the Haar condition in the case of a best approximation problem. If $(H_3)$ is satisfied, a non-zero element $y = \Sigma_{i=1}^k \rho_i y_i$ with $\rho_i > 0$, $1 \leq k \leq n+1$ and $y_i \in \bar{E}$ can only belong to

V° if $k = n+1$ and the $y_i$ are non-zero and not proportional. The condition $(H_3)$ together with $\alpha_0 < \alpha$, where $\alpha_0 = \mathrm{Inf}(f(x)|x\in X)$, implies that the solution $\bar{x}$ of (P) is unique. In the same way, $(H_3)$ with $\alpha_0 < \alpha^1$ implies that the algorithm is iterative. Thus we can obtain the following theorem:

Theorem. If $(H_1)$, $(H_2)$, $(H_3)$ are satisfied and if $\alpha_0 < \alpha^1$, then the algorithm is iterative and $\lim_{\nu\to\infty} \alpha^\nu = \lim_{\nu\to\infty} f(\bar{x}^\nu) = \alpha$. Moreover, the sequence $\bar{x}^\nu$ converges towards $\bar{x}$.

Remarks. (a) In order to start the algorithm, it is sometimes easier to choose, at the first iteration, $f_i^1(x) = \langle x, y_i^1\rangle - r_i^1$, such that $f_i^1(x) \leq f(x)$, for all $x\in X$, i.e., $y_i^1 \in \mathrm{dom}\,(f^*)$ and $r_i^1 \geq f^*\,(y_i^1)$. This modification does not affect the convergence.

(b) The case where there are additional inequality constraints (of the type $g(x) \leq 0$, where g has the same properties as f) can be solved by a natural extension of this algorithm.

## 4. Applications

(a) If X is finite dimensional and f is polyhedral, then the algorithm becomes equivalent to the classical dual simplex algorithm.

(b) Let X be a normed linear space and Y its topological dual. If $f(x) = \|x-x_0\|$, where $x_0$ is a fixed element of X, then the algorithm is the generalization of the Rémès-algorithm (see [3], p. 454).

(c) If A denotes a compact subset of X (for the norm-topology) we consider the following convex functional: $f(x) = \max(\|x-a\| \mid a\in A)$.

The problem consisting in minimizing f over V occurs when we have to approximate an element which is incompletely determined, i.e., precisely a compact set A of elements (see [4]).

All the extremal points of epi $(f^*)$ can be written $[y, f^*(y)]$ with $y\in E(S')$, the set of extreme points of the dual sphere S' of Y. We choose $E = E(S')$. Suppose we have the n+1 elements $y_i^\nu$, $i=1,\ldots,n+1$, the n+1 affine functionals $f_i^\nu(x) = \langle x, y_i^\nu\rangle - f^*(y_i^\nu)$, and the solution $\bar{x}^\nu \in V$. We have to choose the new element $\hat{y}^\nu \in \partial f(\bar{x}^\nu) \cap E$.

By Valadier's theorem ([3], p. 355) we have $\partial f(x) = \overline{co} \bigcup_{a\in\Omega(x)} \partial n(x-a)$, where $n(x) = \|x\|$ and $\Omega(x) = \{\bar{a}\in A \mid \|x-\bar{a}\| = \max_{a\in A}\|x-a\|\}$. Thus, in order to obtain $\hat{y}^\nu$, first we select $\hat{a}^\nu \in \Omega(\bar{x}^\nu)$ and then $\hat{y}^\nu \in E$ such that $\hat{y}^\nu \in \partial n(\bar{x}^\nu - \hat{a}^\nu)$, i.e., $\langle \bar{x}^\nu - \hat{a}^\nu, \hat{y}^\nu\rangle = \|\bar{x}^\nu - \hat{a}^\nu\|$. Since $\hat{a}^\nu \in \Omega(\bar{x}^\nu)$, $f^*(\hat{y}^\nu) = \langle \hat{a}^\nu, \hat{y}^\nu\rangle$. This gives the new affine functional $\hat{f}^\nu(x) = \langle x, \hat{y}^\nu\rangle - \langle \hat{a}^\nu, \hat{y}^\nu\rangle$, which we exchange with one of the $f_i^\nu$.

## References


[1] Carasso, C., Etude de l'algorithme de Rémes en l'absence de conditions de Haar. Numer. Math. 20 (1972), 165-178.

[2] Carasso, C., Convergence de l'algorithme de Rémes. A paraître.

[3] Laurent, P. J., Approximation et optimisation. Hermann, Paris, 1972.

[4] Laurent, P. J., Global approximation of a compact set by elements of a convex set in a normed space. Num. Math. 15 (1970), 137-150.

[5] Rockafellar, R. T., Convex Analysis. Princeton University Press, Princeton, 1970.



Mathématiques appliquées.
Univ. Sci. et méd. de Grenoble.
Cédex 53
38041 - Grenoble, France

THE CONSTANT ERROR CURVE PROBLEM
FOR VARISOLVENT FAMILIES

William H. Ling and J. Edward Tornga

## 1. Introduction

Let $[a,b]$ be a compact real interval, P be a nonempty subset of Euclidean s space, $R^s$, and $C[a,b]$ be the set of real valued continuous functions on $[a,b]$. Our approximating family will be denoted by $V = \{F(A,x) | A = (a_1,a_2,\ldots,a_s) \in P\}$, $x \in [a,b]$, $F(A,x) \in C[a,b]$. For $f \in C[a,b]$, $\|f(x)\|$ will mean $\max_{x\in[a,b]}|f(x)|$.

Property Z. $F(A,x) \in V$ has property Z of degree n on $[a,b]$, (n a positive integer), if for $F(A_1,x) \in V$ with $F(A_1,x) \not\equiv F(A,x)$, $F(A_1,x) - F(A,x)$ has at most n-1 zeros on $[a,b]$.

Local Solvency. $F(A,x) \in V$ is locally solvent of degree n on $[a,b]$ if (1) given n distinct points $a \leq x_1 < x_2 < \ldots < x_n \leq b$ and (2) given $\varepsilon > 0$, there exists a $\delta(A,\varepsilon,x_1,\ldots x_n) = \delta > 0$ such that for any set of real numbers $\{z_i\}_{i=1}^n$ with $|z_i - F(A,x_i)| < \delta$ for $1 \leq i \leq n$, there exists $F(A_1,x) \in V$ with (3) $F(A_1,x_i) = z_i$, $1 \leq i \leq n$ and (4) $\|F(A_1,x) - F(A,x)\| < \varepsilon$.

Definition. $F(A,x) \in V$ is a varisolvent function of degree n at A if (1) $F(A,x)$ has Property Z of degree n at A, (2) $F(A,x)$ is locally solvent of degree n and (3) $F(A,x)$ is not locally solvent of degree n+1. We denote the degree of $F(A,x)$ by $m(A) = n$.

Henceforth V will denote a varisolvent family (a set of varisolvent functions) with $\max_{A\in P} m(A) \leq M < \infty$. We say $m(A^*)$ is maximal if $m(A^*) = M$.

Approximating Problem. Given $f \in C[a,b]$, $f \notin V$, we wish to find $F(A^*,x) \in V$ (if it exists) such that

$$\|f(x) - F(A^*,x)\| = \inf_{F(A,x)\in V} \|f(x) - F(A,x)\|.$$

If such an $F(A^*,x)$ exists, it is called a best approximation (b.a.) to f from V.

Definition. For $F(A,x) \in V$, $E(A,x) \equiv f(x) - F(A,x)$ is called the error curve with respect to f(x). E(A,x) is said to alternate n times on [a,b] if there exist n+1 distinct points $a \le x_1 < x_2 < \ldots < x_{n+1} \le b$ such that (1) $|f(x_i) - F(A,x_i)| = \|f(x) - F(A,x)\|$, $1 \le i \le n+1$, and (2) $[f(x_i) - F(A,x_i)] = -[f(x_{i+1}) - F(A,x_{i+1})]$, $1 \le i \le n$.

In 1961, J. Rice [4] presented the following theorem.

Theorem. $F(A^*,x)$ is a b.a. to $f \in C[a,b]$ from V if and only if $E(A^*,x) \equiv f(x) - F(A^*,x)$ alternates at least $m(A^*)$ times.

However the possiblity that one could have a constant error curve, i.e. have $E(A^*,x) \equiv C$, C a nonzero constant, was not considered in the proof.

## 2. Elimination of Constant Error Curve--Previous Work

Let $f \in C[a,b]$ be given and assume $F(A^*,x)$ is a b.a. to f from V. The constant error curve possibility was eliminated if (1) $m(A^*) = 1$, 2 or 3 by R. Barrar and H. Loeb or if (2) $m(A^*)$ is maximal by D. Braess or if (3) V satisfies the derivative assumptions (local Haar condition) of G. Meinardus and D. Schwedt and R. Barrar and H. Loeb.

## 3. Elimination of Constant Error Curve--Our Approach

For a given $f \in C[a,b]$ and $\varepsilon > 0$, we define $N_\varepsilon(f) = \{g \in C[a,b] \mid \|g(x) - f(x)\| < \varepsilon\}$.

Definition. V is said to permit a constant error on [a,b] if there exists $f \in C[a,b]$ and a b.a. $F(A^*,x)$ to f from V such that $f(x) - F(A^*,x)$ is a nonzero constant.

With these definitions, we state our main Theorem.

Theorem 1 (Global). Assume $B_A = \{f \in C[a,b] |$ a b.a. to f exists from $V\}$ is open. Then V does not permit a constant error on [a,b]. We note that Theorem 1 is in fact a corollary of a more general, more technical Theorem given in our paper.

Outline of Proof of Theorem 1: Assume V does permit a constant error on [a,b], i.e. assume $f \in B_A$ and $F(A^*,x)$ is a b.a. to f from V with $E(A^*,x) = f(x) - F(A^*,x) \equiv C$, C a positive constant. A similar proof holds for $C < 0$. Our proof then proceeds as follows where $m(A^*) \geq 4$.

Step 1. We construct a sequence of functions $\{g_n\} \subset C[a,b]$, $g_n(x) \leq f(x)$ for all x and for all $n \geq 3$, in the following manner. We split [a,b] into 2K closed intervals, $\{I_i\}_{i=1}^{2K}$, consecutively ordered, each with length $(b-a)/2K$. K denotes the smallest integer $\geq \frac{1}{2}(M+1) + 1$ where $M = \max_{A \in P} m(A)$. For each $n \geq 3$, $g_n(x)$ is defined so that on a subinterval of $I_i$, i odd, $g_n(x) \equiv f(x)$, and on a subinterval of $I_i$, i even, $g_n(x) \equiv f(x) - C/n$. With the requirement $\|g_n(x) - f(x)\| = C/n$, one can see that each $g_n$ appears as a "notch" function and $\{g_n\}$ converges uniformly to f on [a,b].

Step 2. Let $F(A_n,x)$ be the b.a. to $g_n$ from V, which exists since $B_A$ is open. Using the "notch" appearance of $g_n$, we show that $E(A_n,x) \equiv g_n(x) - F(A_n,x)$ cannot be a constant.

Step 3. On each $I_i$, i odd, we now construct, using varisolvency, an auxiliary function $F(A^i,x) \in V$. Using the facts that $E(A_n,x)$ alternates for all n and $\{g_n\}$ converges uniformly to f; we find that there exists an n and an i such that $F(A_n,x) - F(A^i,x)$ has at least $m(A_n)$ zeros. This contradicts Property Z for $F(A_n,x)$.

It appears that for most known varisolvent families, the

set $B_A$ is open. For those V which are dense compact, see [2], $B_A$ is open. This includes such families as unisolvent families and the rationals, $R_{n,m}$. A theorem appearing in [1], implies that there are V for which $B_A$ is open and not dense compact. In fact this theorem also guarantees that Braess' result concerning m(A*) maximal is a direct consequence of our Theorem 1.

## 4. Applications and Results

With an application of Theorem 1, we show that the constant error curve possibility can be eliminated for simultaneous approximation, restricted range approximation and approximation on a proper compact subset of [a,b]. As results of our Theorem 1, we obtain a betweeness property for varisolvent families and a uniqueness statement. Lastly we present three examples of varisolvent families to which Theorem 1 is applicable but no previous theory applies.

## References

[1] Dunham, C. B., Continuity of the varisolvent Chebyshev operator. Bull. Amer. Math. Soc. 74 (1968), 606-608.

[2] Dunham, C. B., Existence and continuity of the Chebyshev operator. SIAM Review 10 (1968), 444-446.

[3] Ling, W. H., The constant error curve problem for varisolvent families. Ph.D. Dissertation, Rensselaer Polytechnic Institute, 1972.

[4] Rice, J. R., Tchebycheff approximations by functions unisolvent of variable degree. Trans. Amer. Math. Soc. 99 (1961), 298-302.

[5] Tornga, J. E., Approximation from varisolvent and unisolvent families whose members have restricted ranges. Ph.D. Dissertation, Michigan State University, 1971.

Department of Mathematics
Union College
Schenectady, New York 12308

# SOME MINIMUM PROBLEMS FOR SPLINE FUNCTIONS WITH APPLICATIONS TO QUADRATURE FORMULAS

C. A. Micchelli

## 1. Introduction

The motivation for our interest in the subject of this paper originated with certain problems from the theory of best quadrature formulas. For a thorough account of this subject see the paper of I. J. Schoenberg [5]. For our purposes it suffices to remark that the basic problem in the theory of best quadrature formulas (after an integration by parts is performed) is one of finding a member of some class of piecewise polynomials which has smallest norm, usually taken to be some $L^p$-norm. Quite a variety of constraints on the piecewise polynomials can arise in this fashion depending on what class of quadrature formulas are considered. Whether or not the nodes of the quadrature formula are fixed or variable and what type of function data (function values and/or derivatives) is specified at a given node all lead to a different class of piecewise polynomials. The usual questions that are investigated are those of existence, uniqueness, characterization and construction of best or asymptotically best formulas. There are still many open problems of a fundamental nature; we mention some at the end of the paper.

In this paper, we will be almost exclusively concerned with problems related to quadrature formulas where the nodes are fixed, in fact, equally spaced, and also with the construction of special formulas. By this we mean reducing the problem of finding the best formula to the solution of a polynomial best approximation problem. Our main results are presented in a general setting and then we give applications to the construction of best quadrature formulas and other minimum problems with spline functions. The results presented here represent an extension of the results in our previous paper [2].

## 2. A Minimum Problem

Let n,r be positive integers, $\lambda$ a nonzero complex constant and I some subset of the integers $\{0,1,\ldots,n-1\}$. We define $M_r(I,\lambda) = M_r$ to be the class of all piecewise polynomials M of degree n which satisfy the requirement that in each of the intervals $(\nu,\nu+1)$, $\nu=0,1,\ldots,r-1$ M is a polynomial of degree n with leading coefficient $\lambda^\nu$, $M^{(n)}(x) = n!\lambda^\nu, x \in (\nu,\nu+1)$, $\nu = 0,1,\ldots,r-1$. Furthermore, we assume that the pieces of M are joined together at the knots so that

$$M^{(i)}(\nu^+) = M^{(i)}(\nu^-),\ \nu=1,\ldots,r-1.\quad i \in I \tag{1}$$

and that M satisfies the boundary conditions

$$M^{(i)}(r) = \lambda^r M^{(i)}(0),\quad i \in I. \tag{2}$$

For instance, if $\lambda = 1$ and $I = \{0,1,\ldots,n-2\}$ then $M_r$ is the class of cardinal monosplines satisfying the boundary conditions (2), while $\lambda = -1$ and $I = \{0,1,\ldots,n-1\}$ leads to perfect cardinal splines satisfying (2) (cf. [1]).

The class $M_1(I,\lambda)$ will play an important role in what follows and so we give it the special designation $P$. $P$ is the subspace of polynomials of degree n with leading coefficient one which satisfy the boundary conditions, $p^{(i)}(1) = \lambda p^{(i)}(0)$, $i \in I$. We can give a description of $P$ which is not hampered by these constraints. To do this we follow [4] and introduce the exponential Euler polynomials. These are defined by means of the expression

$$\frac{\lambda-1}{\lambda-e^z} e^{xz} = \sum_{n=0}^{\infty} \frac{A_n(x,\lambda)}{n!} z^n,\quad \lambda \neq 1. \tag{3}$$

$A_n(x,\lambda)$ is a polynomial of degree n with leading coefficient one and for $\lambda = -1$, $A_n(x,-1)$ is the Euler polynomial $E_n(x)$. Similarly, the Bernoulli polynomials are defined by the generating function

$$\frac{z}{e^z - 1} e^{xz} = \sum_{n=0}^{\infty} \frac{B_n(x)}{n!} z^n . \tag{4}$$

Lemma 1. For $\lambda \neq 1$, every element $p \in P$ has a unique representation of the form $p(x) = A_n(x,\lambda) + \Sigma_{j \notin I} c_j A_j(x,\lambda)$. If $\lambda = 1$ then $P \neq \emptyset$ if and only if $I \subseteq \{0,1,\ldots,n-2\}$ and in this case every element has a unique representation of the form

$$p(x) = B_n(x) + \sum_{\substack{j \notin I \\ j \leq n-2}} c_j B_{j+1}(x) + c_{-1} .$$

Proof: When $\lambda \neq 1$ we may verify by using the defining equation for $A_n(x,\lambda)$ that $A_n^{(i)}(1,\lambda) - \lambda A_n^{(i)}(0,\lambda) = n!(1-\lambda)\delta_{in}$, $i=0,1,\ldots$, $n=0,1,\ldots$.

Since the only polynomial of degree n-1 which satisfies the relations $q^{(i)}(1) = \lambda q^{(i)}(0)$, $i=0,1,\ldots,n-1$ is identically zero the lemma follows in a straightforward manner.

If $\lambda = 1$ and $n-1 \in I$ then $P = \emptyset$ because for every polynomial of exact degree n we have $p^{(n-1)}(1) - p^{(n-1)}(0) \neq 0$. However, when $\lambda = 1$ and $n-1 \notin I$ the proof of the lemma proceeds as above after noting that (4) implies $B_n^{(i-1)}(1) - B_n^{(i-1)}(0) = -n!\delta_{in}$, $i=1,2,\ldots,n=0,1,\ldots$ and $\int_0^1 B_n(x)dx = \delta_{0n}$, $n=0,1,\ldots$.

Remark. Lemma 1 tells us that for $\lambda = 1$ $P$ may be empty. It will become clear later that $P \neq \emptyset$ if and only if $M_r \neq \emptyset$, thus we always assume $I \subseteq \{0,1,\ldots,n-2\}$ when $\lambda = 1$.

Let us consider two ways to construct elements of $M_r$. We do this by defining two mappings, the first of which is a 1-1 extension map taking an element of $P$ into $M_r$. We define it by means of the difference relation $(Sp)(x) = \lambda(Sp)(x-1)$ for

$1 \leq x \leq r$ and $(Sp)(x) = p(x)$ for $0 \leq x \leq 1$. Secondly, we define a map $T$ which circulates (from left to right) the pieces of an element $M \in \mathcal{M}_r$ by setting $TM(x) = \lambda^{-1}M(x+1), 0 \leq x \leq r-1$ and $TM(x) = \lambda^{r-1}M(x-r+1), r-1 \leq x \leq r$. Since $T$ circulates the pieces of $M$ it follows that $T$ has order $r$, $T^r = I$. The mappings $T$ and $S$ are related by the fact that $TM = M$ if and only if there is $p \in \mathcal{P}$ such that $Sp = M$. We omit the details.

To complete the picture we need a norm on $\mathcal{M}_r$ which we choose to be the $L^p$-norm on $[0,r], 1 \leq p \leq \infty$. We denote it by $\|\cdot\|_r$ where the subscript refers to the interval of integration.

<u>Theorem 1.</u> Suppose $\lambda$ is a complex constant of modulus one. Then an element of smallest norm in $\mathcal{M}_r$ is given (uniquely when $1 < p < \infty$) by $M_0 = Sp_0$ where $p_0$ is the element of minimum norm in $\mathcal{P}$.

<u>Proof:</u> The proof proceeds by showing that we can always decrease the value of the norm by an element in the range of $S$. Let $\overline{M}$ be the element of minimum norm in $\mathcal{M}_r$, $\|\overline{M}\|_r \leq \|M\|_r$ for all $M \in \mathcal{M}_r$. Define $M_0 = \frac{1}{r}(\overline{M} + T\overline{M} + \ldots + T^{r-1}\overline{M})$, then it follows that $\|M_0\|_r \leq \|\overline{M}\|_r$ since it can be verified that $\|TM\|_r = \|M\|_r$ for all $M \in \mathcal{M}_r$. However, because $T$ has order $r$ we see that $TM_0 = M_0$. Thus $M_0 = Sp_0$ for some $p_0 \in \mathcal{P}$. To identify $p_0$ observe that for any function $f$, $\|Sf\|_r = r^{1/p}\|f\|_1$ for $p < \infty$ while for $p = \infty$, $\|Sf\|_r = \|f\|_1$. Thus it follows that $p_0$ is the element of minimum norm in $\mathcal{P}$.

<u>Corollary 1.</u> Among all perfect spline functions $P$ of degree $n$ which satisfy the boundary conditions $P^{(i)}(r) = (-1)^r P^{(i)}(0)$, $i=0,1,\ldots,n-1$ the Euler spline function $\overline{E}_n(x)$ defined by $\overline{E}_n(x) = E_n(x)$ for $0 \leq x \leq 1$ and extended for all real $x$ by means of the difference equation $\overline{E}_n(x+1) = -\overline{E}_n(x)$ has smallest $L^p$-norm on $[0,r], 1 \leq p \leq \infty$.

For a related result concerning perfect spline functions on the whole real axis see [1].

Corollary 2. Let $\overline{B}_n(x)$ denote the Bernoulli monospline which is defined by setting $\overline{B}_n(x) = B_n(x), 0\leq x\leq 1$ and extending it for all real x by means of the difference equation $\overline{B}_n(x+1) = \overline{B}_n(x)$. Then among all monosplines which satisfy the boundary conditions $M^{(i)}(r) = M^{(i)}(0), i=n-3, n-5, \ldots$, the monospline $M = \overline{B}_n + c_0$ has the smallest $L^p$-norm on $[0,r]$. The constant $c_0$ is defined to give the smallest value to $\|B_n + c\|_1$ for all real values of c.

Proof: The minimum we are searching for can be assumed to satisfy the symmetry relation $M(r-x) = (-1)^n M(x)$. Thus it is necessarily in $M_r$ for $I = \{0,1,\ldots,n-2\}$ and $\lambda = 1$.

As a further application, we give an extremal characterization of the exponential Euler spline, $S_n(x,\lambda)$. $S_n(x,\lambda)$ is defined to be the spline function of degree n (globally $C^{n-1}(-\infty,+\infty)$) which interpolates the power sequence $\lambda^\nu$, that is, $S(\nu,\lambda) = \lambda^\nu, \nu=0,\pm1,\pm2,\ldots$. The existence and uniqueness of this function is assured when $|\lambda| = 1$ but not equal to -1 ([4]).

Corollary 3. Let $|\lambda| = 1$, $\lambda \neq -1$ and $I = \{0,1,\ldots,n-1\}$. Then the element of smallest $L^p$-norm on $[0,r]$ from $M_r(\lambda, I)$ is $S_n(x,\lambda)\cdot A_n(0,\lambda)$.

Proof: Corollary 1 of [4] states that $S_n(x,\lambda)$ can be defined by setting $S_n(x,\lambda) = A_n(x,\lambda)/A_n(0,\lambda)$ for $0\leq x\leq 1$ and extended for all real x by means of the difference relation $S_n(x+1,\lambda) = \lambda S_n(x,\lambda)$.

We give an extension of Theorem 1 which we will use to find best quadrature formulas for integrals with a weight function.

Let g(x) be any function in $L^p$ which satisfies the recurrence relation $\lambda g(x-1) = g(x), 1\leq x\leq r$. We denote by $G_r$ the class of functions for which $G^{(n)}(x) = g(x)$, $x\epsilon(\nu,\nu+1), \nu=0,1, \ldots, r-1$ and (1) and (2) hold. If we choose g = S1 then $G_r$ is just $M_r$.

Theorem 2. Suppose $\lambda$ is a complex constant of modulus one. Then an element of smallest norm in $G_r$ is given (uniquely when $1<p<\infty$) by $G = Sh$ where h is the element of minimum norm in $G_1$.

The application we have in mind for this theorem is the construction of best quadrature formulas of the form

$$(5) \qquad \sum_{k=0}^{r} \sum_{\ell\in J} c_{k\ell} f^{(\ell)}(k)$$

for the $n^{th}$ Fourier coefficient of a function. Here J is some subset of $\{0,1,\ldots,n-1\}$. Thus we desire to find the minimum of the norm of the error functional

$$(6) \qquad \operatorname*{Max}_{\|f^{(n)}\|_r} \left|\int_0^r f(x)\cos\frac{2\Pi Nx}{r}\,dx - \sum_{k=0}^{r} \sum_{\ell\in J} c_{k\ell} f^{(\ell)}(k)\right|.$$

An appropriate integration by parts will show that this problem is equivalent to finding the element of minimum norm among all functions H which satisfy the conditions (a) $H^{(n)}(x) = \cos(2\Pi N/r)x$, $x\epsilon(\nu,\nu+1)$, $\nu=0,1,\ldots r-1$, (b) $H^{(i)}(\nu^+) = H^{(i)}(\nu^-)$, $\nu=1,\ldots,r-1$, $i\epsilon I \equiv \{n-1-j:j\not\epsilon J\}$ and (c) $H^{(i)}(r) = H^{(i)}(0) = 0$, $i\epsilon I$ (cf. [2]). Theorem 2 does not apply directly for two reasons. First we don't have periodic boundary conditions and secondly $\cos(2\Pi N/r)x$ does not satisfy the appropriate recurrence relation. The first difficulty can be overcome by making the assumption that $J\supseteq\{0,1,3,\ldots,\leq n-1\}$.

Thus from the invarience of $\cos(2\Pi N/r)x$ when x is replaced by r-x we conclude that the minimum to the above problem is also the minimum within the larger class of functions satisfying (a), (b) and $H^{(i)}(r) = H^{(i)}(0)$, $i\epsilon I$. To circumvent the second difficulty we consider the class $G_r$ where $g(x) = \exp(2\Pi iNx/r)$ and $\lambda = \exp(2\Pi iNx/r)$. If we let $G_0$ be the element of minimum norm in $G_r$ (which can be constructed by Theorem 2) we would expect since $\lambda^r = 1$ that $\operatorname{Re}G_0$ is the function we seek. This does not always seem to be the case except when $p=2$. Thus we have proved the following.

Corollary 4. Suppose $J$ is some subset of $\{0,1,\ldots,n-1\}$ which contains the set $\{0,1,3,\ldots,\leq n-1\}$. Then the unique best quadrature formula in $L^2[0,r]$ of the form (5) for the $N^{th}$ Fourier coefficient of a function $\int_0^r f(x)\cos(2\pi N/r)x$ is given by

$$(7)\quad \sum_{0<2i\leq n} \frac{\operatorname{Re} h^{(n-2i)}(0)}{n!}\left(f^{(2i-1)}(0) - f^{(2i-1)}(r)\right) - 2\sum_{2i\in J} \frac{\operatorname{Re} h^{(n-2i-1)}(0)}{n!} T_r f^{(2i)} ,$$

where $h$ is the function of minimum norm among all functions satisfying $h^{(n)}(x) = \exp(2\pi iNx/r)$, $0\leq x\leq 1$ and $h^{(n-1-i)}(1) = \lambda h^{(n-1-i)}(0)$, $i\notin J$ and $T_r f = \frac{1}{2}f(0)+f(1)+\ldots+f(r-1)+\frac{1}{2}f(r)$, the trapezoidal sum.

The above result holds for all $L^p$-spaces in the case that $N=0$ ([2]). We also remark the proofs of Theorems 1 and 2 have the advantage over the method employed in [2] in that they apply to certain convex subsets of $M_r$. For instance, in a recent paper [3] finding the minimum element among all nonnegative elements of $M_r(\lambda,I)$ was considered for a special choice of $\lambda$ and $I$. We state one possible corollary which generalizes some results in [3]. The proof is identical with that for Theorem 1.

Corollary 5. Suppose $n$ is an even integer. Then among all monosplines which are nonnegative on $[0,r]$ and satisfy the boundary conditions $M^{(i)}(r) = M^{(i)}(0), i=n-3,n-5,\ldots$, the one of minimum norm is given by $M = \overline{B}_n + c_0$. The constant $c_0$ is defined to give the smallest value of $\|B_n + c\|_1$ for all $c$ which satisfy the constraint $B_n(x) + c \geq 0$, $0\leq x\leq 1$.

Our previous discussion was restricted to nodes equally spaced. Let us now consider nodes $x_\mu$ which are periodic with period s ($\geq 1$), $x_{\mu+s} = x_\mu + 1$, $\mu = 0,1,\ldots,s(r-1)$ and $0 = x_0 < x_1 < \ldots < x_{s-1} < 1$. We define the class $M_{r,s} = M_{r,s}(I_0, I_1, \ldots, I_{s-1}, \lambda)$ to be all functions such that

(1) $M^{(n)}(x) = n!\lambda^\ell$, $x \in (x_\mu, x_{\mu+1})$, $\mu=0,1,\ldots,rs-1$, $\ell = [\frac{\mu}{s}]$,

(2) $M^{(i)}(x_\mu^+) = M^{(i)}(x_\mu^-)$, $i \in I_j$, $\mu=1,2,\ldots,rs-1$, $\mu \equiv j \pmod s$

and

(3) $M^{(i)}(x_{rs}) = \lambda^r M^{(i)}(0)$.

Just as before an element of minimum norm in $M_{r,s}$ is Sp where p is the element of minimum norm from $M_{1,s}$. We can therefore give a class of quadrature formulas with the property that compounding a best quadrature formula always leads to a best formula. This is the class of formula such that the function data at the ends of the interval is the same and includes the value of the function as well as all odd order derivatives $\leq n-1$ where the formula integrates polynomials of degree n-1 exactly.

The question of uniqueness has not been considered here when $p = 1$ or $\infty$. It would be of interest to settle this question. For one result of this type see [3]. In [1] and [6] uniqueness is established in certain cases for the analogous problem on the whole real axis for $p = \infty$. As a final remark we note that an interesting feature of the result in Corollary 4 is that the second sum of (7) is only sensitive to the even integers in J. Thus if we consider formulas which only contain the value of the function and its derivative at each interior node the best formula is one where only the function values occur (just choose $J = \{0,1,3,\ldots,n-1\}$). This has led to the conjecture (at least when $p = 2$ and $N = 0$) that equally spaced nodes are best [5]. The question of finding optimal (variable nodes) quadrature formulas within the class considered here gives rise to nonlinear best approximation problems which are still not completely understood.

## References

[1] Cavaretta, A. S., On Cardinal Perfect Splines of Least Sup Norm on the Real Axis. Doctoral thesis, University of Wisconsin (1970).

[2] Micchelli, C. A., Best Quadrature Formulae at Equally Spaced Nodes. IBM Research Report, RC # (1972).

[3] Schmeisser, G., Optimale Quadraturformeln mit Semidefiniten Kernen. Numer. Math. 20, (1972), 32-53.

[4] Schoenberg, I. J., Cardinal Interpolation and Spline Functions IV. The Exponential Euler Splines, MRC Tech. Sum. Report # 1153 (1971).

[5] Schoenberg, I. J., Monosplines and Quadrature Formulae. In Theory and Applications of Spline Functions, T. N. E. Greville, ed., Academic Press, New York, 1969.

[6] Schoenberg, I. J. and Z. Ziegler, On Cardinal Monosplines of Least $L_\infty$-norm on the Real Axis, J. d'Analyse 23 (1970), 409-436.

Mathematical Sciences Department
IBM Thomas J. Watson Research Center
Yorktown Heights, New York

Remark added in proof: If we choose $I = \{0,1,\ldots,\ell\}, \ell \leq n-1$, in Theorem 1 we are led to similar results for spline functions with multiple knots. We omit the details.

# APPROXIMATIONS OF GENERALIZED INVERSES OF LINEAR OPERATORS IN BANACH SPACES[1]

R. H. Moore and M. Z. Nashed

Let $X$ be a (real or complex) Banach space, and let $[X]$ be the space of all bounded linear operators on $X$ equipped with the uniform norm. Consider $A \in [X]$ with null space $N$ and range $R$. Assume there exist projectors (i.e. continuous linear idempotents) $U: X \to N$, and $E: X \to R$. Let $P = I - U$ and $M = PX$, so that $X = M \oplus N = PX \oplus UX$ and $X = EX \oplus (I - E)X$. Relative to these projectors, A has a unique generalized inverse $A^{\dagger}: X \to M$ such that

$$(1) \quad \begin{array}{ll} \text{(a)}\ A^{\dagger}A = P, & \text{(b)}\ A^{\dagger}AA^{\dagger} = A^{\dagger}, \\ \text{(c)}\ AA^{\dagger} = E, & \text{(d)}\ AA^{\dagger}A = A, \end{array}$$

(see for instance [5], [6]). For each $y \in X$, $A^{\dagger}y$ is a solution of the equation $Ax = Ey$. For $y \notin AX$, $A^{\dagger}y$ may thus be considered an approximate solution of the operator equation $Ax = y$.

If $X$ is a Hilbert space and the range $AX$ is closed, then projectors E, U, as specified above always exist; further if they are chosen to be orthogonal projectors, then $A^{\dagger}$ is the Moore-Penrose inverse and $A^{\dagger}y$ is the least-squares solution of minimal norm of $Ax = y$.

One objective of this paper is to show that if $B \in [X]$ approximates A in a suitable sense then, while $B^{\dagger}$ need not approximate $A^{\dagger}$, still there is an operator $B^{\emptyset}$ mapping $X$ onto $M$ which approximates $A^{\dagger}$ and corresponds to projectors satisfying conditions similar to (1).

Theorem 1. Let $A \in [X]$ with associated projectors E, U, P, and with $M = PX$ as described above. Let $p(\alpha)$ denote a polynomial in $\alpha$ such that $p(0) = 1$. Let $B \in [X]$ and define

$$H := A^{\dagger} (A - B)\, p(EB).$$

Let $H_M$ denote the restriction of H to $M$. If for some polynomial p, $\delta := \|H_M\| < 1$, then there exists a continuous linear operator $B^{\emptyset}$ from $X$ onto $M$ such that

(a) $B^{\emptyset}B = I_M$ so $B^{\emptyset}BB^{\emptyset} = B^{\emptyset}$ and $BB^{\emptyset}B_M = B_M$,

(b) $F := BB^{\emptyset}$, $Q := B^{\emptyset}B$ and $V := I - Q$ are projectors,

(c) $QX = M$, $FX = BM$,

(d) $B_M$ is one-to-one, $(B_M)^{-1} = B^{\emptyset}|_{FX}$,

(e) $N(B) \subset N(FB) = VX$,

(f) $\dim N(B) \leq \dim N(A)$,

(g) $B^{\emptyset}(I - E) = 0$ and $N(A^{\dagger}) = (I - E)X \subset (I - F)X = N(B^{\emptyset})$.

Furthermore, the following estimates hold:

(h) $\|Q - P\| = \|V - U\| \leq \frac{1}{1-\delta} \|HU\| \leq \frac{\delta}{1-\delta} \|U\|$,

(i) $\|B^{\emptyset}x - A^{\dagger}x\| \leq \frac{\|A^{\dagger}\|}{1-\delta} (\delta\|x\| + \|(A - B)q(EB)Ex\|)$, $x \in X$,

(j) $\|Fx - Ex\| \leq \|B\| \, \|B^{\emptyset}x - A^{\dagger}x\| + \|(B - A)A^{\dagger}x\|$, $x \in X$.

The operator $B^{\emptyset}$ is given by

$$B^{\emptyset} := J_M^{-1} A^{\dagger}[I + (A - B)\, q(EB)E],$$

where

$$J := I - H, \text{ and } p(\alpha) = 1 - \alpha q(\alpha).$$

Remark. There is a particular provision for the cases $\dim N(B) \neq \dim N(A)$ or $\operatorname{codim} BX \neq \operatorname{codim} AX$, in which cases $B^{\emptyset}$ is not a generalized inverse of B. The range of $BB^{\emptyset}$ does not coincide with $BX$ and hence the analog of (1c) does not hold.

When is $B^{\emptyset} = B^{\dagger}$ (relative to the projectors Q, F)? According to the definition of the generalized inverse, it is necessary and sufficient that $QX = M$ be complementary to $N(B)$, and $FX = BX$. A more useful form of this criterion is given in the special case $p(\alpha) \equiv 1$ in the following corollary.

Corollary. Suppose $\|A^{\dagger}(A - B)\| < 1$. Then $B^{\phi} = B^{\dagger}$ if and only if $\dim N(B) = \dim N(A)$ and $\operatorname{codim} BX = \operatorname{codim} AX$.

The setting of this paper includes as special cases results on continuous dependence and perturbations of generalized inverses of matrices obtained by Ben-Israel [3] and Stewart [8]. Other specializations of Theorem 1 yield some results of Anselone [1], Anselone and Moore [2], and Ostrowski [7] on inverse operator approximations.

Theorem 1 is applied for $A = I - K$ and $A_n = I - K_n$, where $\|K_n x - Kx\| \to 0$ (but $\|K_n - K\| \not\to 0$) and $\{K_n\}$ is collectively compact (e.g. K is an integral operator, $K_n$ is defined by numerical quadrature). In this case the key estimate shows that $\|(A - B)\, p(EB)\|$ can be made small even when $\|A - B\|$ cannot. Two types of theorems are established (cf. [1, §1.8; p. 10] and [4, Ch. XIV; p. 542]): (1) $A_n^{\phi}$ is obtained from and related to $A^{\dagger}$, which is assumed given; (2) $A^{\phi,n}$ is obtained from and related to $A_n^{\dagger}$, which is assumed given.

These results are applied to least-squares solutions of minimal-norm of Fredholm integral equations of the second kind, and to consideration of the convergence of approximations obtained by means of numerical quadrature in the space of continuous functions. The details will appear elsewhere.

[1]Research supported in part by the Mathematics Research Center, the University of Wisconsin-Madison under contract No. DA-31-124-ARO-D-462.

## References

[1] Anselone, P. M., Collectively Compact Operator Approximation Theory. Prentice-Hall, Englewood Cliffs, New Jersey, 1971.

[2] __________ and R. H. Moore, Approximate solutions of

integral and operator equations. J. Math. Anal. Appl. 9 (1964), 268-277.

[3] Ben-Israel, A., On error bounds for the generalized inverse. SIAM J. Numer. Anal. 3 (1966), 585-592.

[4] Kantorovich, L. V. and G. P. Akilov, Functional Analysis in Normed Spaces. Transl. by D. E. Brown, Pergamon, New York, 1964.

[5] Nashed, M. Z., Generalized inverses, normal solvability, and iteration for singular operator equations, Nonlinear Functional Analysis and Applications. L. B. Rall, ed., Academic Press, New York, 1971, pp. 311-359.

[6] Nashed, M. Z. and G. F. Votruba, A unified approach to generalized inverses of linear operators: algebraic, topological and proximal properties. To appear.

[7] Ostrowski, A. M., General existence criteria for the inverse of an operator. Amer. Math. Monthly 74 (1967), 826-827.

[8] Stewart, G. W., On the continuity of the generalized inverse. SIAM J. Appl. Math. 17 (1969), 33-45.

Department of Mathematics
University of Wisconsin
Milwaukee, Wisconsin 53201
and
School of Mathematics
Georgia Institute of Technology
Atlanta, Georgia 30332

# RECENT RESULTS ON MINIMAL PROJECTIONS

P. D. Morris

Little is known about the problem of characterizing the minimal projections onto finite-dimensional subspaces of Banach spaces. (By a <u>minimal projection</u> onto a given subspace, we mean a bounded linear projection with least norm.) Even in the important special case where the Banach space is $C[a,b]$, $[a,b]$ a real interval, and the subspace is $Y_n$, the space of polynomials with degree at most $n$, this problem is unsolved (when $n \geq 2$).

The difficulty of these problems is partly explained by the following observation. Let $Y$ be an $n$-dimensional subspace of an infinite-dimensional normed space $X$. Let $\{y_1,\ldots,y_n\}$ be a basis of $Y$. Then any set $\{\varphi_1,\ldots,\varphi_n\} \subseteq X^*$ satisfying $\varphi_i(y_j) = \delta_{ij}$ determines a projection onto $Y$ via: $x \to \Sigma\varphi_i(x)y_i$. It is thus seen that the set of all bounded projections of $X$ onto $Y$ is huge, in fact an infinite-dimensional affine manifold in the space of operators from $X$ to $Y$.

In this paper, we shall consider the more tractable problem of characterizing the elements of least norm in a set of projections much smaller than the entire set. This is done with the hope that results obtained concerning the easier problem will yield insight into the general one.

Our setting will be as follows. Let $Y$ be an $n$-dimensional subspace of (real) $C(T)$, where $T$ is a compact Hausdorff space. Let $S = \{t_0,\ldots,t_n\}$ be a set of $n+1$ distinct points of $T$. We assume that the point-evaluation functionals $\{\hat{t}_0,\ldots,\hat{t}_n\}$ are total over $Y$. Thus any $y$ in $Y$ which vanishes on $S$ must vanish everywhere. Let $\mathcal{P}$ denote the set of projections of C(T) onto Y which are supported on S. (A linear operator L, defined on C(T), is <u>supported</u> on a subset A of T if Lx = 0 whenever x in C(T) vanishes on A.) It is easy to see that every member P of $\mathcal{P}$ has the form $P = \Sigma\, \hat{t}_i \otimes y_i$, where $y_0, \ldots, y_n$ are in Y and the tensor notation means: $Px = \Sigma\, x(t_i)y_i$ for any x in C(T). Our assumption that $\{\hat{t}_0, \ldots, \hat{t}_n\}$ is total over Y implies that there exist (uniquely) numbers $\theta_0, \ldots, \theta_n$ such that $\Phi = \Sigma\, \theta_i \hat{t}_i$ annihilates Y, such that $\Sigma\, |\theta_i| = 1$, and such that the first non-zero $\theta_i$ is positive. Then if P is in $\mathcal{P}$, we have $\mathcal{P} = \{P + \Phi \otimes y : y \in Y\}$, so that $\mathcal{P}$ is an n-dimensional affine manifold (see [1]).

Let $P = \Sigma\, \hat{t}_i \otimes y_i$ in $\mathcal{P}$. We define crit $P = \{t \in T : \|\hat{t} \circ P\| = \|P\|\}$. If $\Lambda = \Sigma\, |y_i|$ (the <u>Lebesgue function</u> of P), then it is easy to see that crit $P = \{t \in T : \Lambda(t) = \|\Lambda\| = \|P\|\}$. To state a theorem, we need to define two auxiliary functions: $v(t) = \Sigma \theta_i \operatorname{sgn} y_i(t)$, $u(t) = \Sigma\, \{|\theta_i| : y_i(t) = 0\}$, $t \in T$.

<u>Theorem 1</u>. $P = \Sigma\, \hat{t}_i \otimes y_i$ is minimal in $\mathcal{P}$ if and only if there does not exist y in Y with $v \operatorname{sgn} y > u$ on crit P.

This result was proved by Price and Cheney [2] with more hypotheses on T and Y (viz. T a real interval, Y a Haar subspace of dimension at least 3 which contains the constants). As they point out, their proof of the sufficiency does not require the additional hypotheses.

We sketch the proof of the necessity in Theorem 1. Assume y in Y is such that $v \operatorname{sgn} y > u$ on crit P. Note that, with $\Sigma$ denoting the set $\{z \in C(T) : \|z\| = 1,\ |z(t_i)| = 1,\ 0 \le i \le n\}$, $\|P\| = \max\, \{\|Pz\| : z \in \Sigma\}$. Note also that this maximum is actually taken over a finite set since functions which agree on S are not distinguished by P. Now for $\lambda > 0$, let $Q_\lambda = P - \lambda\Phi \otimes y$. Then $Q_\lambda$ is in $\mathcal{P}$. With the aid of the inequality $v \operatorname{sgn} y > u$, one can show that, given z in $\Sigma$ with $\|Pz\| = \|P\|$, $|(Q_\lambda z)(t)| < \|P\|$ for sufficiently small $\lambda$ and for t in a neighborhood of the set

$\{t\in T: |(Pz)(t)| = \|P\|\}$. Then, for $\lambda$ sufficiently small, $\|Q_\lambda z\| < \|P\|$. For z in $\Sigma$ with $\|Pz\| < \|P\|$, using the observation that, if $0<\lambda_1<\lambda_2$, $Q_{\lambda_1}$ is a convex combination of P and $Q_{\lambda_2}$, one easily establishes again that $\|Q_\lambda z\| < \|P\|$ for sufficiently small $\lambda$. Since, as above, $\Sigma$ is essentially finite, we are done.

For the sequel, we need more assumptions on T and Y. From now on, T is a closed real interval $[a,b]$ and Y is a Haar subspace which contains the constants. It turns out that the way in which S is distributed in T makes a difference. We define $E = \max\{\|y\|: y\in Y \text{ and } |y(t_i)| \leq 1, 0\leq i\leq n\}$. It is easy to see that $\|P\| \geq E$ for each P in $\mathcal{P}$. Hence $\|P\| = E$ implies P is minimal in $\mathcal{P}$. We have

<u>Theorem 2</u>. $P = \Sigma\, \hat{t}_i \otimes y_i$ is minimal in $\mathcal{P}$ if and only if either:

(a) $\|P\| = E$;

or

(b) there exist n+1 points $s_0<s_1<\ldots<s_n$ in crit P with $\operatorname{sgn} v(s_{i+1}) = -\operatorname{sgn} v(s_i)$, $0\leq i\leq n-1$.

The main point of the proof is to verify that if $\|P\| > E$, then $|v| > u$ on crit P.

The condition (a) may or may not hold for a minimal projection in $\mathcal{P}$. Let $T = [-1,1]$, let $Y = Y_2$, the polynomials of degree no greater than 2, and let $S = \{-1,-s,s,1\}$, where $0<s<1$. Then if $s = 1/2$, the minimal projections have norm E, while if $s = 1/4$, the minimal projection has norm greater than E. Verification of these assertions is straightforward but tedious.

As a consequence of Theorem 2, we have:

<u>Corollary 3</u>. If P in $\mathcal{P}$ is minimal and if $\|P\| > E$, then P is the unique minimal projection in $\mathcal{P}$.

To prove this, suppose $Q = P + \Phi \otimes y$ is also minimal. It follows, without much difficulty, that $y(s)v(s) \leq 0$ for s in crit P. Since condition (b) of Theorem 2 holds, this implies that $y = 0$, so that $Q = P$.

The "De La Vallée Poussin Best Approximation Operator" A [2] is an interesting member of $\mathcal{P}$. For x in C[a,b], Ax is defined to be that member y of Y such that

$$\max \{|x(t_i) - y(t_i)| : 0 \leq i \leq n\}$$

is a minimum. In the case where $T = [-1,1]$, $Y = Y_2$, $S = \{-1,-s,s,1\}$, $0<s<1$, Price and Cheney [2] showed that A is minimal if $1/2 \leq s < 1$ and not otherwise. In this setting, it happens that (a) of Theorem 2 always holds when A is minimal. In these cases, it is also not difficult to construct other minimal projections. Thus the conclusion of the Corollary may not hold if the hypothesis "$\|P\| > E$" is dropped.

## References

[1] Cheney, E. W., Projections with Finite Carrier. CNA Report 28 (1971), University of Texas.

[2] Price, K. H. and E. W. Cheney, Extremal Properties of Approximation Operators. CNA Report 54 (1972), University of Texas.

Department of Mathematics
Pennsylvania State University
University Park, Pa. 16802

## SOME REMARKS ON POINTWISE SATURATION

G. Mühlbach

### 1. Introduction

This note can be regarded as a reformulation of the author's note [4] in view of the results obtained by Lorentz and Schumaker [3] and Berens [1].

Saturation is a well-known phenomenon in approximation theory. Roughly speaking, a sequence of approximation operators is said to be saturated if there exists an "optimal" order of approximation, called the "saturation order," such that better approximation occurs only in trivial cases. This will be made precise in two directions corresponding to pointwise and uniform convergence, respectively. Applications will be made to sequences of positive linear operators of Voronovskaja-type and of Bernstein-type, generalizing some results of the author [4] and of Lorentz and Schumaker [3].

### 2. Pointwise and Uniform Saturation

Let I be a nonvoid set and $\mathcal{F} := \{s = (s_n)_1^\infty : 0 \leq s_n \in R^I\}$. For elements of $\mathcal{F}$ we define

$$s \sim t \; [s \approx t] : \Longleftrightarrow s_n = \mathcal{O}(t_n) \text{ and } t_n = \mathcal{O}(s_n) \text{ pointwise}$$
$$[\text{uniformly}] \text{ on } I.$$

Clearly, $\sim$ and $\approx$ are equivalence relations on $\mathcal{F}$, the corresponding quotient sets will be denoted by $F := \mathcal{F}/\sim$ and $F^* := \mathcal{F}/\approx$.

Now let E be a vector subspace of the real vector space $R^I$ of all real functions on I. Let $(L_n)_1^\infty$ be a sequence of operators $L_n: E \to E$ and $R_n f := L_n f - f$. To each $s \in \mathcal{F}$ we assign sets $\mathcal{O}(s)$, $o(s)$ $[\mathcal{O}^*(s), o^*(s)]$

$$\begin{aligned}(1)\quad \mathcal{O}(s) &= \{f : R_n f = \mathcal{O}(s_n) \text{ pointwise}\}\ [\mathcal{O}^*(s) = \{f : R_n f = \mathcal{O}(s_n) \text{ uniformly}\}] \\ o(s) &= \{f : R_n f = o(s_n) \text{ pointwise}\}\ [o^*(s) = \{f : R_n f = o(s_n) \text{ uniformly}\}]\end{aligned}$$

It is easily seen that in fact these classes depend only on the equivalence class, corresponding to $\sim$ or $\approx$, respectively, generated by s, which for convenience in every case is also denoted by s. Indeed, we have

$$t \sim s \Rightarrow \mathcal{O}(t) = \mathcal{O}(s),\ o(t) = o(s)\ [t \approx s \Rightarrow \mathcal{O}^*(t) = \mathcal{O}^*(s),\ o^*(t) = o^*(s)].$$

Evidently, for each $s \in F$

$$\begin{array}{ccc}\mathcal{O}^*(s) & \subset & \mathcal{O}(s) \\ \cup & & \cup \\ o^*(s) & \subset & o(s).\end{array}$$

Now let a sequence of operators $(L_n)$ be fixed and consider

$$A := \{s \in F: o(s) \neq \mathcal{O}(s)\}\quad [A^* := \{s \in F^* : o^*(s) \neq \mathcal{O}^*(s)\}].$$

Each $s \in A$ $[A^*]$ is called a strong approximation order for $(L_n)$ with respect to pointwise [uniform] convergence.

Note that F $[F^*]$ if partially ordered by

$$s \leq t :\Longleftrightarrow s = t \text{ or } s < t,$$

where $s < t$ is defined to hold if and only if $s_n = o(t_n)$ pointwise [uniformly] on I, for at least one and hence by equivalence for all elements of each class compared.

<u>Definition 1</u>. If for a sequence $(L_n)$ of operators $L_n : E \to E$ there exists in A $[A^*]$ a minimal element s with respect to $\leq$, then $(L_n)$ is said to be pointwise [uniformly] <u>saturated</u> with order s and $o|\mathcal{O}$-saturation classes (1).

In general minimal elements are not unique. As is to be

expected it can be shown by examples (compare the example given at the end of this note) that a saturation order need not be unique. But it always can be characterized by the following

Theorem 1. Let $s \in A\ [A^*]$. A sequence $(L_n)$ of operators $L_n : E \to E$ is pointwise [uniformly] saturated with order s if and only if

$$f \in o(s) \Rightarrow \bigwedge_{x\in I} \bigvee_{n_o\in N} \bigwedge_{n\geq n_o} R_n f(x) = 0 \left[ f \in o^*(s) \Rightarrow \right.$$

$$\left. \bigvee_{n_o\in N} \bigwedge_{x\in I} \bigwedge_{n\geq n_o} R_n f(x) = 0 \right].$$

Proof: (i) Suppose $\hat{s} \in F$ and $\hat{s} < s$. Then $\hat{s} \notin A$, for $f \in O(\hat{s})$ implies $R_n f = O(\hat{s}_n)$ pointwise (where $(\hat{s}_n)$ generates $\hat{s}$) and, since $f \in o(s)$, $\bigwedge_x \bigvee_{n_o} \bigwedge_{n\geq n_o} R_n f(x) = 0$, but it is easily seen that this is equivalent to $R_n f = o(R_n f)$ pointwise. Hence $R_n f = o(R_n f) = o(\hat{s}_n)$.

(ii) When $(L_n)$ is saturated with order s and $f \in o(s)$, then for the equivalence class t (generated by $(R_n f)$) we have $t < s$. Since s is a saturation order it follows $t \notin A$, hence $O(t) = o(t)$, which means that $(R_n f)$ is of the kind indicated.

Similarly in the case corresponding to uniform convergence.

In some cases it is easy to see that a given sequence of operators is saturated.

Theorem 2. Let $(L_n)$ be a given sequence of operators. If there exists a function $h \in E$ such that for $s = (|R_n h|)$

1. $s<s$ (in F) is not true,
2. $R_n f = o(|R_n h|)$ pointwise on I, implies that f is an invariant element for all operators $L_n$,

then $(L_n)$ is pointwise and uniformly saturated with order s and $o^*(s) = o(s) = \{f : \bigwedge_{n\in N} L_n f = f\}$.

Proof: Since in F $s<s$ is false, $s \in A$. Suppose $\hat{s} \in F$

and $\hat{s}<s$. Then $o(\hat{s}) \subset O(\hat{s}) \subset o(s) \subset o(\hat{s})$, hence $\hat{s} \notin A$. Similarly $\hat{s} \in F^*$ and $\hat{s}<s$ (in $F^*$) implies $\hat{s} \notin A^*$. Hence $(L_n)$ is pointwise and uniformly saturated with order $s = (s_n)$, $s_n = |R_n h|$.

## 3. Saturation of Sequences of Bernstein- and Voronovskaja-type

To give an application we consider some sequences of positive linear operators. Let $I = [a,b]$ be a compact real interval (not consisting of one point only), $E = C(I)$ and $(f_0,f_1,f_2)$ a complete Čebyšev system [2, p. 1] in $C(I)$.

Definition 2. A sequence $(L_n)$ of operators $L_n\colon C(I) \to C(I)$ is called of Voronovskaja-type with respect to $(f_0,f_1,f_2)$ if

(A) for all $n \in \mathbb{N}$, $L_n$ is linear and positive,

(B) $\bigwedge_{n\in\mathbb{N}} R_n f_2 \geq 0$ and $\bigwedge_{x\in(a,b)} \bigwedge_{m\in\mathbb{N}} \bigvee_{n\geq m} R_n f_2(x) > 0$,

(C) $\bigwedge_{x\in(a,b)} R_n f(x) = o[R_n f_2(x)]$, $n\to\infty$ whenever $f=f_0, f=f_1$ or $f$ vanishes in some neighborhood of $x$.

The sequence $(L_n)$ is called of Bernstein-type with respect to $(f_0,f_1,f_2)$ if in addition

(D) $\bigwedge_{n\in\mathbb{N}} \bigwedge_{k=0,1} L_n f_k = f_k$.

For example, the well-known Bernstein polynomials are of Bernstein-type but there are many other examples, compare [3] and [4]. Two facts should be noted here: 1. Some special cases of sequences of Voronovskaja-type have been investigated earlier by Lorentz and Schumaker [3]. They have used a complete extended Čebyšev system $(f_0,f_1,f_2)$ in the sense of Karlin and Studden ([2], p. 375) and instead of our conditions (B) and (C) they have assumed a Voronovskaja-theorem to be valid.

2. Lorentz and Schumaker have noticed that sequences of this special Voronovskaja-type of "higher orders" than two do

not exist ([3], Theorem 4.2, p. 419). This can easily be extended to the more general case considered here. The following theorem completes investigations of the author [4] and generalizes results obtained by Lorentz and Schumaker [3].

<u>Theorem 3</u>. If $(L_n)$ satisfies (A), (B) and (D), then $(L_n)$ is pointwise and uniformly saturated with order $s = (R_n f_2)$ and $o^*(s) = o(s) = \text{span}(f_0, f_1)$.

If $(L_n)$ is of Bernstein-type with respect to $(f_0, f_1, f_2)$, then $O^*(s) = \text{Lip}\,(f_0, f_1, f_2)$. Moreover, for $M \geq 0$

$$(2) \quad \bigwedge_{n\in\mathbb{N}} \bigwedge_{x\in I} |R_n f(x)| \leq M \cdot R_n f_2(x) \iff f \in \text{Lip}_M(f_0, f_1, f_2).$$

If $(L_n)$ is of Voronovskaja-type with respect to $(f_0, f_1, f_2)$, then $s \in A$ and $s \in A^*$ (i.e., s is a strong approximation order for $(L_n)$ with respect to pointwise and to uniform convergence) with $o^*(s)$, $o(s)$, $O^*(s)$ as before. Moreover (with $o(1) \to 0$ pointwise)

$$(3) \quad \bigwedge_{n\in\mathbb{N}} \bigwedge_{x\in I} |R_n f(x)| \leq [M + o(1)] \cdot R_n f_2(x) \iff f \in \text{Lip}_M(f_0, f_1, f_2).$$

Here we have used the notation:

$$\text{Lip}\,(f_0, \ldots, f_k) := \bigcup_{M>0} \text{Lip}_M(f_0, \ldots, f_k),$$

$$\text{Lip}_M(f_0, \ldots, f_k) := \left\{ f : \left\| \begin{bmatrix} f_0, \ldots, f_k \\ x_0, \ldots, x_k \end{bmatrix} f \right\| \leq M \text{ for all distinct knots } x_i \in I \right\},$$

$$\begin{bmatrix} f_0, \ldots, f_k \\ x_0, \ldots, x_k \end{bmatrix} f := \frac{V\begin{pmatrix} f_0, \ldots, f_{k-1}, f \\ x_0, \ldots, x_{k-1}, x_k \end{pmatrix}}{V\begin{pmatrix} f_0, \ldots, f_{k-1}, f_k \\ x_0, \ldots, x_{k-1}, x_k \end{pmatrix}}, \quad V\begin{pmatrix} g_0, \ldots, g_k \\ x_0, \ldots, x_k \end{pmatrix} := \det g_j(x_i).$$

<u>Proof:</u> The saturation properties will follow by Theorem 2 if it can be shown that

$$R_n f = o(R_n f_2) \text{ pointwise} \iff f \in \text{span}(f_0, f_1).$$

But this is a part of (Theorem 3.13, [4], p. 288). To prove the second statement (about Bernstein-type operators) it suffices to show (2). We need the

<u>Lemma</u>. ([4], p. 290 or [3], p. 418). If $(L_n)$ is of Voronovskaja-type with respect to $(f_0,f_1,f_2)$ then

$$\bigwedge_{x\in(a,b)} \bigwedge_{n\in\mathbb{N}} R_nf(x) \geq o[R_nf_2(x)], n\to\infty \Rightarrow f \text{ is convex with respect to } (f_0,f_1,f_2) \text{ on } [a,b].$$

f is called convex [strictly convex] with respect to a Čebyšev system $(f_0,\dots,f_k)$ on I if $\begin{bmatrix} f_0,\dots f_k \\ x_0,\dots,x_k \end{bmatrix} f \geq 0$ [$> 0$] for all distinct knots $x_i \in I$.

Now the left hand side of (2) is equivalent to $\bigwedge_{x\in I} \bigwedge_{n\in\mathbb{N}} R_n(Mf_2 \pm f)(x) \geq 0$. By the lemma it follows that both functions $Mf_2 \pm f$ are convex with respect to $(f_0,f_1,f_2)$ which in fact is equivalent to $f \in \mathrm{Lip}_M(f_0,f_1,f_2)$. Conversely this implies $\bigwedge_{x\in I} \bigwedge_{n\in\mathbb{N}} |R_nf(x)| \leq MR_nf_2(x)$ by (Theorem 1.5 [4], p. 276).

The proof of the third statement (concerning $o^*(s)$ and $o(s)$ in the case of operators of Voronovskaja-type) can be established by the same arguments used for the first statement (compare Theorem 3.21 in [4], p. 290). "$\Rightarrow$" in (3) again follows by the lemma. To prove the converse we use the fact that to each convex function g and to each $x \in (a,b)$ there exists a "supporting function" $\varphi = a_0f_0 + a_1f_1$ such that $\varphi(x) = g(x)$, $\varphi \leq g$ (see for example [5]). From this "$\Leftarrow$" in (3) follows by property (C).

In some cases where the underlying system $(f_0,f_1,f_2)$ is generated like an extended complete Čebyšev system (compare [2], Theorem 1.2 on p. 379) the functions of $\mathrm{Lip}_M(f_0,f_1,f_2)$ possess characteristic differential properties.

Corollary. Let $c \in [a,b]$ and

$$f_0(x) := w_0(x)$$

$$f_1(x) := w_0(x) \int_c^x w_1(\xi)dv_1(\xi)$$

$$f_2(x) := w_0(x) \int_c^x w_1(\xi) \int_c^\xi w_2(\eta)dv_2(\eta)dv_1(\xi),$$

where the functions $w_0, w_1, w_2 \in C[a,b]$ are strictly positive and $v_1, v_2 \in C[a,b]$ are strictly increasing on $[a,b]$. Then

(4) $f \in Lip_M(f_0,f_1,f_2) \Longleftrightarrow d_1 f \in Lip_M(w_1, d_1 f_2)$

$\left[\Longleftrightarrow |d^2 f(x)| \leq M w_2(x)\right.$ a.e. if in addition $v_2$ is assumed to be absolutely continuous on $\left.[a,b]\right]$.

Here we have used the notation

$$d^2 = d_2 d_1, \quad d_k f = D_k\left(\frac{f}{w_{k-1}}\right), \quad D_k f(x) = \lim_{h\to 0} \frac{f(x+h) - f(x)}{v_k(x+h) - v_k(x)}.$$

An elementary proof can be found in [5].

## 4. Saturation of Čebyšev Splines of Order 1

Consider a complete Čebyšev system $(f_0,f_1,f_2)$ in $C[a,b]$ where for all $x$, $f_0(x) = 1$. For each sequence of subdivisions $(Z_n)$ of $[a,b]$

$$Z_n = \{x_k^{(n)}: a = x_0^{(n)} < x_1^{(n)} < \ldots < x_n^{(n)} = b\}$$

and for each $f \in C[a,b]$ let $L_n f$ on $[x_k^{(n)}, x_{k+1}^{(n)}]$ be the unique linear combination of $f_0, f_1$ that interpolates $f$ in $x_k^{(n)}$ and $x_{k+1}^{(n)}$ $(k=0,\ldots,n-1)$. Then $L_n f$ is the unique continuous Čebyšev spline of order 1 with respect to $(f_0,f_1)$, interpolating $f$ in the knots of $Z_n$.

Evidently for each $n$, $f_0$ and $f_1$ are invariant elements of $L_n$. Now consider a sequence $(Z_n)$ of subdivisions such that

$$\max_k (x_{k+1}^{(n)} - x_k^{(n)}) \to 0$$

as $n \to \infty$ and such that for each $x \in (a,b)$ there are infinitely many n such that $x \notin Z_n$. Then $(L_n)$ is of Bernstein-type with respect to $(f_0, f_1, f_2)$ and hence by Theorem 3 pointwise and uniformly saturated with order $s = (s_n)$, $s_n = R_n f_2$.

Now consider a function $g \in C[a,b]$, strict convex with respect to $(p_0, p_1, p_2)$ $[p_i(x) = x^i]$, which is not differentiable at least at one point of $(a,b)$. Then according to Theorem 3, $(L_n)$ is uniformly saturated both with order $(R_n g)$ and $(R_n p_2)$ but $(R_n g) \approx (R_n p_2)$ does not hold. According to the corollary, from the contrary it would follow $d_1 g = g' \in \mathrm{Lip}\,(1, 2p_1)$ which contradicts the choice of g. Thus we conclude that a saturation order need not be unique.

## References


[1] Berens, H., Pointwise saturation of positive operators. J. Approx. Theory, 6 (1972), 135-147.

[2] Karlin, S. and W. Studden, Tshebycheff systems. Interscience, New York, 1966.

[3] Lorentz, G. G. and L. L. Schumaker, Saturation of positive operators. J. Approx. Theory, 5 (1972), 413-424.

[4] Mühlbach, G., Operatoren vom Bernsteinschen Typ. J. Approx. Theory, 3 (1970), 274-292.

[5] Mühlbach, G., Čebyšev -Systeme und Lipschitzklassen. J. Approx. Theory. In print.



Institut für Mathematik
Technische Universität Hannover
3000 Hannover
West Germany

# A LINEAR BEST APPROXIMATION OPERATOR

K. H. Price and E. W. Cheney[1]

In an effort to elucidate some of the long outstanding problems associated with approximation by linear projections, recent papers [1], [2], [3] have focused their attention on projections having finite dimensional range. Motivated by these investigations, we solve in a general setting a classical problem of de La Vallée Poussin [4]. In order to state the problem and the subsequent theorems, we adopt the following notations:

(1) Let $X$ be a normed linear space and $Y$ an $n$-dimensional subspace of $X$.

(2) Let $\{f_0,\ldots,f_n\}$ be a subset of $X^*$ total over $Y$, i.e. $y \in Y$ and $f_i(y) = 0$ for $0 \le i \le n$ implies $y = 0$.

(3) Define a seminorm on $X$ by

$$\Delta(x) = \max_{0 \le i \le n} |f_i(x)| .$$

The problem is to find those elements of $Y$ which best approximate an arbitrary element of $X$ with respect to this seminorm. The solution is given in terms of a functional $\Phi$ having the following form and properties:

$$(4) \quad \Phi = \sum_{i=0}^{n} \theta_i f_i, \quad \Phi \in Y^{\perp}, \quad \sum_{i=0}^{n} |\theta_i| = 1.$$

The existence of $\Phi$ follows from the observation that $Y^*$ is of dimension $n$, and hence, the restricted functionals $f_i|Y$ form a linearly dependent set.

<u>Theorem 1</u>. Let the set of functionals (2) be independent over X and total over Y. Select $z \epsilon X$ such that $\Phi(z) = 1$, and define $Z = Y \oplus z$. Select $q_0, \ldots, q_n$ in Z such that $f_i(q_j) = \delta_{ij} - \theta_j \operatorname{sgn}\theta_i$. Then the operator A whose value at x is

$$\Sigma\, f_i(x) q_i, \text{ denoted } A = \Sigma\, f_i \otimes q_i,$$

is a linear projection of X onto Y and solves de La Vallée Poussin's problem: i.e. $\Delta(x-Ax) \leq \Delta(x-y)$ for all $x \epsilon X$ and $y \epsilon Y$.

<u>Proof</u>: First, we have that $q_i \epsilon Y$ for $0 \leq i \leq n$, because $\Phi(q_i) = 0$, and as a subspace of Z, Y is the null space of $\Phi|Z$. Hence, the range of A is a subspace of Y.

Second, since $\{f_0, \ldots f_n\}$ is total over Y and $f_i \circ (A-I) = -(\operatorname{sgn}\theta_i)\Phi$ for $0 \leq i \leq n$, $A|Y = I|Y$.

Third, for all $x \epsilon X$ and $y \epsilon Y$,

$$|\Phi(x)| = |\Phi(x-y)| = |\Sigma\, \theta_i f_i(x-y)| \leq \Sigma |\theta_i| \max_j |f_j(x-y)|$$

$$= \Delta(x-y).$$

Fourth, for all $x \epsilon X$ and $y \epsilon Y$,

$$\Delta(x-Ax) = \max_j |f_i(x-Ax)| \leq |\Phi(x)| \leq \Delta(x-y),$$

by the last two parts above.

We now relate this best approximation operator A to analogues of the classical Lagrange interpolation projections. Select an index $i \epsilon \{0,1,\ldots,n\}$, and suppose that the set of functionals $\{f_0, \ldots f_{i-1}, f_{i+1}, \ldots f_n\}$ is total over Y. This is equivalent to the assumption that $\theta_i \neq 0$. Now select elements $w_{ij} \epsilon Y$, such that $f_\nu(w_{ij}) = \delta_{\nu j}$ for $\nu \neq i$ and $j \neq i$. For convenience, define $w_{ii} = 0$ and let

$$(5) \quad L_i = \sum_j f_j \otimes w_{ij}.$$

By using the properties of $\Phi$ and the above definitions, we can verify

(6) $f_j \circ L_i = f_j$ if $j \neq i$.

(7) $f_i(w_{ij}) = -\theta_j \theta_i^{-1}$ if $i \neq j$ and $\theta_i \neq 0$.

(8) $L_j = L_i - \theta_j^{-1} \Phi \otimes w_{ij}$ if $\theta_i \neq 0$ and $\theta_j \neq 0$.

Theorem 2. The operator A is a convex linear combination of the operators $L_i$ (those that exist). In fact,

$$A = \sum_{\theta_i \neq 0} |\theta_i| L_i.$$

Proof: Let $B = \Sigma_{\theta_i \neq 0} |\theta_i| L_i$. To show that A = B, it is straightforward to prove that $f_i \circ A = f_i \circ B$ for $i=0,\ldots,n$, and then use totality of the set $\{f_0,\ldots,f_n\}$ over Y.

Theorem 3. If P is any projection of the form $\Sigma f_i \otimes y_i$ from X onto Y, then P and A are related by the equation

$$A = P - \Phi \otimes Pq \qquad q \in X,\ f_i(q) = \operatorname{sgn} \theta_i.$$

Proof: By Lemma 2 of [1], the operator P-A is of the form $\Phi \otimes u$ for some $u \in Y$. Then $(P-A)(q) = \Phi(q)u = u$. From the properties of A it follows that $f_i(Aq) = 0$ for $0 \leq i \leq n$. Hence, $Aq = 0$ and thus $u = Pq$.

---

[1]The first author was supported by a Faculty Research Grant from Stephen F. Austin State University. The second author was supported by the Air Force Office of Scientific Research, Office of Aerospace Research.

## References

[1] Cheney, E. W., Projections with finite carrier. Proceedings of a Conference at Oberwolfach June 1971. Series ISNM, Birkhauser-Verlag, Basel. To appear. CNA Report 28, The University of Texas at Austin.

[2] Morris, P. D. and E. W. Cheney, On the existence and characterization of minimal projections. CNA Report 37, The University of Texas at Austin, May 1972. To appear in

J. Reine und Angewandte Math.

[3] Morris, P. D. and E. W. Cheney, Stability properties of trigonometric interpolating operators. CNA Report 51, The University of Texas at Austin, August 1972. To appear in Math. Z.

[4] de La Vallée Poussin, C., Sur les polynomes d'approximation et la representation approchée d'un angle. Acad. Royale de Belgique, Bulletin de la Classe de Sciences, 1910, No. 12, 808-844.

Mathematics Department
Stephen F. Austin College
Nacogdoches, Texas 75961

Department of Mathematics
The University of Texas
Austin, Texas 78712

# APPROXIMATION THEORY AND ABSOLUTE CONVERGENCE OF FOURIER SERIES ON COMPACT LIE GROUPS

David L. Ragozin

Let G be a compact connected Lie group. Any f in $L_1(G)$ has a Fourier series $f \sim \Sigma_{\hat{G}} d(\lambda)\mathrm{Tr}(\hat{f}(\lambda)U^{\lambda}(g))$, where $\hat{G}$ is the set of equivalence classes of irreducible representations of G, $U^{\lambda}$ is a unitary representation in the class $\lambda$ with $d(\lambda) = \dim U^{\lambda}$, and $\hat{f}(\lambda) = \int f(g)U^{\lambda}(g^{-1})dg$. Several authors have considered the problem of finding smoothness criteria on f which insure some form of absolute convergence of the Fourier series of f (see [4], [5], [7], [8]). This paper concerns two aspects of this problem. First we apply approximation theoretic tools to derive several smoothness criteria which imply absolute convergence. This part closely resembles Bernstein's approach to his theorem that each f in Lip $(1/2 + \varepsilon)$ on the circle has an absolutely convergent Fourier series. Second, and more interestingly, we combine techniques of random Fourier series and approximation theory to produce examples which show that most of our criteria in the first part are best possible.

To summarize our results we need to recall several facts and introduce some additional notation.

1. The usual definition of absolute convergence of the Fourier series for f in $L_1(G)$ is the condition $\Sigma_{\hat{G}} d(\lambda)\mathrm{Tr}(|\hat{f}(\lambda)|) < \infty$, where $|\hat{f}(\lambda)|$ is the positive square root of the matrix $\hat{f}(\lambda)\hat{f}(\lambda)^*$. We shall also consider the conditions $\Sigma_{\hat{G}}\{d(\lambda)\mathrm{Tr} \cdot (|\hat{f}(\lambda)|^p)\}^{1/p} < \infty$ and $\Sigma_{\hat{G}} \|f_{\lambda}\|_p < \infty$, where $f_{\lambda}(g) = d(\lambda)\mathrm{Tr} \cdot (\hat{f}(\lambda)U^{\lambda}(g))$, as representing some form of absolute convergence. All these conditions are equivalent when G is abelian.

2. The Lie group G contains a maximal torus $T^{\ell}$ and the set $\hat{G}$ is parameterized (and will be identified with) a certain subsemigroup of the integer lattice $Z^{\ell} = \hat{T}^{\ell}$, which generates $Z^{\ell}$ as a group. We choose some norm, $| \; |$, on $R^{\ell}$. If $m = (\dim G - \ell)/2$, then we have the estimate $d(\lambda) \leq C|\lambda|^m$. In fact on a subset of $\hat{G}$, which still generates $Z^{\ell}$, $d(\lambda) \approx |\lambda|^m$ as $|\lambda| \to \infty$.

3. We write $P_n$ for the space of f in $L_1(G)$ with $\hat{f}(\lambda) = 0$ for $|\lambda| > n$, and we let $E_n(f)$ denote the error in the best uniform approximation to f by elements in $P_n$.

4. $C^{k,\alpha}(G)$ is the space of all f in $C^k(G)$ such that any $k^{th}$ derivative of f is in Lip $\alpha$, $\alpha<1$, or in Lip* $\alpha$, $\alpha=1$.

Now our first result is

Theorem 1. Suppose $\Sigma_{n=1}^{\infty} n^{r-1}E_n(f) < \infty$. If $r \geq \ell/2 + m(2/p - 1)$ for some p with $1 \leq p \leq 2$, then

(1.i) $\Sigma_{\hat{G}}\{d(\lambda)\mathrm{Tr}(|\hat{f}(\lambda)|^p)\}^{1/p} < \infty$ and

(1.ii) $\Sigma_{\hat{G}}\|f_\lambda\|_{p'} < \infty$, where $1/p + 1/p' = 1$.

The proof of (1.i) is straightforward given the estimate for $d(\lambda)$ in 2, while (1.ii) follows from (1.i) by use of the Hausdorff-Young inequality.

In view of the extension of the Jackson theorems to G (see [2], [6]) we have the following smoothness criteria for absolute convergence.

Theorem 2. If f is in $C^{k,\alpha}(G)$ and $k + \alpha > \ell/2 + m(2/p - 1)$, for some p, $1 \leq p \leq 2$, then (1.i) and (1.ii) hold.

Note that for p=1 our condition is $k + \alpha > \dim G/2$ while for p=2 we merely need $k + \alpha > \ell/2$.

Now we turn to the construction of examples which prove

Theorem 3. For each p, $1 \leq p \leq 2$, there exist f in $C^{k,\alpha}(G)$, $k + \alpha = \ell/2 + m(2/p - 1)$, with $\Sigma_{\hat{G}}\{d(\lambda)\mathrm{Tr}(|\hat{f}(\lambda)|^p)\}^{1/p} = \infty$.

To indicate the method of construction we introduce some notation. Form the product $\Omega = \Pi_{\hat{G}} U(d(\lambda))$ of the unitary groups $U(d(\lambda))$ and let $\mu_\Omega$ denote its normalized Haar measure. For any $f$ in $L_2(G)$ and $(W_\lambda) = W$ in $\Omega$, let $f^W$ in $L_2(G)$ be such that $f^W \sim \Sigma_{\hat{G}} d(\lambda) \mathrm{Tr}(\hat{f}(\lambda) W_\lambda U^\lambda(g))$.

Now a fundamental estimate due to Rider (see [1, (29.12)]) leads to the result that for any $d(\lambda)$ by $d(\lambda)$ matrix $A$,

$$\int_{U(d(\lambda))} \exp\{\mathrm{Re}(\mathrm{Tr}(AW_\lambda))\} dW_\lambda \leq \exp\{(2\mathrm{Tr}(AA^*))/d(\lambda)\}$$

When combined with the Bernstein inequality for $P_n$ and certain techniques from random Fourier series (compare [3, Ch. VI]) this inequality leads to

Lemma 4. There exist constants $C_1, C_2$ such that for $f$ in $P_n$

$$\mu_\Omega\{W: \|f^W\|_\infty \geq C_1(\log n)^{1/2}\|f^W\|_2\} \leq (C_2/n).$$

Now suppose $r>0$ is given and let

$$F^{r,W}(g) = \Sigma_{|\lambda|>0} d(\lambda)|\lambda|^{-(r+\dim G/2)} \log^{-1/2}|\lambda| \mathrm{Tr}(W_\lambda U^\lambda(g)).$$

Then we have

Theorem 5. Suppose $r = k + \alpha$, $\alpha \leq 1$. Then for $\mu_\Omega$-almost all $W$, $F^{r,W}$ is in $C^{k,\alpha}(G)$.

Outline of proof: Write $P_j^W$ for the orthogonal projection of $F^{r,W}$ on $P_{2^{j+1}} \cap P_{2^j}^\perp$. Then Lemma 4 and the Borel-Cantelli lemma show that for almost all $W$, $\|P_j^W\|_\infty = O(j^{1/2}\|P_j^W\|_2)$ as $j \to \infty$. But using the estimates in 2 we get $j^{1/2}\|P_j^W\|_2 = O((2^j)^{-r})$. Now the converse Bernstein theorems for $G$ give the desired result.

When $r = \ell/2 + m(2/p - 1)$ the functions in Theorem 5 suffice to prove Theorem 3 in light of

Lemma 6. $\Sigma_{\hat{G}}\{d(\lambda)|\lambda|^{-p(r+\dim G/2)} \log^{-(p/2)}|\lambda| \mathrm{Tr}(W_\lambda W_\lambda^*)\}^{1/p}$

$$\geq C \Sigma_{n=1}^\infty n^{-(r+1-\ell/2-m(2/p-1))}.$$

This is proved using the last estimate mentioned in 2.

Remark. Theorem 3 shows that with respect to absolute convergence of type (1.i) and the classes $C^{k,\alpha}(G)$, Theorem 2 is best possible. We expect the same is true with respect to type (1.ii). However, for non-commutative G and $p<2$ the functions in Theorem 5 do not show this. In fact for any $r>0$ Lemma 4 shows that $\|F_\lambda^{r,W}\|_\infty = O(|\lambda|^{-(r+\ell/2)})$ for almost all W. From this it follows that for any $r > \ell/2$, $\Sigma_{\hat{G}}\|F_\lambda^{r,W}\|_\infty < \infty$. So, a fortiori, for $r = \ell/2 + m(2/p - 1)$, $\Sigma_{\hat{G}}\|F_\lambda^{r,W}\|_{p'} < \infty$.

## References


[1] Hewitt, E., and K. Ross, Abstract Harmonic Analysis. Vol. II, Springer-Verlag, Berlin, 1970.

[2] Johnen, H., Sätze vom Jackson-Typ auf Darstellungsräumen kompakter, zusammenhängender Liegruppen. In Linear Operators and Approximation, P. L. Butzer, J. P. Kahane, B. Sz.-Nagy, eds., Birkhäuser-Verlag, Basel, 1972.

[3] Kahane, J. P., Some Random Series of Functions. D. C. Heath and Co., Lexington, Mass., 1968.

[4] Peetre, J., Absolute convergence of eigenfunction expansions. Math. Ann., 169 (1967), 307-314.

[5] Ragozin, D. L., Approximation theory on compact manifolds and Lie groups with applications to harmonic analysis. Ph.D. Dissertation, Harvard, 1967.

[6] Ragozin, D. L., Polynomial approximation on compact manifolds and homogeneous spaces. Trans. Amer. Math. Soc., 150 (1970), 41-53.

[7] Sugiura, M., Fourier series of smooth functions on compact Lie groups. Osaka J. Math. 8 (1971), 33-47.

[8] Taylor, M. E., Fourier series on compact Lie groups. Proc. Amer. Math. Soc., 19 (1968), 1103-1105.


Department of Mathematics
University of Washington
Seattle, Washington 98195

# ON THE COMPUTATIONAL COMPLEXITY OF APPROXIMATION OPERATORS

John R. Rice[1]

## 1. Introduction

Computational complexity is a measure of the number of operations that some abstract machine requires to carry out a task. The task considered here is the approximation of a real function $f(x)$ in $C^p[0,1]$ and the only operations counted are evaluations of $f(x)$. We denote the approximation by $P_n(x)$, the error $\|f - P_n\|_\infty$ by $\varepsilon$ and the approximation operator by $T: f(x) \rightarrow P_n(x)$, where n is the dimension (or polynomial order) of the approximation space. The approximation $P_n(x)$ is estimated to within $\mathcal{O}(\varepsilon)$ by a computation algorithm A using M evaluations of $f(x)$ and, with $\bar{\varepsilon} = -\log_{10}\varepsilon$, we let $d(M,T,A)$ measure the computational complexity of the process. We are most interested in $d(M,T) = \inf_A d(M,T,A)$ which measures the computational complexity of the approximation operator T. We conjecture that $d(M,T)$ is at most $p \log M + c$ (c is a constant) for common operators such as $L_1$, $L_2$, and $L_\infty$ approximation. The purpose of this paper is to exhibit algorithms A whose complexity is close to this conjecture. This conjectured value is established here only for the spline approximation projector of deBoor [1], close results are obtained for $L_1, L_2, L_\infty$ approximation.

## 2. Discretization

The following procedure is occasionally presented for the computation of approximations: Choose a uniformly spaced finite set $X \in [0,1]$, compute the best approximation to $f(x)$ on X and use this as the approximation to f(x). Specifically, set $P_n(x) = \Sigma_{i=0}^{n} a_i x^i$ and let $a_i^*$ and $a_i(X)$ be the best approximation coefficients (in either the $L_2$ or $L_\infty$ norms) on [0,1] and X, respectively. If $X = \{x_i | i=1,2,\ldots,M\}$ is uniform (or nearly so) in [0,1] then Rice [7] has established that, for some constant K,

$$|a_i^* - a_i(X)| < \frac{K}{M}$$

for both $L_2$ and $L_\infty$ approximation and this bound cannot be improved. These results are elaborated upon by Cheney [2]. The inadequacy of this procedure is seen by observing that $\varepsilon = \mathcal{O}(cn^{-p})$ and thus M must be $\mathcal{O}(n^p)$ to obtain a reasonable approximation. Let D be the operator defined by this procedure and we have

Theorem 1. $d(M,D) = \log M + c$.

Note the result is independent of p. This indicates that D is almost useless for the approximation of smooth functions and experience supports this conclusion.

## $L_1$ Approximation

Let $T_1$ denote the best approximation operator in the $L_1$ norm. The general computation algorithms for this norm are not particularly well understood, but there is one striking, though specialized, possibility. One may obtain the best $L_1$ approximation for some functions f(x) simply by interpolating f(x) at the canonical points. A sufficient condition for this approach is that f(x) be convex with respect to the approximating functions (i.e., $f(x) - \Sigma_{i=1}^{n} a_i \phi_i(x)$ has at most n zeros). If we consider the case of polynomial approximation, then we have

Theorem 2. Suppose f(x) is convex with respect to polynomials of all degrees. Then

$$d(M,T_1) = p \log M + c \ .$$

The class of functions satisfying the hypothesis of Theorem 2 is rather narrow, see [4] for further discussion. It is plausible that the hypothesis in this theorem is somewhat stronger than necessary and it is of considerable interest to determine the class of functions where this interpolation process produces the best $L_1$ approximation.

## 4. Least Squares Approximation

We consider least squares approximation by the Legendre Polynomials $\phi_i(x)$, $i=0,1,2,\ldots,n$. The coefficients $a_i^*$ of the best $L_2$ approximation are given by

$$a_i^* = \int_0^1 f(x)\phi_i(x)dx \ .$$

The obvious computational approach is to evaluate this integral by a quadrature formula. Assume that $f(x) \in C^p[0,1]$ and that we use Gauss quadrature of order n to obtain $a_i'$. We know that $\varepsilon = O(n^{-p})$ and, also [3], that

$$|a_i^* - a_i'| = O(\varepsilon) = O(n^{-p}).$$

However, the fact that $|a_i^* - a_i'| = O(n^{-p})$ does not imply that we have calculated the best approximation to within $\varepsilon$. In fact we have

$$|\Sigma(a_i^* - a_i')\phi_i(x)| \leq \sum_{i=0}^{n} |a_i^* - a_i'|\max|\phi_i(x)|$$

or

$$|\Sigma(a_i^* - a_i')\phi_i(x)| < \max|a_i^* - a_i'| \sum_{i=0}^{n} |\phi_i(x)| \ .$$

Both of these estimates indicate that the effect of the quadrature error in the approximation is $O(n^{-p+3/2})$.

Note that we may (and should) economize on f(x) evaluations by using the same formula (and hence the same f(x) values) for all the quadratures. Let $\Gamma_2$ denote the best least squares approximation operator and then a manipulation establishes

Theorem 3. $d(M,T_2) \leq p \log M - 3/2 \log M + c$ .

It seems unlikely that this result can be improved by a more sophisticated analysis of quadrature errors. However, it is plausible that a better result may be obtained by using specialized quadrature formulas (e.g., Filon's rule and generalizations) or methods for computing Fourier transforms It is easy to conjecture that the constant 3/2 may be replaced in this way by 1 or 1/2, but not so easy to see why it would be replaced by zero.

## 5. Tchebycheff Approximations

Let $T_\infty$ denote the best Tchebycheff approximation operator for polynomials. The most efficient common algorithm for computing an estimate of the best approximation is the Remes algorithm. The process has three steps as follows: Step 1: determine an initial guess, Step 2: apply the Remes algorithm until the error in determining the best approximation is "small," Step 3: continue to apply the Remes algorithm until sufficient accuracy is obtained. The dividing line between steps 2 and 3 is that quadratic convergence takes place in step 3. To quantify things, we observe that step 1 requires somewhere between 0 and n function evaluations, depending on the scheme chosen. Each iteration of the Remes algorithm requires Kn function evaluations where K is a small fixed integer. Thus, in order to obtain the extrema of the error function one uses the approach of Murnaghan and Wrench [5] and obtains K=3. The number of iterations required in step 3 is seen to be bounded by $K \log \log \varepsilon$ for some constant K. The estimation of the number of iterations for step 2 is more difficult. It is clear that this number is not bounded as a function of $f(x) \in C^p[0,1]$ for $\varepsilon$ and n fixed.

To analyze this situation recall [8], [9] that the Remes

algorithm (in conjunction with the Murnaghan and Wrench perturbation) is Newton's method for the system of equations

$$\frac{d}{dx}\left[f(x) - P_n(x)\right]\Big|_{x=x_i} = 0 \qquad i=0,1,2,\dots,n+1$$

$$f(x_i) - P_n(x_i) = (-1)^i \varepsilon \qquad i=0,1,2,\dots,n+1$$

in the 2n+2 unknowns $x_i$ (critical points), $a_i^*$ (best approximation coefficients) and $\varepsilon$. Let J denote the Jacobian of this system and $H_i$ (the Hessian) denote the matrix of the second partial derivatives for each equation. A fairly standard analysis shows that $J^{-1}H_i$ governs the rate of convergence of the algorithm. A straightforward manipulation exhibits the structure (which is fairly simple) of these matrices. They involve diagonal submatrices of derivatives (first, second and third) of the error function and Vandermonde type submatrices. Bounds on $J^{-1}H_i$ may then be estimated using assumptions on the smoothness of f(x) and the fact that the critical points are separated [6]. The detailed results become complicated, but we may present them in the following qualitative form

<u>Theorem 4.</u> Assume $p \geq 3$ and let $C_0^p[0,1]$ denote those f(x) where the second derivatives of the error curve at the critical points is bounded from zero independent of f(x) and n. Then there exists a nested sequence of compact sets $F_k$ whose union is the interior of $C_0^p[0,1]$ and such that in each $F_k$ we have

$$d(M,T_\infty) \leq p \log M - p \log \log \log M + c$$

The restriction about the second derivative at the error curve is due to properties of Newton's Method. Other methods exist which are also quadratically convergent and which do not have this property. Thus it is reasonable to believe that this restriction can be removed. The intuitive content of this theorem is that the computational complexity of the $T_\infty$ operator is almost p log M on large subsets of $C^p[0,1]$ which are difficult to define precisely in a simple manner.

## 6. Spline Approximation.

Let $T_s$ denote the spline projector operator introduced by deBoor [1]. He showed that

$$\|f - T_s f\|_\infty \le K\|f - T_s^* f\|_\infty$$

where $T_s^*$ is the best approximation operator for splines of degree q and n prescribed knots. The constant K depends on q, but not f(x). Furthermore, the operator $T_s$ is linear and involves a fixed number of function values between each pair of knots. Furthermore, it is well known that if p=q and h is the maximum knot separation then $\|f - T_s^* f\|_\infty$ is $\mathcal{O}(h^{-p})$. These observations may be combined to establish

Theorem 5. $d(M, T_s^*) \le p \log M + c$ .

In comparing the computational use of splines and polynomials, it is noteworthy that $T_s f$ may be evaluated with $\mathcal{O}(\log n)$ arithmetic operations while $L_2 f$ and $L_\infty f$ require $\mathcal{O}(n)$ arithmetic operations.

## 7. Conclusions

The most interesting conclusion is that it appears to be as easy to compute $L_\infty$ approximations as $L_2$ approximations for a wide class of functions. Similar results may also hold for $L_1$ approximation. The next conclusion is that spline approximations are probably less complex to compute than polynomial approximations of comparable accuracy, again for certain wide classes of functions. The most interesting open questions are: (a) is $d(M,T) = p \log M + c$ a lower bound on the computational complexity and (b) are there processes for $L_1$, $L_2$ and $L_\infty$ that have computational complexity $p \log M + c$.

---

[1]This work was partially supported by NSF grant GP-11695.

## References

[1] deBoor, Carl, On uniform approximation by splines. J. Approx. Theory 1 (1968), 219-235.

[2] Cheney, E. W., Introduction to Approximation Theory. McGraw-Hill, New York, 1966 (Ch. 3).

[3] Davis, P. J. and P. Rabinowitz, Numerical Integration. Blaisdell, New York, 1967 (Ch. 4).

[4] Karlin, S. J. and W. J. Studden, Tchebysheff Systems. Interscience, New York, 1966 (Ch. 11).

[5] Murnaghan, F. D. and J. W. Wrench, The determination of the Chebyshev approximating polynomial of a differentiable function. MTAC, 13 (1958), 185-193.

[6] de la Vallée-Poussin, Ch., L'approximation des functions d'une variable réelle. Gauthier-Villars, Paris, 1919.

[7] Rice, J. R., On the convergence of an algorithm for best Tchebycheff approximations. J. Soc. Indust. Appl. Math. 7 (1959), 133-142.

[8] _______, The approximation of functions, Vol. II. Addison Wesley, Reading, Mass. 1969 (Ch. 9).

[9] Werner, H., Die construktive Ermittlung der Tschebyscheff-Approximierenden im Bereich der rationalen Funktionen. Arch. Rat. Mech. Anal. 11 (1962), 368-384.

Mathematics Department
Purdue University
Lafayette, Indiana 46907

# MERGELYAN SETS AND THE MODULUS OF CONTINUITY

L. A. Rubel,[1] A. L. Shields, B. A. Taylor

For f a continuous function on the closed unit disc, that is analytic in the interior, let

$$\omega(\delta,f) = \sup \{|f(z_1)-f(z_2)| : |z_1-z_2| \le \delta, |z_1| \le 1, |z_2| \le 1\}$$

and

$$\tilde{\omega}(\delta,f) = \sup \{|f(z_1)-f(z_2)| : |z_1-z_2| \le \delta, |z_1| = |z_2| = 1\}$$

denote, respectively, the modulus of continuity of f on the closed unit disc, and the modulus of continuity of the restriction of f to the boundary $\{|z| = 1\}$.

<u>Theorem 1.</u> There exists a positive absolute constant C such that

$$\omega(\delta,f) \le C\, \tilde{\omega}(\delta,f)$$

for all $\delta>0$ and all f analytic for $|z|<1$ and continuous for $|z|\le 1$. An example shows that one cannot take C=1.

This theorem answers a question posed by W. E. Sewell in 1942. It would be interesting to see whether $\lim \frac{\omega(\delta,f)}{\tilde{\omega}(\delta,f)} = 1$ as $\delta \to 0$, and whether the analogue of Theorem 1 holds for other regions. It is well known that Theorem 1 is false for harmonic functions.

<u>Definition.</u> A relatively closed subset F of the open unit disc is called a Mergelyan set if every function g that is analytic on the open unit disc and uniformly continuous on F can be uniformly approximated by polynomials on each set of the form $F \cup \{|z| \le r\}$ for each r with $0<r<1$.

Theorem 2. There exist two Mergelyan sets whose intersection is not a Mergelyan set. There exist two Mergelyan sets whose union is not a Mergelyan set.

The proof of Theorem 1 can be reduced to the following result.

Lemma 1. Let u be harmonic for $|z|<1$ and continuous for $|z|\leq 1$, and let v be a harmonic conjugate of u. There exists an absolute constant C such that, with $\delta = 1-r\leq 1/2$

$$|[u(1)-u(r)] - [v(re^{i\delta/2})-v(re^{-i\delta/2})]| \leq C\,\tilde{\omega}(\delta,u).$$

This is a discrete analogue of the Cauchy-Riemann equation

$$\frac{\partial u}{\partial r} = \frac{1}{r}\frac{\partial v}{\partial \theta}\,.$$

The proof of Lemma 1 is obtained by making direct estimates of the Poisson and conjugate Poisson kernels:

$$(1)\quad P(r,t) = \frac{1-r^2}{(1-r)^2+4r\,\sin^2(t/2)}\,,$$

$$(2)\quad Q(r,t) = \frac{2r\,\sin\,t}{(1-r)^2+4r\,\sin^2(t/2)}\,.$$

Lemma 2. For $0<\delta\leq 1/2$, $\delta\leq|t|<\pi$, $r=1-\delta$, we have

$$|P(r,t)+\{Q(r,\tfrac{\delta}{2}-t)+Q(r,\tfrac{\delta}{2}+t)\}| = O(\delta + \frac{\delta^2}{|t|^3}).$$

Example. To show that we may not take C=1 in Theorem 1, let

$$F(z) = \frac{1-\sqrt{1-z^2}}{z}$$

and take $\delta=1$. Somewhat complicated calculations show that

$$\tilde{\omega}(1,F) = |F(e^{i\pi/3})-F(1)| = .9332\ldots < 1 = |F(1)-F(0)| \leq \omega(1,F).$$

We turn now to Mergelyan sets.

<u>Definition.</u> Let F be a relatively closed subset of the open unit disc D. Then U(F) is the space of all functions that are analytic in D and uniformly continuous on F, in the topology of uniform convergence on every set of the form $K \cup F$, as K ranges over the compact subsets of D. We then say that F is a Mergelyan set if and only if the polynomials are dense in U(F).

<u>Definition.</u> A bullseye is a closed subset of D that contains circles $\{z: |z|=r\}$ for values of r arbitrarily close to 1.

It follows from Theorem 1 that every bullseye is a Mergelyan set. Now choose two bullseyes whose intersection is a Blaschke sequence that is dense on $\{|z|=1\}$. It is easy to see that this intersection cannot be a Mergelyan set.

<u>Definition.</u> The polynomial hull of a set $G \subset D$, $H_p(G)$, is the set of points z for which

$$|p(z)| \leq \sup\{|p(w)|: w \in G\}$$

for all polynomials p.

<u>Definition.</u> The uniformly continuous analytic hull of $G \subset D$ with respect to F, $H_U(G:F)$ is the set of all points $z \in D$ for which

$$|f(z)| \leq \sup\{|f(w)|: w \in G\}$$

for all $f \in U(F)$.

<u>Proposition 1.</u> If F is a Mergelyan set, then

$$H_U(K \cup F:F) = D \cap H_P(K \cup F)$$

for every compact subset K of D.

This proposition is quite easy to prove. We do not know whether its converse is true.

<u>Proposition 2.</u> Let J be a simple closed Jordan curve in $\overline{D}$ that intersects $\partial D$ only at z=1 and let $J' = J \setminus \{1\}$. Then

$H_U(J':J') = J'$.

The proof of Proposition 2 uses Arakelian's Theorem on uniform approximation to continuous functions by entire functions. Using Propositions 1 and 2 it is easy to see that if $F_1$ and $F_2$ are two disjoint short circular arcs with 1 as common endpoint and that are orthogonal to $\partial D$ at their endpoints, then $F_1 \cup F_2$ is not a Mergelyan set. But $F_1$ (and likewise $F_2$) is a Mergelyan set since it is carried onto a radial segment by a map $\varphi$ that maps D conformally one:one onto D and that is analytic in $\overline{D}$. But the standard argument of replacing f(z) by f(rz) shows that a radial segment is a Mergelyan set.

Full details will appear elsewhere.

---

[1]The first author acknowledges partial support from the National Science Foundation.

Mathematics Department
University of Illinois
Urbana, Illinois 61801

Mathematics Department
University of Michigan
Ann Arbor, Michigan 48104

Mathematics Department
University of Michigan
Ann Arbor, Michigan 48104

Note added in proof: A. Straÿ has informed us, since this talk was given, that he has obtained a geometrical characterization of Mergelyan sets.

# ON SEMI-CARDINAL QUADRATURE FORMULAE

I. J. Schoenberg and S. D. Silliman

This is an expanded abstract of the paper [3]. We take this opportunity to sketch its contents in the wider context of semi-cardinal quadrature formulae (Q.F.) with general boundary conditions of which the Euler-Maclaurin Q.F. is the prototype. For the more fundamental discussion of the underlying <u>interpolation</u> methods see Lectures 7 and 8 of the forthcoming monograph [5]. As early as 1950 some conjectures of Meyers and Sard [1], [2, 60-61], were pointing towards semi-cardinal approximation methods. In describing them we may as well start with the Q.F. of Newton and Cotes

$$(1)\quad \int_0^n f(x)dx = \sum_{\nu=0}^{n} H_{\nu,n} f(\nu) + R_n f,$$

characterized by the property that $Rf = 0$ if $f(x) \in \pi_n$, where $\pi_n$ is the class of polynomials of degree not exceeding n. In 1949 A. Sard generalized the Q.F. (1) as follows: Let m be a natural number, $m \leq n+1$, and let us require that the Q.F.

$$(2)\quad \int_0^n f(x)dx = \sum_{\nu=0}^{n} H_{\nu,n}^{(m)} f(\nu) + R_{n,m} f$$

be exact if $f(x) \in \pi_{m-1}$. This condition allows to write the remainder in the form

$$(3)\quad R_{n,m} f = \int_0^n K_{n,m}(x) f^{(m)}(x)dx,$$

where the kernel K still depends on n-m+1 free parameters. These are so chosen as to minimize the value of the integral

$$(4)\quad J_n^{(m)} = \int_0^n (K_{n,m}(x))^2 dx.$$

The parameters being so chosen, (2) becomes Sard's best Q.F.

In [1] Meyers and Sard keep m fixed in (2) and let $n\to\infty$. On numerical evidence they formulated the following conjectures:

Conjecture 1. The limits

$$(5)\quad \lim_{n\to\infty} H^{(m)}_{\nu,n} \qquad (\nu=0,1,\ldots)$$

exist.

Conjecture 2.

$$(6)\quad \lim_{n\to\infty} H^{(m)}_{[n/2]+k,n} = 1 \qquad \text{for each integer } k.$$

Conjecture 3. The limits

$$(7)\quad \lim_{n\to\infty} (J^{(m)}_{n+1} - J^{(m)}_{n})$$

exist.

## Semi-cardinal Quadrature Formulae

In order to understand the problems raised by the Meyers-Sard conjectures, the results that establish them, as well as the purpose of the paper [3], we digress and discuss semi-cardinal Q.F. Let

$$(8)\quad \mathcal{S}^+_n = \{S(x)\}$$

denote the class of functions $S(x)$ such that

$$(9)\quad \begin{cases} \text{(i)} & S(x) \in C^{n-1}(R^+) \\ \text{(ii)} & S(x) \in \pi_n \text{ in each of the intervals } [0,1],[1,2],\ldots. \end{cases}$$

Such $S(x)$ may be called semi-cardinal splines.

A function $K(x)$ of the form

$$(10)\quad K(x) = \frac{x^{2m}}{(2m)!}+S_{2m-1}(x), \text{ where } S_{2m-1}(x) \in \mathcal{S}^+_{2m-1}$$

is called a semi-cardinal monospline of degree 2m. It is well known that quadrature formulae and monosplines are equivalent (dual) subjects that are related by successive integrations by parts. This brings up the subject of boundary conditions. A useful type of such conditions is conveniently described as follows.

Let m be an integer, $m \geq 2$, and let us partition the numbers $1,2,\ldots,2m-2$ in $m-1$ pairs as follows

(11) $(1, 2m-2), (2, 2m-3), (3, 2m-4),\ldots,(m-1,m)$.

Notice that in each pair the sum of the two elements is $2m-1$. Let I be a set of $m-1$ numbers

(12) $I = \{i_1,i_2,\ldots,i_{m-1}\}, \quad (1 \leq i_1 < i_2 < \ldots < i_{m-1} \leq 2m-2)$,

obtained by choosing one and only one element from each of the pairs (11). We call such sets I admissible. Simple examples of admissible sets are

$$I_1 = \{1,2,3,\ldots,m-1\},$$

(13) $$I_2 = \{1,3,5,\ldots,2m-3\},$$

$$I_3 = \{m,m+1,\ldots,2m-2\}.$$

We refer to [5, Lecture 8] for proofs of the following results. To each admissible set (12) there corresponds a Q.F.

(14) $$\int_0^\infty f(x)dx = \sum_0^\infty H_\nu f(\nu) + \sum_{i\in I} A_i f^{(i)}(0) + \int_0^\infty K(x)f^{(2m)}(x)dx.$$

Here K(x) is a monospline of the form (10) which is uniquely characterized by the following three conditions

(15) $K(\nu) = 0 \qquad (\nu=0,1,\ldots)$,

(16) $K^{(i)}(0) = 0 \qquad (i \in I)$,

(17) $K(x) = \frac{1}{(2m)!}(\overline{B}_{2m}(x) - B_{2m}) + O(1)$ as $x \to +\infty$.

In fact it suffices to ask that K(x) should grow at most like a power of x, as $x\to\infty$, and then (17) follows. In (17) $\overline{B}_{2m}(x)$ is the usual periodic extension of the Bernoulli polynomial $B_{2m}(x)$ and $B_{2m}$ is the Bernoulli number. In (14) we assume that the derivatives $f^{(\nu)}(x)$, $(\nu=0,1,\ldots,2m)$ are in $L_1(R^+)$, that they all vanish at $\infty$, and that

$$(18)\quad \sum_{\nu=0}^{\infty} |f(\nu)| < \infty.$$

The Q.F. (14) then arises by successive integrations by parts of its remainder term and using the relations (15), (16).

In (14) we may write the remainder term in the form of a Stieltjes integral

$$(19)\quad \int_0^{\infty} K(x)\,df^{(m-1)}(x).$$

It is then shown that the Q.F. (14) is exact, i.e., its remainder term vanishes, whenever

$$(20)\quad f(x) \in S_{2m-1}^{+} \cap L_1(R^+).$$

In this connection we observe that the $H_\nu$ of (14) have the property $H_\nu \to 1$ as $\nu \to \infty$, and that (20) implies that the derivatives $f^{(\nu)}(x)$ have the required behavior, including the validity of (18). That this condition of exactness characterizes the Q. F. (14) is stated by

<u>Theorem 1.</u> If the Q.F.

$$(21)\quad \int_0^{\infty} f(x)dx = \sum_0^{\infty} \tilde{H}_\nu f(\nu) + \sum_{i\in I} \tilde{A}_i f^{(i)}(0) + \tilde{R} f,$$

satisfies

$$(22)\quad \tilde{H}_\nu = O(1) \text{ as } \nu\to\infty$$

and has the property that

$$(23)\quad \tilde{R} f = 0 \text{ if } f(x) \in S_{2m-1}^{+} \cap L_1(R^+)$$

then (21) must reduce to the Q.F. (14), hence $\tilde{H}_\nu = H_\nu$, $\tilde{A}_i = A_i$.

The three admissible sets (13) give rise to the special Q.F.

$$(24)\quad \int_0^{\infty} f(x)dx = \sum_{\nu=0}^{\infty} H_\nu^{[m]} f(\nu) + \sum_{i=1}^{m-1} A_i^{[m]} f^{(i)}(0) + \int_0^{\infty} K_1(x) f^{(2m)}(x)dx,$$

$$(25)\quad \int_0^\infty f(x)dx = \tfrac{1}{2} f(0) + f(1) + f(2) + \ldots + \sum_{\nu=1}^{m-1} \frac{B_{2\nu}}{(2\nu)!} f^{(2\nu-1)}(0)$$

$$+ \int_0^\infty K_2(x) f^{(2m)}(x)dx,$$

where $K_2(x) = (\overline{B}_{2m}(x) - B_{2m})/(2m)!$, and

$$(26)\quad \int_0^\infty f(x)dx = \sum_{\nu=0}^{\infty} H_\nu^{(m)} f(\nu) + \sum_{i=m}^{2m-2} A_i^{(m)} f^{(i)}(0) + \int_0^\infty K_3(x) f^{(2m)}(x)dx,$$

respectively. The Q.F.(24) may be called the <u>complete Q.F.</u> (25) is, of course, the classical <u>Euler-Maclaurin Q.F.</u> Finally, (26) may be called the <u>natural Q.F.</u> with forcing terms. By previous remarks concerning the general Q.F. (14), all these three formulae are exact if (20) holds.

Our paper [3] is mainly concerned with the Q.F. (26), the reason being its connection with the Meyers-Sard conjectures. The following results were established in [4].

<u>Theorem 2.</u> The limits (5) in Conjecture 1 are identified as the coefficients of (26), hence

$$(27)\quad \lim_{n\to\infty} H_{\nu,n}^{(m)} = H_\nu^{(m)} \qquad \text{for } \nu=0,1,\ldots.$$

Also the relations (6) are there established and finally

<u>Theorem 3.</u> The limits (7) of Conjecture 3 have the values

$$(28)\quad \lim_{n\to\infty} (J_{n+1}^{(m)} - J_n^{(m)}) = (-1)^{m-1} B_{2m}/(2m)! \quad .$$

In [4] the derivation of these results required the use of the so-called eigensplines of the class $S_{2m-1}$ of cardinal splines. The objectives of [3] are much more limited. Its first aim is to establish directly and independently of Theorem 1, of which it is a special case, the following

<u>Theorem 4.</u> If the Q.F.

$$(29)\quad \int_0^\infty f(x)dx = \sum_0^\infty H_\nu f(\nu) + Rf,$$

subject only to the condition that

(30) $H_\nu = O(1)$ as $\nu \to \infty$,

has the property that $Rf = 0$ for all elements of $S^+_{2m-1} \cap L_1(R^+)$ which satisfy the boundary conditions

(31) $f^{(m)}(0) = f^{(m+1)}(0) = \ldots = f^{(2m-2)}(0) = 0,$

then (29) must reduce to the Q.F.

(32) $$\int_0^\infty f(x)dx = \sum_0^\infty H_\nu^{(m)} f(\nu) + R_3 f$$

formed with the first sum of (26).

The second purpose of [3] is to do for the semi-cardinal Q.F. (32) what Cotes did for (1) and Sard did for (2): To determine accurate values of the coefficients $H_\nu^{(m)}$. The method of generating functions used in [3] to establish Theorem 4 also allows to compute the $H_\nu^{(m)}$. These are determined to nine decimal places for m=2,3,4,5,6, and 7. One reason why we went as far as m=7 is that for the first time a negative coefficient appeared: $H_4^{(7)} = -.300\ 751\ 517$.

Since the general Q.F. (14) is exact if f(x) satisfies (20), the same property is shared by the three special cases (24), (25), and (26). These three formulae are therefore of entirely comparable "strength." However, on adapting them in an obvious way to an arbitrary step h, and applying them to such standard functions like $f(x) = e^{-x}$, or $f(x) = (1+x)^{-2}$, we find that the results furnished by (24) are best, while those due to (26) are worst, especially if it is used in the simplified form (32). The reason for this seems fairly obvious: The complete Q.F. (24) uses the earliest consecutive derivatives $F^{(i)}(0)$ (i=1,2,...,m-1), while those used by (25) or (26), while equal in number, are more remote, hence less likely to furnish significant information on f(x). Besides, except for $f(x) = e^{-x}$, they are also harder to compute. A more sophisticated reason for the preeminence of (24) will be found on inspecting the boundary conditions (16) satisfied by their respective kernels.

For this reason, in a subsequent paper, the second-named author proposes to do for the complete Q.F. (24) what was done in [3] for (26): To determine accurately the coefficients $H_\nu^{[m]}$ and $A_i^{[m]}$ for m=2,3,4,5,6, and 7.

---

Sponsored by the United States Army under Contract No. DA-31-124-ARO-D-462.

## References


[1] Meyers, L. F. and A. Sard, Best approximate integration formulas. J. of Math. and Phys., 29 (1950), 118-123.

[2] Sard, A., Linear Approximation. American Math. Society, Providence, R. I., 1963.

[3] Schoenberg, I. J. and S. D. Silliman, On semi-cardinal quadrature formulae. MRC T.S.R. #1300, October 1972. To be submitted to Math. of Computation.

[4] Schoenberg, I. J., Cardinal interpolation and spline functions VI. Semi-cardinal interpolation and quadrature formulae. MRC T S.R. #1180, December 1971. To appear in J. d'Analyse Mathématique.

[5] ________, Cardinal spline interpolation. To appear as the 12th Regional Conference Monograph published by SIAM.


Mathematics Research Center
University of Wisconsin
Madison, Wisconsin 53706

Department of Mathematics
Cleveland State University
Cleveland, Ohio 44115

# CONSTRUCTIVE ASPECTS OF DISCRETE POLYNOMIAL SPLINE FUNCTIONS

Larry L. Schumaker

## 1. Introduction

There are two basic approaches to developing spline functions--the variational approach wherein splines are defined as the solutions of certain constrained minimization problems, and the constructive approach wherein they are defined by piecing together classes of functions at certain "knots." Both approaches are well represented in the literature for problems dealing with functions defined on intervals in R. Here we are interested in functions on discrete sets.

The variational approach to discrete splines was initiated in [5], where natural discrete polynomial splines were defined as functions minimizing $\|\Delta^m f\|_{\ell^2}$ over an appropriate subset of sequences. They are analogs of the well-known natural polynomial splines and are useful for solving (approximately) certain (continuous) spline interpolation problems with inequality constraints. These same discrete splines play a fundamental role in connection with certain best and optimally best summation formulae--see [6].

In the present paper we concentrate on the constructive approach to discrete splines.

## 2. Definition and Basic Properties

Let Z be the set of integers and let $F$ be the set of all real-valued functions on Z. We begin with a set $\Delta = \{x_1 < x_2 < \dots < x_k\} \subset Z$. Such a set partitions Z into intervals

$I_0 = \{j \in Z\colon j \le x_1\}$, $I_k = \{j \in Z\colon j > x_k\}$ and $I_i = (x_i, x_{i+1}]$, $i = 1, 2, \ldots, k-1$. Suppose $M = \{m_1, \ldots, m_k\}$ is a set of integers with $0 \le m_i \le n-1$ and $x_i + m_i < x_{i+1} + m_{i+1}$.

Definition 2.1. Let

(2.1) $DSp(\pi_n;\Delta;M) = \{s \in F\colon$ there exist $s_0, \ldots, s_k \in \pi_n$ such that $s_i(\nu) = s(\nu)$ for $\nu \in I_i$ and $s_{i-1}(x_i+j) = s_i(x_i+j)$, $j = 0,1,\ldots,m_i;\ i = 1,2,\ldots,k\}$.

We call this the class of discrete polynomial splines of degree n with knot intervals $[x_1,\ x_1+m_1], \ldots, [x_k,\ x_k+m_k]$.

$DSp(\pi_n;\Delta;M)$ is a class of bi-infinite sequences which reduce to $n^{th}$ degree polynomials in certain intervals such that the polynomials are tied together in the sense that adjacent polynomials coincide on the knot intervals. It is clearly a linear subspace of $F$. Moreover, $\pi_n \subset DSp$.

We now show that $DSp(\pi_n;\Delta;M)$ is finite dimensional and exhibit some explicit bases for it. To this end we associate integers

$$\ell_i = \min(n-m_i,\ x_{i+1}+m_{i+1}-x_i-m_i), \qquad i = 1,2,\ldots,k-1$$
$$\ell_k = n-m_k$$

with each knot interval. We need the factorial function (see [1]) $\nu^{(j)} = \nu(\nu-1)\ldots(\nu-j+1)$ and the notation

$$\nu_+^{(j)} = \begin{cases} \nu^{(j)} & \text{if } \nu \ge 0 \\ 0 & \text{otherwise} \end{cases}.$$

Theorem 2.2. Every $s \in DSp(\pi_n;\Delta;M)$ has a unique representation of the form

$$(2.2) \qquad s(\nu) = \sum_{j=0}^{n} a_j \nu^{(j)} + \sum_{i=1}^{k} \sum_{j=1}^{\ell_i} c_{ij} (\nu - x_i)_+^{(m_i+j)}.$$

Thus the dimension of $DSp(\pi_n;\Delta;M)$ is $n + 1 + K$, where $K = \Sigma_{i=1}^{k} \ell_i$.

Associated with the original set of knots $\{x_i\}_1^k$ and

integers $\{m_i\}_1^k$ we define the sets

$$X_1 = \{x_1 - n + m_1 + 1, \ldots, x_1\}$$

$$X_{i+1} = \{\max(x_i+1,\ x_{i+1} - n + m_i + 1), \ldots, x_{i+1}\},$$

$i = 1, 2, \ldots, k-1$. Let $\{y_i\}_1^{\overline{K}} = \bigcup_{i=1}^{k} X_i$.

<u>Theorem 2.3.</u> We have $K = \overline{K}$. The functions

$$(2.3) \qquad \{\nu^{(j)}\}_{j=0}^{n} \cup \{(\nu-y_i)_+^{(n)}\}_{i=1}^{K}$$

span $DSp(\pi_n;\Delta;M)$.

Theorem 2.3 can be stated as follows: $DSp(\pi_n;\Delta;M) = DSp(\pi_n;\overline{\Delta};N)$, where $\overline{\Delta} = \{y_1 < \ldots < y_K\}$ and $N = (n-1,\ldots,n-1)$. We can think of the class $DSp(\pi_n;\overline{\Delta};N)$ as $n^{th}$ degree polynomial splines with simple knots at the $y_i$; the succeeding polynomial pieces tie together in a maximal way (without coinciding)--namely at n points. The original class could be considered as $n^{th}$ degree polynomial splines with knots of multiplicity $\ell_i$. What we now see is that for discrete splines these classes are the same; multiple knots can be replaced by an appropriate set of simple knots, where some are <u>consecutive</u>. For the remainder of the paper we consider only distinct (simple) knots.

## 3. <u>Discrete B-splines</u>

In this section we develop local-support discrete splines which are the analogs of the classical B-splines (cf. [2]). They will be defined as linear combinations of evaluations of the kernel $G_{n+1}(\nu,x) = (\nu-x)_+^{(n)}$.

Suppose that

$$(3.1) \qquad \ldots\ x_{-1} < x_0 < x_1 < \ldots$$

is some bi-infinite sequence of integers.

Theorem 3.1. Let $s(\nu) = \Sigma_{i=0}^{m} c_i(\nu - x_i)_+^{(n)}$. If $m \leq n$ and $s(\nu) \equiv 0$ for $\nu \geq x_m$, then $s(\nu) \equiv 0$.

This result asserts that an $n^{th}$ degree discrete polynomial spline which vanishes outside an interval (we call it a confined spline) must have at least n+2 knots in the interval if it is non-trivial.

We now construct the discrete B-splines with a minimal number of knots. For any $j \geq 0$, $k \geq 1$ let

$$Q_{i,j}^{k}(\nu) = (-1)^j G_k[\nu;\, x_i, \ldots, x_{i+j}], \tag{3.2}$$

where $G_k(\nu;x) = (\nu - x)_+^{(k-1)}$ and the bracket notation denotes the $j^{th}$ divided difference over $x_i, \ldots, x_{i+j}$.

Theorem 3.2. For $k \geq 2$, $j \geq 1$, the functions (3.2) satisfy the recurrence

$$Q_{i,j}^{k}(\nu) = \frac{(\nu - x_i - k + 2)Q_{i,j-1}^{k-1}(\nu) + (x_{i+j} - \nu + k - 2)Q_{i+1,j-1}^{k-1}(\nu)}{(x_{i+j} - x_i)}. \tag{3.3}$$

Theorem 3.3 Let $k \geq 1$. Then

(i) $Q_{i,k}^{k}(\nu) \in DSp(\pi_{k-1}; \{x_i, \ldots, x_{i+k}\};\ )$

(ii) $Q_{i,k}^{k}(\nu) > 0$ for $\nu = x_i + k - 1, \ldots, x_{i+k} - 1$

(iii) $Q_{i,k}^{k}(\nu) = 0$ otherwise

(iv) $f[x_i, \ldots, x_{i+k}] = \sum_{\nu = x_i}^{x_{i+k} - k} \frac{Q_{i,k}^{k}(\nu + k - 1)}{(k-1)!} \Delta^k f(\nu)$

(v) $\sum_{\nu=-\infty}^{\infty} k\, Q_{i,k}^{k}(\nu) \equiv 1.$

Theorem 3.4. Let $\Delta$ denote the partition (3.1).

(i) If $s \in DSp(\pi_n; \Delta; N)$ and $s(\nu) = 0$ for $\nu \notin [x_\ell, x_r]$, then s has the representation $s(\nu) = \Sigma_{j=\ell-n}^{r-n} c_j Q_{j,n+1}^{n+1}(\nu)$.

(ii) If $s \in DSp(\pi_n;\Delta;N)$ and $s(\nu) = 0$ for $\nu < x_\ell$ then $s(\nu) = \Sigma_{j=\ell-n}^{\infty} c_j Q_{j,n+1}^{n+1}(\nu)$.

(iii) If $s \in DSp(\pi_n;\Delta;N)$ then s has a unique representation $s(\nu) = \Sigma_{j=-\infty}^{\infty} c_j Q_{j,n+1}^{n+1}(\nu)$.

## 4. Total Positivity Properties

In this section we study the kernel $(\nu-x)_+^{(n)}$.

Theorem 4.1. For $n \geq 1$ the kernel $G_{n+1}(\nu,x) = (\nu-x)_+^{(n)}$ is totally positive on $Z \times Z$, i.e.,

(4.1) $$\det[(\nu_i - x_j)_+^{(n)}] \geq 0$$

for all choices of $\nu_1 < \ldots < \nu_r$ and $x_1 < \ldots < x_r$. Moreover, strict inequality holds if and only if

(4.2) $$\nu_{i-n-1} < x_i < \nu_i - (n-1), \qquad i = 1, 2, \ldots, r.$$

It would be possible to extend this result to allow confluent $\nu_i$'s or $x_j$'s (cf. [3]). However, in view of the multiple knot--simple knot equivalence (cf. §2), in the discrete case we do not need it. Instead we prove only

Theorem 4.2. Let $n \geq 1$. For any $x_1 < x_2 < \ldots < x_k$ and $\nu_1 < \nu_2 < \ldots < \nu_{n+k+1}$

$$D\begin{pmatrix} 0,\ldots,0, x_1,\ldots,x_k \\ \nu_1,\ldots\ldots\ldots\ldots,\nu_{n+k+1} \end{pmatrix}$$

$$= \begin{vmatrix} 1 & \nu_1 & \ldots\nu_1^{(n)} & (\nu_1-x_1)_+^{(n)} & \ldots & (\nu_1-x_k)_+^{(n)} \\ 1 & \nu_2 & \ldots\nu_2^{(n)} & (\nu_2-x_1)_+^{(n)} & \ldots & (\nu_2-x_k)_+^{(n)} \\ \ldots & & & & & \\ 1 & \nu_{n+k+1} & \ldots & & & (\nu_{n+k+1}-x_k)_+^{(n)} \end{vmatrix} \geqq 0 .$$

Strict inequality prevails if and only if

(4.3) $\nu_i < x_i < \nu_{i+n+1}-(n-1), \quad i = 1, 2,\ldots,k.$

Corollary 4.3. Fix $n \geq 1$, and let $\nu_1 < \nu_2 < \ldots < \nu_{n+k+1}$ and real numbers $d_1,\ldots,d_{n+k+1}$ be prescribed. Suppose $\Delta = \{x_1 < \ldots < x_k\}$ satisfies (4.3). Then there exists a unique discrete spline $s \in DSp(\pi_n;\Delta;N)$ satisfying

(4.4) $s(\nu_j) = d_j, \quad j = 1, 2,\ldots,n+k+1.$

## 5. Discrete Natural Splines

Let $\Delta = \{x_1 < x_2 < \ldots < x_k\}$.

An important subclass of the discrete polynomial splines of odd degree is

(5.1) $DNSp(\pi_{2m-1};\Delta;N) = \{s \in DSp(\pi_{2m-1};\Delta;N): s_0, s_k \in \pi_{m-1}\}.$

(cf. (2.1).) Following Greville [2] we define a local-support basis for this class. We suppose $k > 2m$.

We need further notation. Let $z_i = x_k - x_i$, $i = 1, 2, \ldots,k$ and let $\overline{\Delta} = \{z_1 < z_2 < \ldots < z_k\}$. Suppose $\overline{Q}$ is defined in the same way as $Q$, only with respect to $\overline{\Delta}$.

Theorem 5.1. Let

(5.2) $$B_i(\nu) = \begin{cases} \overline{Q}^{2m}_{k-m-i+1,m+i-1}(x_k-\nu) & i = 1,\ldots,m \\ Q^{2m}_{i-m,2m}(\nu) & i = m+1,\ldots,k-m \\ Q^{2m}_{i-m,k-i+m}(\nu) & i = k-m+1,\ldots,k \,. \end{cases}$$

Then $\{B_i(\nu)\}_1^k$ form a basis for $DNSp(\pi_{2m-1};\Delta;N)$.

The basis splines $\{B_i\}_{m+1}^{k-m}$ can be computed by the convenient recurrences of §3. Following [4] we may also derive recurrences for the end-bases which permit computations involving only positive combinations of positive quantities.

Theorem 5.2. For $j \geq 2$ and $k \geq 3$,

$$(5.3)\quad Q^{k}_{i,j}(\nu) = Q^{k-2}_{i,j-2}(\nu) - (x_{i+j} + x_{i+j-1} + 2k - 5 - 2\nu)Q^{k-2}_{i,j-1}(\nu) + (\nu - x_{i+j} - k + 3)^{(2)}Q^{k-2}_{i,j}(\nu) \qquad (j \leq k-2).$$

$$(5.4)\quad Q^{k}_{i,k-1}(\nu) = Q^{k-2}_{i,k-3}(\nu) - \left[x_{i+k-1} + x_{i+k-2} - 2\nu + 2k - 5 - \frac{(\nu - x_{i+k-1} - k + 3)^{(2)}}{(x_{i+k-1} - x_i)}\right] Q^{k-2}_{i,k-2}(\nu) - \frac{(\nu - x_{i+k-1} - k + 3)^{(2)}}{(x_{i+k-1} - x_i)} Q^{k-2}_{i+1,k-2}(\nu) .$$

Corollary 5.3. Let $\{B_i\}$ be as in (5.2). Then

(i) $B_i(\nu) > 0$ for $\nu = \ldots, x_{m+i} - 1$, $\quad i = 1, 2, \ldots, m$

(ii) $B_i(\nu) > 0$ for $\nu = x_{i-m} + 2m - 1, \ldots,$ $\quad i = k-m+1, \ldots, k$.

Further results and complete proofs of the theorems quoted here can be found in [7].

---

Supported in part by AFOSR-69-1812D.

## References


[1] Fröberg, C.-E., Introduction to Numerical Analysis. 2nd ed., Addison Wesley, Reading, 1969.

[2] Greville, T. N. E., Introduction to spline functions. In Theory and Application of Spline Functions, T. N. E. Greville, ed., Academic Press, New York, 1969, 1-35.

[3] Karlin, S., Total Positivity Vol. I. Stanford Press, Stanford, 1968.

[4] Lyche, T. and L. L. Schumaker, Computation of smoothing and interpolating natural splines via local bases. SIAM J. Numer. Anal., to appear.

[5] Mangasarian, O. L. and L. L. Schumaker, Discrete splines via mathematical programming. SIAM J. Control 9 (1971), 174-183.

[6] Mangasarian, O. L. and L. L. Schumaker, Best summation formulae and discrete splines. SIAM J. Numer. Anal., to appear.

[7] Schumaker, L. L., Constructive aspects of discrete polynomial spline functions. C.N.A. 62, Center for Numerical Analysis, The University of Texas, Austin, 1973.

Department of Mathematics
The University of Texas at Austin
Austin, Texas 78746

# TWO THEOREMS ON SUNS IN CONTINUOUS FUNCTION SPACES

Paul Schwartz

The concept of a "sun" was introduced into abstract approximation theory by Efimov and Stechkin [5]. Informally, a set V is a <u>sun</u> if the condition that v (in V) is a best approximation to f(outside V) implies that v is also a best approximation to all elements on the ray from v through f. This concept is meaningful in any linear metric space. Making a local definition, we say that v is a <u>solar</u> point of V if it has the property just described for all f which have v as a best approximation.

In [2] Brosowski proved that a set V in a normed linear space is a sun if and only if best approximations from V (when they exist) can be characterized by the <u>Kolmogorov Criterium</u>. We say that a pair (v,f) satisfies the Kolmogorov Criterium if $v \in V$, $f \notin V$, and

(K) For each $u \in V$ there exists a continuous linear functional L such that $\|L\| = 1$, $L(f-v) = \|f - v\|$, and $L(u-v) \leq 0$.

Here we are concerned with these concepts in the particular space $C = C(Q,X)$, where Q is a compact Hausdorff space and X is an arbitrary real normed linear space. In C one uses the norm

$$\|f\| = \max \{\|f(q)\|: q \in Q\}.$$

Two theorems are established concerning suns in the space C. These theorems extend results of [1], [3] in which X was required to be an inner-product space. On the other hand, these results seem not to be readily derivable from the abstract results of [3].

In the case that $V \subset C$, the Kolmogorov Criterium can be stated in the following form:

(K') For each $u \in V$ there exists a point $q \in Q$ and a continuous linear functional L on X such that $\|L\| = 1$, $L(f(q) - v(q)) = \|f - v\|$, and $L(u(q) - v(q)) \leq 0$

In both (K) and (K') the concept is not changed if L is required to be an extreme point of the relevant unit sphere. As before, we can make the definition local by saying that v is a <u>Kolmogorov point</u> of V if (K) or (K') holds whenever f is outside V and has v for a best approximation. In the following theorem, the critical set of an f in C is the set crit $(f) = \{q \in Q: \|f(q)\| = \|f\|\}$.

<u>Unicity Theorem</u>. Let V be a sun in C, and suppose that Q has at least two points. In order that best approximations out of V be unique (when they exist at all) it is necessary that whenever some v in V is a best approximation to some $f \in C\backslash V$, v must be uniquely determined among the elements of V by its values on the set crit (f-v). If the space X is strictly convex, this condition is also sufficient.

<u>Proof of necessity</u>: Suppose that v is a best approximation to f, that $f \neq v$, and that v is <u>not</u> uniquely determined by its values on S = crit (f - v). We shall construct a function h having two best approximations. First, select an element w in V, different from v, which agrees with v on S. Define h by

$$h(q) = v(q) + \{\|w - v\| - \|w(q) - v(q)\|\} \frac{f(q) - v(q)}{\|f - v\|}.$$

If $q \in S$, then $\|h(q) - v(q)\| = \|w - v\|$ and $h(q) - v(q)$ is a multiple of $f(q) - v(q)$. If $q \notin S$, then $\|h(q) - v(q)\| < \|w - v\|$. Hence crit $(h - v) = S$. From these facts, one sees that (v,h) satisfies (K') since (v,f) does. Hence, v is a best approximation to h and $\|h - v\| = \|w - v\|$. A short calculation reveals, however, that $\|h(q) - w(q)\| \leq \|w - v\|$ for all q. Thus w is also a best approximation to h.

Proof of sufficiency: Assume that X is strictly convex, and that some $f \in C$ has two best approximations, v and w, in V. We will show that although w is a best approximation to 2f - v, w is not uniquely determined in V by its values on crit (2f - v - w).

Since (2f - v) - v is a multiple of f - v and (v,f) satisfies (K'), we see that (v,2f - v) also satisfies (K'). Hence v is a best approximation to 2f - v.

Since $\|2f - v - w\| \leq \|f - v\| + \|f - w\| = 2\|f - v\|$, we see that w is also a best approximation to 2f - v.

Now for $q \in$ crit (2f - v - w), we have

$$2\|f - v\| = \|f - v\| + \|f - w\| \geq \|f(q) - v(q)\| + \|f(q) - w(q)\| \geq \|2f(q) - v(q) - w(q)\| = \|2f - v - w\| = 2\|f - v\|.$$

By the strict convexity of X, this implies that f(q) - v(q) = f(q) - w(q), so that w(q) = v(q) for all $q \in$ crit (2f - v - w).

Special cases of the Unicity Theorem are Satz 3.10 of [1] and Theorem 10 of [4]. Theorem 3 of [6] is a related unicity theorem for arbitrary subspaces. For finite-dimensional subspaces, more concrete unicity theorems are available in [8]. That the sufficiency of our condition fails when X is not strictly convex can be shown by taking $X = \ell_\infty^{(2)}$, Q = [0,1], f(t) = (0,t) for $0 \leq t \leq 1$, and $V = \{v_{a,b}: a,b \in \mathbb{R}\}$, where $v_{a,b}(t) = (a + bt, b)$.

Theorem Characterizing Solar Points. In order that a point v of V be a solar point it is necessary and sufficient that for each pair (u,f) having property (1) below there exist a sequence $v_n \in V$ such that $v_n \to v$ and $\|f(q) - v_n(q)\| < \|f - v\|$ for all $q \in$ crit (f - v).

Property (1). $L(u(q) - v(q)) > 0$ whenever $q \in Q$, $L \in X^*$, $\|L\| = 1$ and $L(f(q) - v(q)) = \|f - v\|$.

The proof, while not difficult, is too long to be included here. This theorem extends Satz 11 of [3], and is comparable to the regularity property given there (p. 379) in an abstract normed linear space. In a certain technical sense, the characterization of solar points given here is "intrinsic" since it does not refer explicitly to an approximation problem. However, an approximation problem is actually present since the characterization can be reformulated as follows. If v is locally a best approximation to f on the critical point set of f - v, then the pair (v,f) satisfies the Kolmogorov Criterium (K').

## References

[1] Brosowski, B., "Nicht-Lineare Tschebysheff-Approximation." Mannheim (1968).

[2] Brosowski, B., Nichtlineare Approximation in normierten Vektorräumen. ISNM 10 (1969), 140-159.

[3] Brosowski, B. and R. Wegmann, Charakterisierung bester Approximation in normierten Vektorräumen. J. Approx. Theory 3 (1970), 369-397.

[4] Cheney, E. W. and D. E. Wulbert, The Existence and Unicity of Best Approximations. Math. Scand. 24 (1969), 113-140.

[5] Efimov, N. V. and S. V. Stechkin, "Some Properties of Chebyshev Sets." Doklady Akad. Nauk 118 (1958), 17-19.

[6] Johnson, L. W., Unicity in the Uniform Approximation of Vector-Valued Functions. Bull. Austral. Math. Soc. 3 (1970), 193-198.

[7] Schwartz, P., A Report on Kolmogoroff-Type Characterization of Best Chebyshev Approximation in Spaces of Continuous Vector-Valued Functions. Preprint.

[8] Singer, I., Sur la meilleure approximation des fonctions abstraites continue à valeurs dans un espace de Banach. Rev. Roumaine Math. Pures Appl. 2 (1957), 245-262.

Department of Mathematics
The University of Texas at Austin
Austin, Texas 78712

# APPROXIMATION BY FUNCTIONS WITH RESTRICTED RANGES

Wilhelm Sippel

Let $C[a,b]$ be the space of all real valued continuous functions defined on the real closed interval $[a,b]$ and let $C[a,b]$ be normed by

$$\|f\| = \max\{|f(x)| : x \in [a,b]\} \quad \text{for } f \in C[a,b].$$

For a given subspace $V$ of $C[a,b]$ we define a subset $W$ of $V$ by

$$W = \{v \in V : \ell(x) \leq v(x) \leq u(x) \quad \text{for all } x \in [a,b]\}$$

where $\ell$ and $u$ are two fixed functions in $C[a,b]$ satisfying $\ell(x) \leq u(x)$ for all $x \in [a,b]$. In the following we assume that $V$ is an $n$-dimensional extended Haar subspace of order $n$ of $C[a,b]$. (For the definition of an extended Haar subspace see [1] and [2].)

If $W \neq \emptyset$ then we have for a given $f \in C[a,b]$ that there exists a function $v_0 \in W$ such that

$$\|f - v_0\| = \min\{\|f - v\| : v \in W\}.$$

The function $v_0$ is said to be a best restricted approximation to $f$.

G. D. Taylor [2] gave a complete treatment of this approximation problem in the case that the functions $u$ and $\ell$ fulfill special conditions. In this paper we shall use other restrictions to these two functions and in this way we shall find a new type of alternant points.

Now let $T$ be the set of points of equality of $u$ and $\ell$. We assume that $T$ is a finite set and we denote the elements of $T$ by $y_1, y_2, \ldots, y_s$. Also we assume that there exist $s$ positive integers $\omega_\sigma$ $(1 \leq \sigma \leq s)$ such that

$$u^{(\rho)}(y_\sigma) = \ell^{(\rho)}(y_\sigma) \qquad \text{for } 0 \leq \rho \leq \omega_\sigma - 1$$

and

$$u^{(\omega_\sigma)}(y_\sigma) \neq \ell^{(\omega_\sigma)}(y_\sigma).$$

(We suppose that all these derivatives exist in a neighborhood of $y_\sigma$ and are continuous there.) In addition we assume that

$$\sum_{\sigma=1}^{s} \omega_\sigma < n.$$

In order to develop an alternation theorem it is necessary to define the sets of upper and lower extremal points, $E_u$ and $E_\ell$, in a generalized sense. For given $f \in C[a,b]$ and $v \in W$ let

$$E_u = \{x \in [a,b] \sim T\colon\ v(x) = u(x) \text{ or } f(x) - v(x) = -\|f-v\|\} \cup \{y_\sigma \in T\colon\ v^{(\omega_\sigma)}(y_\sigma) = u^{(\omega_\sigma)}(y_\sigma)\},$$

$$E_\ell = \{x \in [a,b] \sim T\colon\ v(x) = \ell(x) \text{ or } f(x) - v(x) = \|f-v\|\} \cup \{y_\sigma \in T\colon\ v^{(\omega_\sigma)}(y_\sigma) = \ell^{(\omega_\sigma)}(y_\sigma)\}.$$

Finally we define a subspace $V_o$ of $V$ by

$$V_o = \{v \in V\colon\ v^{(\rho)}(y_\sigma) = 0 \text{ for } 0 \leq \rho \leq \omega_\sigma - 1 \text{ and } 1 \leq \sigma \leq s\}.$$

Theorem. Let $f \in C[a,b]$, $v_o \in W$ and suppose

(1) $f(x) + \|f-v_o\| > \ell(x)$ and $f(x) - \|f-v_o\| < u(x)$ for all $x \in [a,b]$.

Let the integer $k$ be defined by $k = n+1 - \Sigma_{\sigma=1}^{s} \omega_\sigma$.

Then the following statements are equivalent:

(a) $v_o$ is a best restricted approximation to $f$.

(b) There exist $k$ points $x_1 < x_2 < \ldots < x_k$ in $[a,b]$ which are elements of $E_u$ and $E_\ell$ alternately.

(c) There exists no function $v \in V_o$ such that

$$v(x) > 0 \text{ for } x \in E_u \sim T, \quad v(x) < 0 \text{ for } x \in E_\ell \sim T,$$

$$v^{(\omega_\sigma)}(y_\sigma) > 0 \text{ for } y_\sigma \in E_u \cap T, \quad v^{(\omega_\sigma)}(y_\sigma) < 0 \text{ for } y_\sigma \in E_\ell \cap T.$$

Remark. If condition (1) of the theorem is violated, i.e. if there exists $x \in [a,b]$ such that

$$f(x) + \|f-v_o\| = \ell(x) \text{ or } f(x) - \|f-v_o\| = u(x),$$

then $v_o$ is a best restricted approximation to f.

Proof: (a) $\Rightarrow$ (c). This and the following two parts are proved by contradiction. Let $v \in V_o$ be a function which fulfills the conditions (c). Then there exists a positive constant d such that $v_o - dv \in W$ and $\|f-(v_o-dv)\| < \|f-v_o\|$. Therefore $v_o$ is not a best restricted approximation to f.

(c) $\Rightarrow$ (b). We assume that there exist at most m, $m<k$, consecutive points $x_1<x_2<\ldots<x_m$ in $[a,b]$, which are elements of $E_u$ and $E_\ell$ alternately. Because we have $m - 1 + \Sigma_{\sigma=1}^{s} \omega_\sigma < n$ we are able to construct a function $v \in V_o$ having exactly m-1 simple zeros at the points $z_1, z_2, \ldots, z_{m-1} \in [a,b] \sim T$ and zeros of order $\omega_\sigma$ at the points $y_\sigma$ $(1 \leq \sigma \leq s)$. Let the points $z_1, z_2, \ldots, z_{m-1}$ be in (a,b). Then they determine m subintervals of [a,b]. We assume that each of these subintervals contains either no point of $E_u$ or no point of $E_\ell$ and that $z_\mu \notin E_u \cup E_\ell$ for $1 \leq \mu \leq m-1$. Then one of the functions v and -v satisfies the conditions of (c).

(b) $\Rightarrow$ (a). If under the hypothesis of (b) there exists a function $w \in W$ such that $\|f-w\| \leq \|f-v\|$ then we find in the same way as in [2] that w=v.

Note that from part "(b) $\Rightarrow$ (a)" of the proof there follows also that if $v_o$ is a best restricted approximation to f satisfying (1) then $v_o$ is unique.

Now let us consider the following example where a point of T occurs as an alternant point. Take $[a,b] = [-1,1]$, $u(x) = 2x^2$ and $\ell(x) = 0$ for $x \in [-1,1]$. Let V be the space of all polynomials of degree $\leq 4$ and let $f \in C[-1,1]$ be defined by

$$f(x) = \begin{cases} -\frac{5}{8}x - \frac{5}{8} & \text{for } x \in [-1,\frac{1}{2}] \\ \frac{47}{8}x - \frac{31}{8} & \text{for } x \in [\frac{1}{2},1] \end{cases} .$$

Then we have $T = \{y_1\} = \{0\}$ and $\omega_1 = 2$ and for $v_o(x) = x^4$ we find that $E_u = \{-1,\frac{1}{2}\}$ and $E_\ell = \{0,1\}$. Because $k = n+1-\omega_1 = 4$ the four points $x_1 = -1$, $x_2 = 0$, $x_3 = \frac{1}{2}$, $x_4 = 1$ are characterizing $v_o$ as a best restricted approximation to f.

## References

[1] Karlin, S. and W. J. Studden, Tchebycheff systems with applications in analysis and statistics. Interscience, New York, 1966.

[2] Taylor, G. D., Approximation by functions having restricted ranges: equality case. Numer. Math. 14 (1969), 71-78.

Institut für Angewandte Mathematik I
Universität Erlangen-Nürnberg
8520 Erlangen
West Germany

# CHARACTERIZATION OF THE FUNCTION CLASS $W^{m,p}$

P. W. Smith

## 1. Introduction

In 1964 Schoenberg [4] obtained the following characterization for functions in $H^{m,2}(R) \equiv H^{m,2}$, $m=1,2,\ldots$, the space of functions $x$ such that $x^{(m-1)}$ is absolutely continuous on R, the real line, and $x^{(m)} \in L^2(R) \equiv L^2$.

Theorem 1.1 (Schoenberg). Let $x \in C(R)$. Then $x \in H^{m,2}$ if and only if there is a constant K independent of h such that

$$(1.1) \qquad h \sum_{i=-\infty}^{\infty} \left( \frac{\Delta_h^m x(ih)}{h^m} \right)^2 \leq K$$

for all $h > 0$, where $\Delta_h^m$ denotes the $m^{th}$ successive difference.

Jerome and Schumaker in [2] then generalized this result to the spaces $H^{m,p}$, $m=1,2,\ldots$, and $1<p<\infty$, and in [3] included the $p=\infty$ case. When we specialize the results to uniform meshes, we get

Theorem 1.2 (Jerome-Schumaker). Let $x \in C(R)$. Then $x \in H^{m,p}$, $m=1,2,\ldots$, $1<p\leq\infty$ if and only if there is a constant K independent of h such that

$$(1.2) \qquad \left\{ h \sum_{i=-\infty}^{\infty} \left| \frac{\Delta_h^m x(ih)}{h^m} \right|^p \right\}^{1/p} \leq K$$

for all $h > 0$.

These results were derived by making use of the $H^{m,p}$-splines which are solutions to norm minimization problems in $H^{m,p}$. We note that $H^{m,p}$, $1<p<\infty$, is a uniformly rotund Banach space when normed by

$$(1.3)\qquad \|x\|_{H^{m,p}} = \left\{ \sum_{i=1}^{m} |x(t_i)|^p + \int_R |D^m x|^p \right\}^{1/p} .$$

The $H^{m,p}$-splines are solutions to

$$(1.4)\qquad \inf_{\substack{x \in H^{m,p} \\ x|_E = f}} \|D^m x\|_{L^p} ,$$

where $E \subset R$ and $f:E \to R$. Golomb showed in [1] that (1.4) has a solution if and only if there is an $x \in H^{m,p}$ so that $x|_E = f$. The author considered a similar problem to (1.4) in [5] except that $H^{m,p}$ was replaced by $W^{m,p} \equiv H^{m,p} \cap L^p$. Solutions to these problems were called $W^{m,p}$-splines. In particular if $E = \{t_i\}_{i=-\infty}^{\infty}$ is quasi-uniform and there is an $x \in W^{m,p}$ so that $x|_E = f$ then

$$(1.5)\qquad \inf_{\substack{x \in W^{m,p} \\ x|_E = f}} \|D^m x\|_{L^p}$$

has a unique solution. We will present a new proof of this result and then using estimates already derived in [5] prove the following analogue to Theorems 1.1 and 1.2.

Theorem 1.3. Let $x \in C(R)$. Then $x \in W^{m,p}$, $m=1,2,\ldots$, $1<p\leq\infty$ if and only if there is a constant K independent of h such that

$$(1.6)\qquad \left( h \sum_{i=-\infty}^{\infty} \left\{ |x(ih)|^p + \left| \frac{\Delta_h^m x(ih)}{h^m} \right|^p \right\} \right)^{1/p} \leq K ,$$

for all $1 \geq h > 0$.

## 2. Equivalent Norms on $W^{m,p}$

Throughout this paper E will denote a quasi-uniform partition of R, i.e., $E = \{t_i\}_{i=-\infty}^{\infty}$ and $0<\delta<t_{i+1}-t_i<\frac{1}{\delta}$, and $\mu$ will be an integer-valued function defined on E with $0\leq\mu(t_i)\leq m-1$ for all $t_i \in E$.

Lemma 2.1. If $x \in H^{m,p}$, $1 \leq p \leq \infty$ and $m=1,2,\ldots$, and $x(t_i) = 0$ for all $t_i \in E$, then $x \in W^{m,p}$.

Proof: This follows from the proof of Lemma 1.4.2 [5] (see especially 1.4.30). Lemma 1.4.2 was proven only for $1<p<\infty$, but the modifications for p=1 and $\infty$ are trivial.

Lemma 2.2. If $x \in H^{m,p}$, $1 \leq p \leq \infty$, $m=1,2,\ldots$, and $\{x(t_i)\}_{t_i \in E} \in \ell^p$, then $x \in W^{m,p}$.

Proof: It is easy to see that there is a $y \in W^{m,p}$ so that $y(t_i) = x(t_i)$ for all $t_i \in E$. Now Lemma 2.1 implies that $y - x \in W^{m,p}$ and hence $y - (y-x) = x \in W^{m,p}$.

Lemma 2.3. $W^{m,p}$ is a Banach space when normed by

$$(2.1) \qquad \|x\|_{E,\mu,p} = \left( \sum_{t_i \in E} \sum_{k=0}^{\mu(t_i)} |D^k x(t_i)|^p \right)^{1/p} + \|D^m x\|_{L^p} .$$

Proof: Since $\|\cdot\|_{E,\mu,p}$ is clearly a norm on $W^{m,p}$ by [5], Corollary 1.4.4, it remains to check whether $W^{m,p}$ is complete under this norm. Let $x_n$ be a Cauchy sequence in $\|\cdot\|_{E,\mu,p}$ and $x_n \in W^{m,p}$. Then $x_n$ converges to $x_*$ in $H^{m,p}$, see (1.3). Furthermore, because convergence in $H^{m,p}$ implies local uniform convergence, convergence in $\|\cdot\|_{E,\mu,p}$ implies $\{x_n(t_i)\}_{t_i \in E} \overset{\ell^p}{\to} \{x_*(t_i)\}_{t_i \in E}$. Thus $x_* \in W^{m,p}$ by Lemma 2.2 and $x_n \to x_*$ in the $\|\cdot\|_{E,\mu,p}$ norm. This is true for any Canchy sequence so $W^{m,p}$ is a Banach space when normed by $\|\cdot\|_{E,\mu,p}$.

Now we have the tools to prove the following theorem.

Theorem 2.1. The following norms are equivalent on $W^{m,p}$, $1 \leq p \leq \infty$ and $m=1,2,\ldots$;

$$(2.2) \qquad \|x\|_{W^{m,p}} = \|x\|_{L^p} + \|D^m x\|_{L^p}$$

$$(2.3) \qquad \|x\|_{E,\mu,p} .$$

Proof: It was shown in [6], Lemma 4.1, for $1<p<\infty$ that there is a constant $K > 0$ so that for all $x \in W^{m,p}$

$$(2.4) \qquad \|x\|_{W^{m,p}} \geq K\|x\|_{E,\mu^*,p} \quad ,$$

where $\mu^*(t_i) \equiv m-1$ for all $t_i \in E$. It follows that

$$(2.5) \qquad \|x\|_{W^{m,p}} \geq K\|x\|_{E,\mu,p}$$

for any $\mu$ satisfying $0\leq\mu(t_i)\leq m-1$. The open mapping theorem can now be invoked to conclude that $\|\cdot\|_{W^{m,p}}$ is equivalent to $\|\cdot\|_{E,\mu,p}$. Although the arguments used in [6] were for $1<p<\infty$, only a slight modification is necessary to deduce the results for $p=1$ and $\infty$.

We note the similarity of $\|\cdot\|_{E,\mu,p}$ with the $H^{m,p}$ norm. In fact we can use this similarity to deduce the existence and uniqueness of $W^{m,p}$-splines. Let $E = E_0 \subset E_1 \subset \ldots \subset E_{m-1}$ be given subsets of $R$, $E_i$ may be empty for $i\neq 0$, and suppose we have function $f_i$ so that

$$(2.4) \qquad \text{(i)} \quad f_i : E_i \to R$$

$$\text{(ii)} \quad \{f_i(t_j)\}_{t_j\in E_i} \in \ell^p \; , \quad i=0,1,\ldots,m-1.$$

Then we may state

Corollary 2.1. With $f_i$ and $E_i$ as above and $1<p<\infty$ there exists a unique solution to

$$(2.5) \qquad \inf_{\substack{x \in W^{m,p} \\ \{D^i x|_{E_i} = f_i\}_{i=0}^{m-1}}} \|D^m x\|_{L^p}$$

Proof: Let us choose $\bar{\mu}$ which maps $E$ into the integers to satisfy

$$(2.6) \qquad \bar{\mu}(t_j) = \max\{i: t_j \in E_i\} \; .$$

Then we see that (2.5) is equivalent to

$$(2.7)\qquad \inf_{\substack{x \in W^{m,p} \\ \{D^i x|_{E_i} = f_i\}_{i=0}^{m-1}}} \|x\|_{E,\bar{\mu},p} \quad .$$

Thus we have reduced our problem to a norm minimization problem over a uniformly rotund Banach space so that the solution exists and is unique if and only if the set $x \in W^{m,p}$ such that $\{D^i x|_{E_i} = f_i\}_{i=0}^{m-1}$ is non-empty. But this fact follows trivially from (2.4) and the fact that $\mathbf{E}$ is quasi-uniform.

## 3. Characterization of $W^{m,p}$-functions, $1<p\leq\infty$

We are now in a position to prove Theorem 1.3. If (1.6) holds then Theorem 1.2 implies that $x \in H^{m,p}$. When h=1 we note that $\{x(i)\}_{i=-\infty}^{\infty} \in \ell^p$. From Lemma 2.2 we may conclude that $x \in W^{m,p}$.

Conversely, if $x \in W^{m,p}$, $1<p<\infty$, we denote by $x_h$ the extremal $W^{m,p}$ extension of $x|_{\{ih\}_{i=-\infty}^{\infty}}$ . Corollary 1.4.1 of [5] implies that there is a constant $\eta>0$ independent of h so that

$$(3.1)\qquad \eta\|x_h\|_{L^p}^p \geq h \sum_{i=-\infty}^{\infty} |x(ih)|^p \quad .$$

Theorem 3.2.1 of [5] tells us that there is a constant $C>0$ independent of h so that

$$(3.2)\qquad \|x_h\|_{L^p} \leq C\|x\|_{W^{m,p}} \qquad \text{for } 0 < h \leq 1 \quad .$$

It follows using Theorem 1.2 again that (1.6) holds. This completes the proof of Theorem 1.3 for $1<p<\infty$. If $x \in W^{m,\infty}$ then clearly (1.6) holds for $p=\infty$.

## References

[1] Golomb, M., $H^{m,p}$-Fxtensions by $H^{m,p}$-Splines. J. Approximation Theory 5 (1972), 238-275.

[2] Jerome, J. W. and L. L. Schumaker, Characterization of Functions with Higher Order Derivatives in $L^p$. Trans. Amer. Math. Soc. 143 (1969), 363-371.

[3] Jerome, J. W. and L. L. Schumaker, Characterization of Absolute Continuity and Fssential Boundedness for Higher Order Derivatives. Center For Numerical Analysis, The University of Texas at Austin, 1972.

[4] Schoenberg, I. J., Spline Interpolation and the Higher Derivatives. Proc. Nat. Acad. Sci. U. S. A. 51 (1964), 24-28.

[5] Smith, P. W., $W^{r,p}(R)$-Splines. Dissertation, Purdue University, Lafayette, Indiana, June 1972.

[6] Smith, P. W., $W^{r,p}$-Splines. J. Approximation Theory. To appear.

Mathematics Department
Texas A&M University
College Station, Texas 77843

# APPROXIMATION OF A CLASS OF UNBOUNDED FUNCTIONS

J. J. Swetits and B. Wood

## 1. Introduction

Approximating unbounded functions has been considered by a number of authors, including Müller and Walk [4], Hoischen [2], Hsu [3], and Eisenberg and Wood [1]. We shall use certain entire functions to approximate the following class of unbounded functions: For $I \subset [0,\infty)$ and g defined on I, let

(1) $w_I(g,\delta) = \sup\{|g(x) - g(x')| : |x - x'| \leq \delta,\ x,x' \in I\}$.

Let $A > 0$, r a nonnegative integer be given. Denote by $\theta_r(A)$ the class of all functions, f, such that $f^{(r)}(t)$ exists on $[0,\infty)$ and for each $\delta > 0$ and finite closed interval $I \subset [0,\infty)$

(2) $w_I\{f^{(r)}(t),\delta\} \leq w_I\{e^{At},\delta\}$.

## 2. Preliminaries

Let $\{\alpha_n\}$ be a positive sequence increasing monotonically to $\infty$. For $0<x<\infty$ let $\delta_n(x) = \sqrt{x/\alpha_n}$, $\Delta_n(x) = \max.(1/\alpha_n,\delta_n)$, $p_{nk}(x) = e^{-\alpha_n x}(\alpha_n x)^k/k!$, $k=0,1,\ldots,n=1,2,\ldots$, and $T_{ns}(x) = \Sigma_{k=0}^{\infty} (k - \alpha_n x)^s\, p_{nk}(x)$, $s=0,1,\ldots,n=1,2,\ldots$. We shall make use of the well-known Szasz-Hille positive linear operators defined by

(3) $$S_{\alpha_n}(f,x) = \sum_{k=0}^{\infty} f\left(\frac{k}{\alpha_n}\right) p_{nk}(x)$$

if $f(t) = O(e^{\alpha t})(t \to \infty)$ for some $\alpha > 0$ (see, e.g. [1]).

<u>Lemma</u>. Let $f \in \theta_0(A)$. Then, for $0\leq x\leq a<\infty$ and $n=1,2,\ldots$,

(4) $|f(x) - S_{\alpha_n}(f,x)| \le C_0/\alpha_n^{1/4}$,

where $C_0$ is a constant depending only on a,A.

Outline of Proof: For $N_n > 0$ write $I_n = [0, x + N_n/A]$,

$$k_n(t) = \begin{cases} 0, & 0 \le t \le x + N_n/A, \\ 1, & t > x + N_n/A, \end{cases}$$

and

$$\tilde{k}_n(t) = \begin{cases} 1, & 0 \le t \le x + N_n/A, \\ 0, & t > x + N_n/A. \end{cases}$$

Therefore $k_n(t) \le e^{A(t-x)}/e^{N_n}$ for $0 \le t < \infty$ and, using the fact that $S_{\alpha_n}$ is positive on $[0,\infty)$,

$$(5) \quad |f(x) - S_{\alpha_n}(f,x)| \le |S_{\alpha_n}\{k_n(t)\{f(x)-f(t)\},x\}|$$

$$+ |S_{\alpha_n}\{\tilde{k}_n(t)\{f(x)-f(t)\},x\}|$$

$$\le e^{-N_n} S_{\alpha_n}\{|f(x)|e^{A|t-x|} + |f(t)|e^{A|t-x|},x\}$$

$$+ S_{\alpha_n}\{w_{I_n}(f,|t-x|),x\}.$$

Inequality (2) implies

$$(6) \quad w_{I_n}\{f,\delta_n(x)\} \le \delta_n(x)\, e^{A(x+N_n/A)}[A - \frac{A^2\delta_n}{2!} + \ldots] \le C\delta_n e^{N_n},$$

where C(x,A) is a constant depending only on x and A. It may be shown that there exists a number F = F(x,A) such that

$$(7) \quad S_{\alpha_n}\left\{\{|f(x)| + |f(t)|\}\, e^{A|t-x|},x\right\} \le F, \quad n=1,2,\ldots .$$

After choosing $N_n = \ln(\alpha_n^{1/4})$, the result follows from (5), (6), (7), and well-known estimates involving the modulus of continuity.

## 3. Main Result

Theorem. Let $0<a<\infty$, $A>0$, $r=0,1,\ldots$ be given. Then there exists a constant $C_r(a,A)$ such that for each $f \in \theta_r(A)$ we can find a sequence $\{I_{nr}^f\}$, $n=1,2,\ldots$, of entire functions with

(8) $|f(x) - I_{nr}^f(x)| \leq C_r[\Delta_n(a)]^r/\alpha_n^{1/4}$, $0\leq x\leq a$, $n=1,2,\ldots$.

The $I_{nr}^f$ have the form:

(9) $I_{nr}^f(z) = e^{-\alpha_n z} \sum_{k=0}^{\infty} \beta_k(n,r)z^k$,

where there exist constants $G(r)$, $H(r)$, $M(r,A)$ such that

(10) $|\beta_k(n,r)| \leq \frac{\alpha_n^k e^{Mk/\alpha_n}}{k!} [1 + \frac{G}{\alpha_n^2}]$, $k=0,1,\ldots,r-1$; $n=1,2,\ldots$,

and

(11) $|\beta_k(n,r)| \leq \frac{\alpha_n^k e^{Mk/\alpha_n}}{k!} [1 + \frac{k(k-1)\ldots(k-r+1)H}{\alpha_n^2}]$, $k \geq r$;

$n=1,2,\ldots$.

Outline of Proof: If $r=0$ we may let $I_{nr}^f(z) = S_{\alpha_n}(f,z)$ and use the Lemma to obtain (8). For $r=1$ we also choose $I_{nr}^f(z) = S_{\alpha_n}(f,z)$ and obtain (8) in a manner similar to the proof of the Lemma.

When $r=2$ let

$$I_{nr}^f(x) = S_{\alpha_n}(f,x) - \frac{T_{n2}(x)}{2!\alpha_n^2} S_{\alpha_n}(f^{(2)},x)$$

and use Taylor's expansion of $f$ and techniques similar to those of the Lemma to obtain (8). Now suppose (8) holds for $2\leq i<r$ and define

$$I_{nr}^f(x) = S_{\alpha_n}(f,x) - \sum_{i=2}^{r} \frac{T_{ni}(x)}{i!\alpha_n^i} I_{n,r-i}^{f^{(i)}}(x).$$

We employ Taylor's expansion of $f$ and the Lemma to establish (8) for $i=r$.

Using induction it may be shown that, for $r \geq 2$ and $f \in \theta_r(A)$,

$$(12) \quad I_{nr}^f(z) = e^{-\alpha_n z} \sum_{k=0}^{\infty} \left\{ f\left(\frac{k}{\alpha_n}\right) + \sum_{i=2}^{r} f^{(i)}\left(\frac{k}{\alpha_n}\right) \frac{\tau_{ri}(z,n)}{\alpha_n^i} \right\} \frac{(\alpha_n z)^k}{k!},$$

where $\tau_{ri}(z,n)$ are some polynomials in $w = \alpha_n z$, independent of f, of degree $[i/2]$ (greatest integer) Estimates (10) and (11) on the coefficients $\beta_k(n,r)$ now follow from (12) and elementary calculations.

It is natural to ask if the Theorem is best possible. Namely, if f is defined on $[0,\infty)$ and may be approximated there by entire functions of the form (9), (10), (11), with error (8), does $f \in \theta_r(A)$ for some $A > 0$? The authors have not been able to answer this question.

## References

[1] Eisenberg, S. and B Wood, On the order of approximation of unbounded functions by positive linear operators. S.I.A.M. J. Num. Anal. 9(2) (1972), 266-276.

[2] Hoischen, L., Asymptotische Approximation stetiger Funktionen durch ganze Dirichlet-Reihen. J. Approx. Theory 3 (1970), 293-299.

[3] Hsu, L. C., Approximation of nonbounded continuous functions by certain sequences of linear positive operators or polynomials. Studia Math. 21 (1961-62), 37-43.

[4] Müller, M. and H. Walk, Konvergenz-und Güteaussagen für die Approximation durch Folgen linearer positiver Operatoren. Math. Inst of the Bulgarian Acad. Sciences, Varna, Bulgaria (1970).

Mathematics Department
Old Dominion University
Norfolk, Virginia 23508

Mathematics Department
University of Arizona
Tucson, Arizona 85721

# UNIFORM APPROXIMATION WITH SIDE CONDITIONS

Gerald D. Taylor[1]

## 1. Introduction

Let $W \subset C[0,1]$ be a class of approximants and $K \subset W$ be a subset of W that is determined by certain constraints. Approximating with elements of K is referred to as approximating from W with side conditions and, as usual, given $f \in C[0,1]$, one says $p \in K$ is a best approximation to f if and only if

$$\|f-p\| = \inf\{\|f-q\|: q \in K\}$$

where

$$\|h\| = \sup\{|h(t)|: t \in [0,1]\}.$$

In what follows we shall survey the work of various authors on this topic.

## 2. Equality Constraints

In 1956, S. Paszkowski [1] considered Lagrange interpolatory constraints imposed on W a Haar subspace. Thus, setting

(1) $\tilde{C}[0,1] = \{f \in C[0,1]: f(t_i) = a_i,\ i=1,\dots k, 0 \le t_1 < \dots < t_k \le 1\}$,

where $\{a_i\}_{i=1}^{k}$ is a fixed set of real numbers, $k < \dim W$ and

(2) $L_n = \{p \in W: p(t_i) = a_i,\ i=1,\dots,k, 0 \le t_1 < \dots < t_k \le 1\}$,

his results included existence, alternation-type characterization and uniqueness theorems for approximating $\tilde{C}[0,1]$ with $L_n$. Uniqueness actually holds for some set of functions strictly between $\tilde{C}[0,1]$ and $C[0,1]$. However, the set $\tilde{C}[0,1]$ is a "natural" class for which uniqueness holds and this is indicative of many side condition problems. He also proved for $W = \pi_n$, the set of all algebraic polynomials of degree $\leq n$, that

$$(3) \quad \text{dist}(f,L_n) \leq C \, \text{dist}(f,\pi_n)$$

for $f \in \tilde{C}[0,1]$, and C a constant independent of n, one of the few times that an estimate of the cost of the side conditions has been given. (Zero in the convex hull and Kolmogorov-type characterizations hold for this and most every other side condition problem. See H. K. Hoffmann [2] and L. Wuytack [3].)

In 1968, F. Deutsch [4] gave a modern treatment of this problem including a generalization of the Remes algorithm. In 1969, this was extended to allow Hermite interpolatory constraints by H. L. Loeb, D. G. Moursund, L. L. Schumaker and G. D. Taylor [5] and recently, D. Platte [6] studied this problem with Hermite-Birkhoff interpolatory constraints of certain type and $W = \pi_n$. In both of these studies, existence, alternation and uniqueness results were proved for the set of continuous functions satisfying only the Lagrange part of the constraints. A Remes-type algorithm was given in the Hermite study and in the Hermite-Birkhoff study it was shown that

$$(4) \quad \text{dist}(f,B_n) \leq A_\varepsilon \{\text{dist}(f,\pi_n)\}^{1-\varepsilon}$$

for f satisfying all the interpolatory constraints, where $A_\varepsilon$ is a constant independent of n with $A_\varepsilon \to \infty$ as $\varepsilon \to 0$, $B_n$ is the set of all polynomials satisfying the constraints and that there exists f satisfying all the interpolatory constraints of a specific problem for which

$$(5) \quad \overline{\lim_{n\to\infty}} \frac{\text{dist}(f,B_n)}{\text{dist}(f,\pi_n)} = \infty.$$

(See D. Johnson [7] for a related study.)

In 1967, C. Gilormini [8] announced existence and alternation results for Lagrange interpolatory constraints imposed on $R_n^m[0,1]$. However, H. L. Loeb [9] showed that his existence claim was incorrect. In 1970, A. Perrie [10] extended the results of [5] to the rational setting. In 1969, R. Barrar and H. L. Loeb [11] studied Lagrange interpolatory constraints on varisolvent families. Recently, J. Williams [12] has studied the problem of approximating with rationals all having the same fixed numerator. This study is primarily numerical, although existence was incorrectly claimed. A future paper by G. D. Taylor and J. Williams will treat this problem.

In 1965, B. Brosowski [13] studied approximation from

$$(6) \quad M = \{p(x) = \sum_{i=1}^{n} a_i\varphi_i(x): W = \langle\varphi_1,\ldots,\varphi_n\rangle \text{ is Haar,}$$

$$A(a_1,\ldots,a_n)^T = (b_1,\ldots,b_m)^T\}$$

where A is a fixed $m \times n$ matrix and $\{b_i\}_{i=1}^m$ fixed real numbers. Existence, a sufficient alternation condition and some other general results were given. No natural class for which uniqueness holds is known. In 1967, C. Gilormini [14] announced some results for the rational analog of this problem. Recently, a somewhat less general study of linear approximation with side conditions treating existence, characterization and uniqueness has been given by B. L. Chalmers [15] which includes most of the studies mentioned in this paper. Even with these two studies much remains to be done in this direction. For example, approximation from $\{p\in\pi_n: \int_0^1 p(x)dx = 0\}$ is not treated by [15] and the results of [13] are not specific enough.

## 3. Inequality Constraints

Early studies of this type are one-sided approximation (1959, W. Kammerer [16]) and a general study by J. R Rice [17] in 1962. In 1968, G. D. Taylor [18, 19] studied restricted range approximation:

(7) $K_n = \{p\in W$: W is Haar, $\ell(x) \leq p(x) \leq u(x)$ and $\ell(x) < u(x)$ for

all $x \in [0,1]\}$

with some additional restrictions on the functions $\ell$ and u. This was extended to $\ell(x) \leq u(x)$ by L. L. Schumaker and G. D. Taylor [20] and to $R_n^m[0,1]$ by H. L. Loeb, D. G. Moursund, and G. D. Taylor [21] and H. L. Loeb and D. G. Moursund [22]. Related studies have been done by P. J. Laurent [23], R. J. Duffin and L. A. Karlovitz [24], K. Taylor [25], E. Tornga [26] and C. B. Dunham [27]. Existence, alternation-type characterization and uniqueness for all f belonging to

(8) $C^*[0,1] = \{f \in C[0,1]: \ell(x) \leq f(x) \leq u(x)$ for all $x \in [0,1]\}$

have been established as well as a Remes-type algorithm (R. G. Jones and L. A. Karlovitz [28], for nonnegative approximants; G. D. Taylor and M. J. Winter [29], a single point exchange; H. S Hersey, D. W. Tufts and J. T. Lewis [30] and D. R. Gimlin, R. K. Cavin and M. C. Budge [31], multiple exchange). Error estimates remain open for this problem. This theory has been used for designing nonrecursive digital filters [30], [31].

In 1965, O. Shisha [32] studied the error of approximation with monotone polynomials and in 1968 these results were extended by J. A. Roulier [33]. Starting in 1968, G. G. Lorentz and K. L. Zeller [34, 35] and G. G. Lorentz [36] studied best approximation questions for this problem. Thus, setting

(9) $M_n = \{p \in \pi_n : \varepsilon_i p^{(k_i)}(x) \geq 0,\ 0 < k_1 < \ldots < k_\nu, k_i$ an integer and

$\varepsilon_i = \pm 1\ \forall i\}$,

existence, characterization and partial uniqueness results were given. Error estimates for approximating $f \in C^1[0,1]$ with $f'(x) \geq 0$ from $M_n' = \{p \in \pi: p'(x) \geq 0\}$ of Jackson-type were also given and a function f with $f'(x) \geq 0$ on [0,1] was constructed for which (5) holds with $B_n$ replaced by $M_n'$. Error estimates of the form (4) and for more general $M_n$ remain open. In 1971, R. A. Lorentz [37] proved that uniqueness of best approximation from $M_n$ holds for all $f \in C[0,1]$ using a theorem of K. Atkinson and A. Sharma [38] on Hermite-Birkhoff interpolation. In 1972, J. T. Lewis [39] gave an algorithm for finding best monotone approximations with only one derivative constraint present. This theory has been generalized independently by B. L. Chalmers [15] and J. A. Roulier and G. D. Taylor [40] to that of polynomials and their derivatives having restricted ranges. These results can probably be generalized to extended Chebyshev systems (S. Karlin and J. M. Karon [41]). Questions concerning monotone rational approximation are open. A reasonable first attempt would be to use $\{r \in R_n^0[0,1]: r'(x) \geq 0 \ \forall\, x \in [0,1]\}$.

In 1968, C. Geiger [42] considered approximation in $C[-1,1]$ by rational functions in $R_n^n[-1,1]$ of the form $p(x)/p(-x)$. Existence, characterization and error considerations were discussed. Finally, in 1971, J. A. Roulier and G. D. Taylor [43] considered approximating by

$$(10) \quad C_n = \{p(x) = \sum_{i=0}^{n} a_i x^i \in \pi_n : \alpha_i \leq a_{k_i} \leq \beta_i,\ 0 \leq k_1 < \ldots < k_\nu \leq n,\ k_i \text{ an integer } \forall i\}$$

where $\{a_i\}_{i=1}^{\nu}$ and $\{\beta_i\}_{i=1}^{\nu}$ are fixed sets of real numbers. An existence result, a necessary alternation-type characterization and a "natural" class of functions for which uniqueness holds are given. It does not appear that the Remes algorithm can be modified to treat this problem; however, a linear programming routine of M. I. Zaldak [44] can be used. Extension of this problem to either rationals or an extended Chebyshev system remains open.

In closing we would like to mention an expository paper on this subject by J. T. Lewis [45] and a study by W. Gearhart [46] of this subject using semi-infinite programming techniques.

---

[1]Supported in part by AFSOR 72-2271.

## References

[1] Paszkowski, S., On approximation with nodes. Razprawy Mat. 14 (1957), 1-62.

[2] Hoffmann, K. H., Zur Theorie der nichtlinearen Tschebyscheff-Approximation mit Neberbedingungen. Numer. Math. 14 (1969), 24-41.

[3] Wuytack, L., Kolmogoroff's criterion for constrained rational approximation. J. Approx. Theory 4 (1971), 120-136.

[4] Deutsch, F., On uniform approximation with interpolatory constraints. J. Math. Anal. Appl. 24 (1968), 62-79.

[5] Loeb, H. L., D. G. Moursund, L. L. Schumaker and G. D. Taylor, Uniform generalized weight function polynomial approximation with interpolation. SIAM J. Numer. Anal. 6 (1969), 283-293.

[6] Platte, D., Approximation with Hermite-Birkhoff interpolatory constraints and related H-set theory. Thesis, Michigan State University, 1972.

[7] Johnson, D., Jackson type theorems for approximation with side conditions. Preprint.

[8] Gilormini, C., Approximation rationelle de Tchebycheff avec des noeuds. C. R. Acad. Sci. Sér. A-B 264 (1967), A359-A360.

[9] Loeb, H. L., Un Contre-exemple à un resultat de M. Claude Gilormini. C. R. Acad. Sci. Sér. A-B 266 (1968), A237-A238.

[10] Perrie, A., Uniform rational approximation with osculatory interpolation. J of Comp. and Sys. Sci. 4 (1970), 509-522.

[11] Barrar, R and H. L. Loeb, Best non-linear uniform approximation with interpolation. Arch. Rational Mech. Anal. 33 (1969), 231-237.

[12] Williams, J., Numerical Chebyshev approximation by interpolating rationals. Math. Comp. 26 (1972), 199-206.

[13] Brosowski, B., Approximationen mit linearen Nebenbedingungen. Math. Z. 88 (1965), 105-128.

[14] Gilormini, C., Sur l'approximation par fractions rationelles généralisées avec constraints sur les coefficients. C. R. Acad. Sci Sér. A-B 265 (1967), A235-A236.

[15] Chalmers, B. L., A unified approach to uniform real

approximation by polynomials with linear restrictions. Trans. Amer. Math. Soc. 166 (1972), 309-316.

[16] Kammerer, W., Optimal approximations of functions; one-sided approximation and extrema preserving approximations. Thesis, Univ. of Wisconsin, 1959.

[17] Rice, J. R., Approximation with convex constraints. J. SIAM 11 (1963), 15-32

[18] Taylor, G. D., Approximation by functions having restricted ranges III. J. Math Anal. Appl. 27 (1969), 241-248.

[19] Taylor, G D., Approximation by functions having restricted ranges: equality case. Numer. Math. 14 (1969), 71-78.

[20] Schumaker, L. L. and G. D. Taylor, On approximation by polynomials having restricted ranges II. SIAM Numer. Anal. 6 (1969), 31-36.

[21] Loeb, H. L., D. G. Moursund and G. D Taylor, Uniform rational generalized weight function approximation having restricted ranges. J. Approx. Theory 1 (1968), 401-411.

[22] Loeb, H. L. and D. G Moursund, Continuity of the best approximation operator for restricted range approximation. J. Approx. Theory 1 (1968), 391-400.

[23] Laurent, P. J., Approximation uniforme de fonctions continues sur un compact avec constraints de type inegalite. R.I.P.O. 1 (1967), 81-95.

[24] Duffin, R. J. and L. A. Karlovitz, Formulation of linear programs in analysis I. Approximation theory, SIAM J. Appl. Math. 16 (1968), 662-675.

[25] Taylor, K., Contribution to the theory of restricted polynomial and rational approximation. Thesis, Michigan State University, 1970.

[26] Tornga, E., Approximation from varisolvent and unisolvent families whose members have restricted ranges. Thesis, Michigan State University, 1971.

[27] Dunham, C. B., Alternating minimax approximation with unequal restraints. Preprint, to appear J. Approx. Theory.

[28] Jones, R. C. and L. A. Karlovitz, Iterative construction of constrained Tchebycheff approximation of continuous functions. SIAM J. Numer. Anal. 5 (1968), 574-585.

[29] Taylor, G. D. and M. J. Winter, A modified Remes algorithm for finding best restricted approximations. SIAM Numer. Anal. 7 (1970), 248-255.

[30] Hersey, H. S., D. W. Tufts and J. T. Lewis, Interactive

minimax design of linear-phase nonrecursive digital filters subject to upper and lower function constraints. IEEE Trans. of Audio and Electroacoustics, AU-20 (June 1972), 171-173.

[31] Gimlin, D. R., R. K. Cavin and M. C. Budge, Non-recursive filter design via best restricted approximations. To appear IEEE, Trans of Audio and Electroacoustics.

[32] Shisha, O., Monotone approximation. Pacific J. Math. 15 (1965), 667-671.

[33] Roulier, J. A., Monotone approximation of certain classes of functions. J. Approx Theory 1 (1968), 319-324.

[34] Lorentz, G. G and K. L. Zeller, Degree of approximation by monotone polynomials II. J. Approx. Theory 2 (1969), 265-269.

[35] Lorentz, G. G. and K. L. Zeller, Monotone approximation by algebraic polynomials, Trans. Amer. Math. Soc. 149 (1970), 1-18.

[36] Lorentz, G. G., Monotone approximation. Proceedings of the 3rd Symposium on Inequalities, ed. O. Shisha, Academic Press Inc., 1969, 201-215.

[37] Lorentz, R. A., Uniqueness of best approximation by monotone polynomials. J. Approx Theory 4 (1971), 401-418.

[38] Atkinson, K. and A. Sharma, A partial characterization of posed Hermite-Birkhoff interpolation problems. SIAM J. Numer. Anal. 6 (1969), 230-235.

[39] Lewis, J. T., Computation of best monotone approximations, Math. Comp. 26 (1972), 737-747.

[40] Roulier, J. A. and G. D. Taylor, Approximation by polynomials with restricted ranges on their derivatives. J. Approx Theory 5 (1972), 216-227.

[41] Karlin, S. and J. M. Karon, On Hermite-Birkhoff interpolation. J. Approx. Theory 6 (1972), 90-115.

[42] Geiger, C., Über eine Klasse rationaler Tschebyscheff-Approximationen mit Nebenbedingungen. J. Approx. Theory 1 (1968), 340-354.

[43] Roulier, J. A and G. D. Taylor, Uniform approximation by polynomials having bounded coefficients. Math. Abhandlung. 36 (1971), 126-135.

[44] Zaldak, M. I., Chebyshev approximation of a continuous function by a polynomial with restrictions on its coefficients. Soviet Math. Dokl. 5 (1964), 1515-1518.

[45] Lewis, J. T., Approximation with convex constraints. To appear in SIAM Rev. January, 1973.

[46] Gearhart, W., Analysis of Chebyshev-type approximation through semi-infinite programming. Preprint.

Computer Science Department
Stanford University
Stanford, California 94305

# FOURIER MULTIPLIERS ON $L^p(R^n)$ IN CONNECTION WITH BOUNDED RIESZ MEANS

Walter Trebels

In this note we would like to substantiate the remark of the survey paper of P. L. Butzer [1; Sec. 4] that the techniques outlined there can be applied to Fourier multiplier theory on $R^n$, the Euclidean n-space.

Denote the elements of $R^n$ by $v,x,y,\dots,x = (x_1,\dots,x_n)$, the inner product by $v\cdot x = \Sigma_{k=1}^n v_k x_k$, and the absolute value by $|x| = (x\cdot x)^{1/2}$. Let S be the Schwartzian space of infinitely differentiable, rapidly decreasing functions, and let $L^p$, $1\le p\le\infty$, be the usual Lebesgue space with norm

$$\|f\|_p = (\int_{R^n} |f(x)|^p dx)^{1/p},\ 1\le p<\infty,\ \text{or}\ \|f\|_\infty = \operatorname{ess\,sup}_x |f(x)|.$$

Introduce on $L^p$ the Riesz means of order $\alpha$ by

$$R_\alpha(\rho)f(x) = (2\pi)^{-n/2} \int_{R^n} \rho^n r_\alpha(\rho y) f(x-y)dy \equiv \rho^n r_\alpha(\rho\cdot)*f(x),$$

where $r_\alpha$ is given by its (distributional) Fourier transform

$$r_\alpha^\wedge(v) = \begin{cases} (1-|v|)^\alpha, & |v|\le 1 \\ 0 & , |v|\ge 1 \end{cases},\quad \phi^\wedge(v) = (2\pi)^{-n/2} \int_{R^n} \phi(x) e^{-iv\cdot x} dx.$$

Then it is well-known (see [5], [6; p. 114]) that

$$(*)\quad \|R_\alpha(\rho)f\|_p \le C_\alpha \|f\|_p \qquad (\alpha>(n-1)|\tfrac{1}{p}-\tfrac{1}{2}|;\ 1\le p\le\infty)$$

uniformly in $\rho > 0$; in particular, in case $p = 1$ and $p = \infty$ the function $r_\alpha$ is integrable over $R^n$.

Analogously to [8] let us derive sufficient multiplier criteria for radial functions from property (*). To this end, introduce $BV_1$ as the usual class of functions of bounded variation on $[0,\infty)$ and for $\alpha>0$

$$BV_{\alpha+1} = \{e \in C[0,\infty]; e^{(\gamma)},\ldots,e^{(\alpha-1)} \in AC_{loc}(0,\infty), e^{(\alpha)} \in BV_{loc}(0,\infty),$$

$$\|e\|_{BV_{\alpha+1}} = \frac{1}{\Gamma(\alpha+1)} \int_0^\infty \xi^\alpha |de^{(\alpha)}(\xi)| + \lim_{\xi\to\infty}|e(\xi)| < \infty\},$$

where $\gamma = \alpha - [\alpha]$, $[\alpha]$ being the largest integer less than or equal to $\alpha$; $C[0,\infty]$ is the set of all bounded, uniformly continuous functions for which $\lim_{\xi\to\infty} e(\xi) = e_\infty$ exists. Thus, for integer $\alpha$ the above definition of $BV_{\alpha+1}$ is immediately clear. In order to explain the strictly fractional case consider the fractional integral of order $1-\gamma$, $0<\gamma<1$,

$$I_\omega^{1-\gamma}[e](\xi) = \frac{1}{\Gamma(1-\gamma)} \int_\xi^\omega (\zeta-\xi)^{-\gamma} e(\zeta)d\zeta,$$

and following Cossar (see [8; p. 31]) define the fractional derivative of order $\gamma$ by

$$e^{(\gamma)}(\xi) = -\lim_{\omega\to\infty} (d/dx) I_\omega^{1-\gamma}[e](\xi).$$

Usual differentiation of $e^{(\gamma)}$ yields pure fractional derivatives of order $\alpha = [\alpha] + \gamma$, i.e.,

$$e^{(\alpha)}(\xi) = (d/d\xi)^{[\alpha]} e^{(\gamma)}(\xi).$$

Thus the sets $BV_{\alpha+1}$ are defined for all $\alpha \geq 0$ and note that $BV_{\alpha+1} \subset BV_{\beta+1}$ in the sense of continuous embedding for all $\beta$, $0\leq\beta\leq\alpha$. In order to formulate our first result, let $M^p(=M^p(R^n))$ be the set of all multipliers $m(v)$ acting from $L^p$ in $L^p$. Then for $p=1$ or $p=\infty$ a function $m$ is a multiplier iff it is the Fourier (-Stieltjes) transform of a suitable bounded Borel measure on $R^n$. In case $1<p<\infty$ one has $m(v) \in M^p$ iff there exists a constant $D_m$ such that

$$\|F^{-1}[m\phi^\wedge]\|_p \leq D_m \|\phi\|_p \qquad (\phi \in S),$$

$F^{-1}$ being the inverse Fourier transform.

<u>Theorem 1</u>. If $e \in BV_{\alpha+1}$, then $e(|v|) \in M^p$ provided

$\alpha > (n-1)|1/p-1/2|$, $1\le p\le\infty$.

Proof: Note that for arbitrary $e \in BV_{\alpha+1}$ one has ($\alpha>0$)

$$e(\xi) = H_\alpha \int_\xi^\infty (\zeta-\xi)^\alpha de^{(\alpha)}(\zeta) + e_\infty, \quad H_\alpha = \frac{(-1)^{[\alpha]+1}}{\Gamma(\alpha+1)}$$

(see [8; pp. 25, 33, 36]), and that for arbitrary $\alpha > 0$ the $\alpha^{th}$ derivative of a constant is zero. Then, in case $1<p<\infty$,

$$F^{-1}[e(|v|)\phi^\wedge(v)](x)$$

$$= (2\pi)^{-n/2} \int_{R^n} H_\alpha \int_{|v|}^\infty (\zeta-|v|)^\alpha de^{(\alpha)}(\zeta)\phi^\wedge(v)e^{ix\cdot v}dv + e_\infty\phi(x)$$

for any $\phi \in S$. Since the double integral converges absolutely, we may change the order of integration. Taking norms and applying the generalized Minkowski inequality in connection with (*) we obtain

$$\|F^{-1}[e(|v|)\phi^\wedge(v)]\|_p$$

$$= \|H_\alpha \int_0^\infty \zeta^\alpha de^{(\alpha)}(\zeta)(2\pi)^{-n/2} \int_{|v|\le\zeta} (1 - \frac{|v|}{\zeta})^\alpha \phi^\wedge(v)e^{ix\cdot v}dv\|_p$$

$$+ |e_\infty| \|\phi\|_p \le C_\alpha \|e\|_{BV_{\alpha+1}} \|\phi\|_p .$$

In case $p=1$ and $p=\infty$ consider $F^{-1}[e(|v|)r_\alpha^\wedge(v)]$; since $r_\alpha \in L^1$ for $\alpha > (n-1)/2$, a direct evaluation along the same lines proves e to be a Fourier (-Stieltjes) transform of a suitable bounded measure.

For $p=1$, $n=1$ this theorem is well-known as quasi-convexity theorem (cf. [2; p. 251]); for integer $\alpha$ and $n>1$, $p=1$ see [7].

Note that the $BV_{\alpha+1}$-norm is invariant with respect to dilations; indeed, $\|e(\xi/\rho)\|_{BV_{\alpha+1}} = \|e\|_{BV_{\alpha+1}}$ uniformly in $\rho > 0$, a property standard in multiplier theory. But the property that the multipliers be of Fejér's type, i.e., $e_\rho(\xi) = e(\xi/\rho)$, may be considerably relaxed.

Theorem 2. Let $\psi(\rho)$ be positive for $\rho > 0$, $\chi(\xi)$ be continuous.

strictly increasing on $[0,\infty)$ with $\lim_{\xi\to 0+}\chi(\xi) = 0$ and $\lim_{\xi\to\infty}\chi(\xi) = \infty$. Let $\chi$ possess $([\alpha]+2)$ derivatives on $(0,\infty)$ with

$$\xi^k|\chi^{(k+1)}(\xi)| \le C\chi'(\xi) \qquad (0 \le k \le [\alpha]+2),$$

$C$ being independent of $\xi$, and let $\chi'$ be monotone for all $\xi \ge \xi_0 \ge 0$. Then, for each $e \in BV_{\alpha+1}$ one has uniformly in $\rho>0$

$$e(\chi(|v|)/\psi(\rho)) \in M^p \qquad (\alpha > (n-1)|\frac{1}{p} - \frac{1}{2}|;\ 1\le p\le\infty).$$

In particular $\chi(\xi) = \xi^\beta, \beta > 0$, $\log(1+\xi)$ are admitted.

The proof follows along the same lines as for Theorem 1 since

$$\begin{cases} (1 - \chi(\xi)/\chi(\rho))^\alpha, & 0 \le \xi \le \rho \\ 0 & , \quad \xi \ge \rho \end{cases}$$

belongs to $BV_{\alpha+1}$ uniformly in $\rho > 0$ (see [8; p. 28, 47]). The theorem itself is an appropriate modification of Hardy's "Second theorem of consistency" (1916).

To estimate the $BV_{\alpha+1}$-norm in the pure fractional case is mostly a quite difficult task. A modification of a theorem of Weyl (1917) provides us with a sufficient, easy to verify Lipschitz condition.

Theorem 3. Let $e$ have compact support and be $([\alpha]+1)$-fold differentiable on $(0,\infty)$. If $\int_0^\varepsilon \xi^\alpha|e^{([\alpha]+1)}(\xi)|d\xi$ is finite for some $\varepsilon > 0$ and

$$\int_0^\infty \xi^\alpha|e^{([\alpha]+1)}(\zeta + \xi) - e^{([\alpha]+1)}(\xi)|d\xi = \mathcal{O}(\zeta^\delta),$$

where $0 < \zeta < 1/4$ and $\delta > \alpha - [\alpha]$, then $e(|v|) \in M^p$, $1\le p\le\infty$, provided $\alpha > (n-1)|1/p - 1/2|$.

For the proof we remark that the above hypotheses imply $e \in BV_{\alpha+1}$ (see [8; p. 46]).

Remark. (a) It is clear that any improvement of (*) will give sharper multiplier criteria. In this respect let us mention two recent results.

(i) Fefferman [4] has shown that for $n \geq 2$

$$\|\rho^n b_\alpha(\rho\cdot)* f\|_p \leq C_\alpha \|f\|_p, \quad b_\alpha\hat{\ }(v) = \begin{cases} (1-v^2)^\alpha, & |v| \leq 1 \\ 0 & , |v| \geq 1 \end{cases}$$

provided $1<p<4n/(3n+1)$ and $\alpha > n/p - (n+1)/2$.

(ii) Carleson-Sjölin [3] improved (*) for n=2 in another way:

$$\|\rho^n b_\alpha(\rho\cdot)* f\|_p \leq C'_\alpha \|f\|_p \qquad (\alpha > 0;\ 4/3 < p < 4).$$

Finally, according to a written communication of P. Sjölin, Fefferman's above result may be slightly improved to $1<p\leq 4/3$.

Since $(1+\xi)^{-\alpha} \in BV_{\beta+1}$ for <u>all</u> $\beta \geq 0$ and $r_\alpha\hat{\ }(v) = (1 + |v|)^{-\alpha} b_\alpha\hat{\ }(v)$, it is clear that (*) also holds with $\alpha$ and p as in (i) and (ii).

(b) The above theorems yield additional information (in the case of radial multipliers) to a problem of Stein [6; p.110]: Give sufficient conditions for a multiplier to belong to some $M^p$, $p \neq 2$, without implying that it also belongs to $M^p$, for all $1<p<\infty$. Another approach not restricted to radial multipliers--is given by Hirschman [5]; however, to compute there the $\beta$-variation of e seems to be more complicated than to verify $e \in BV_{\alpha+1}$.

## <u>References</u>


[1] Butzer, P. L., A survey of work on approximation at Aachen, 1968-1972. These proceedings.

[2] Butzer, P. L. and R. J. Nessel, Fourier Analysis and Approximation, Vol. I. Birkhäuser, Basel and Academic Press, New York, 1971.

[3] Carleson, L. and P. Sjölin, Oscillatory integrals and a multiplier problem for the disc. Studia Math. <u>44</u> (1972), 287-299.

[4] Fefferman, C., Inequalities for strongly singular convolution operators. Acta Math. <u>124</u> (1970), 9-36.

[5] Hirschman, Jr., I. I., On multiplier transformations. Duke Math. J. 26 (1959), 221-242.

[6] Stein, E. M., Singular Integrals and Differentiability Properties of Functions. Princeton Univ. Press, Princeton, N. J., 1970.

[7] Trebels, W., On a Fourier-$L^1(E_n)$-multiplier criterion. To appear in Acta Sci. Math. (Szeged) .

[8] Trebels, W., Multipliers for $(C,\alpha)$-bounded Fourier expansions in Banach spaces and approximation theory. In print.

Lehrstuhl A für Mathematik
RWTH Aachen
D-51-Aachen, Templergraben 55
West Germany

# THE APPROXIMATION OF FUNCTIONS BY CERTAIN TRIGONOMETRIC INTERPOLATION POLYNOMIALS

A. K. Varma

Let $f(x)$ be a $2\pi$ periodic continuous function defined on the real line. Throughout this paper, M is assumed to be a fixed even positive integer. Also, we set

$$(1)\quad x_{kn} = \frac{2k\pi}{n}, \quad k = 0,1,\ldots,n-1.$$

In earlier work [2] we have introduced the trigonometric polynomials

$$(2)\quad A_n(f,x) = \sum_{k=0}^{n=1} f(x_{kn})H_n(x - x_{kn}),$$

where $H_n(x)$ is defined by

$$(3)\quad H_n(x) = \frac{1}{n}\left\{1 + 2\sum_{j=1}^{n-1} \alpha_j \cos jx\right\}, \quad \alpha_j = \frac{(n-j)^M}{(n-j)^M + j^M}$$

Here $A_n(f,x)$ has the interpolatory property

$$(4)\quad A_n(f,x_i) = f(x_i) \quad i=0,1,\ldots,n-1.$$

The object of this paper is to improve the earlier estimates of $|A_n(f,x) - f(x)|$ considerably. Our result, is somewhat analogous to Zygmund's theorem [3] concerning the approximation of functions by typical means of their Fourier series. More precisely we aim to prove the following

<u>Theorem 1.</u> If f is a $2\pi$-periodic continuous function and $w_{M-1}(\frac{1}{n}, f)$ is the modulus of continuity of $f(x)$ of $(M-1)^{th}$ order, then we have

$$(5)\quad |A_n(f,x) - f(x)| \le c\, w_{M-1}\left(\frac{1}{n}, f\right).$$

It is well-known that if $k<v$ then $w_v(f,t) \leq 2^{v-k} w_k(f,t)$ for any function $f(x)$. Therefore, from (5) we may conclude that $A_n(f,x)$ gives best order of approximation to $f(x)$ among continuously differentiable function of orders $1,2,\ldots M-1$. The next theorem deals with the order of approximation of $f(x)$ by $A_n(f,x)$ when $f(x)$ is M times continuously differentiable on the real line.

Theorem 2. Let $f(x)$ be $2\pi$-periodic and M times continuously differentiable. Then we have

$$(6)\quad |f(x) - A_n(f,x)| \leq \frac{c_2 \log n}{n^M} \|f^{(M)}\|.$$

Proof of Theorem 1. It is easy to verify that

$$(7)\quad A_n(\cos it,x) - \cos ix = \alpha_{n-i}(\cos (n-i)x - \cos ix)$$

$$A_n(\sin it,x) - \sin ix = -\alpha_{n-i}(\sin ix + \sin (n-i)x)$$

$$i=1,2,\ldots n-1$$

and

$$(8)\quad A_n(1,x) \equiv 1.$$

Let $T_{n-1}(x)$ be any arbitrary trigonometric polynomial of order $n-1$ given by $T_{n-1}(x) = e_0 + \Sigma_{i=1}^{n-1} e_i \cos ix + f_i \sin ix$.

On using (7), (8) and (2) we obtain

$$(9)\quad T_{n-1}(x) - A_n[T_{n-1},x] = (1 - \cos nx) \sum_{i=1}^{n-1} \alpha_{n-i}(e_i \cos ix + f_i \sin ix) + \sin nx \sum_{i=1}^{n-1} \alpha_{n-i} \cdot (e_i \sin ix - f_i \cos ix)$$

we set (M - even)

$$(10)\quad X_n^M(x) = 1/2\, a_0 + \sum_{v=1}^{n-1} (a_v \cos vx + b_v \sin vx) \left(1 - \frac{v^M}{n^M}\right)$$

where $a_k$, $b_k$ are Fourier coefficients of $f(x)$. $X_n^M(x)$ are called typical means of the Fourier series of $f(x)$.

On using the well-known theorem of A.Zygmund [3] and a theorem of S. B. Steckin [1] it follows that

$$(11)\quad |f(x) - X_n^M(x)| \le c_3 w_{M-1}(\tfrac{1}{n}, f)$$

and

$$(12)\quad |f(x) - X_n^M(x)| \le c_4 w_M(\tfrac{1}{n}, f).$$

For our purpose we choose $T_{n-1}(x)$ in particular $X_n^M(x)$ as defined in (10). From (9) we obtain

$$(13)\quad X_n^M(x) - A_n(X_n^M, x) = (1 - \cos nx)P'(x) + (\sin nx)\tilde{P}'(x)$$

where

$$(14)\quad P(x) = \frac{1}{\pi}\int_0^{2\pi} f(t+x)\left(\sum_{i=1}^{n-1} i^{M-1}\delta_i \sin it\right)dt$$

$$(15)\quad \delta_i = \frac{(1 - i^M/n^M)}{(n-i)^M + i^M} \qquad i=1,2,\dots,n-1$$

(we set $\delta_0 = \frac{1}{n^M}$, $\delta_n = 0$).

Let f(x) be (M-1) times continuously differentiable function of x. Integrating by parts (M-1) times we obtain

$$(16)\quad P(x) = \frac{-1}{\pi}\int_0^{2\pi} f^{(M-1)}(t+x)\left(\sum_{i=1}^{n-1} \delta_i \cos it\right)dt.$$

Now, it remains to estimate the expression in the bracket inside the integrand. For this we set

$$(17)\quad 2\cos iu = (i+1)t_{i+1}(u) - 2it_i(u) + (i-1)t_{i-1}(u),$$

where $t_i(u)$ denotes the Fejer kernel

$$(18)\quad t_i(u) = 1 + \frac{2}{i}\sum_{j=1}^{i-1}(i-j)\cos ju, \qquad i > 1,$$

$$=1, \qquad\qquad i = 1,$$

on using (16), (17) and positivity of Fejer kernel together with

(19) $\int_0^{2\pi} |t_i(u)| du = 2\pi$

we obtain the desired estimate of P(x)

$$|P(x)| = O(1/n^M) \max |f^{(M-1)}(x)|.$$

Since P(x) is a trigonometric polynomial of order $\leq n-1$ we have

(20) $|P'(x)| = O(1/n^{M-1}) \max |f^{(M-1)}(x)|.$

and

(21) $|\tilde{P}'(x)| = O(1/n^{M-1}) \max |f^{(M-1)}(x)|.$

On using (13), (20) and (21) we have

(22) $|X_n^M(x) - A_n(X_n^M, x)| = O(1/n^{M-1}) \max |f^{(M-1)}(x)|.$

From earlier results [2] we know that

(23) $|A_n(f,x)| \leq L\|f\|$ if $f \in C_{2\pi}$.

Next we note that

(24) $f(x) - A_n(f,x) = f(x) - X_n^M(x) + X_n^M(x) - A_n(X_n^M, x)$
$+ A_n(X_n^M, x) - A_n(f,x).$

On using (22) to (24) it follows that if f(x) has (M-1) continuous derivatives then $|f(x) - A_n(f,x)| \leq (c_5/n^{M-1}) \cdot \max |f^{(M-1)}(x)|$.

On using Stečkin's theorem once more, we have for $f \in C_{2\pi}$ $|f(x) - A_n(f,x)| \leq c_6 w_{M-1}(\frac{1}{n}, f)$.

This completes the proof of Theorem 1. The proof of Theorem 2 needs similar ideas.

For completeness we state the theorem of the Stečkin which we have used in the proof of our theorem.

Theorem (S. B. Stečkin). Let p be a natural number and $u_n (n=1,2,\ldots)$ be a linear method of approximation of functions, having the following properties

(i) for any function $f(x) \in C_{2\pi}$ $\|u_n(f)\| \leq M_0\|f\|$;

(ii) For any function $f(x) \in C_{2\pi}$ for which $f^{(p)}(x) \in C_{2\pi}$
$\|f - u_n(f)\| \leq M_p\|f^p\|/n^p$ $n=1,2,\ldots$

Then for any function $f(x) \in C_{2\pi}$ we have

$$\|f - u_n(f)\| \leq B_p(M_0 + M_p)\, w_p(\tfrac{1}{n}, f).$$

## References

[1] Stečkin, S. B., The approximation of periodic functions by Fejer sums. Amer. Math. Soc. Translations (2) 28 (1963), 269-282.

[2] Varma, A. K., On a new interpolation process. J. Approx Theory 4 (1971), 159-164.

[3] Zygmund, A., The approximation of functions by typical means of their Fourier series. Duke Math. J. 12 (1945) 695-704.

Department of Mathematics
University of Florida
Gainesville, Florida 32601

# A GENERAL THEOREM OF KOROVKIN TYPE FOR VECTOR LATTICES

Manfred Wolff

A well-known theorem by Korovkin allows to derive the pointwise convergence of a sequence of operators from the convergence of their restrictions to a certain finite subset. In his famous paper [4] J. A. Šaškin proved, that such a test-set or Korovkin-set exists in a space $C(X)$ (i.e., the space of all real valued continuous functions on a compact space X) iff X is homeomorphic to a subspace of $R^n$ for a suitable n or, equivalently, iff $C(X)$ is finitely generated, considered as a Banach algebra over R.

Various authors carried over especially Korovkin's result to certain other Banach lattices (see the references in [7]). But a general theorem characterizing completely all those Banach lattices which possess a Korovkin set has not yet been formulated. It is the main purpose of this paper to give such a result and to sketch its proof. The result itself will be published in [8], but the proof given there differs from our proof here which is in part due to E. Scheffold (oral communication).

Let us call a locally convex vector lattice (lcvl) E finitely generated, if there exists a finite subset A, such that E is the smallest closed vector sublattice containing A (see [5], V. Sections 7 and 8 for the general theory of lcvls).

A complete description of the class of finitely generated Banach lattices will be found in [7], Section 1. Well-known examples are $C(X)(X \subset R^n)$, $L^1(X,\Sigma,\mu)$ (($X,\Sigma,\mu$) being a separable measure space) and $\ell^p(N)$ ($1 \leq p < \infty$). We proved in [7] that all separable Banach lattices with order continuous norm, especially all reflexive ones, are generated by two elements.

We give now our main

Definition. A subset M of a lcvl E is called Korovkin set, if the following holds:

for all lcvls F, for all equicontinuous sequences $(T_n)_{n \in N}$ of positive linear maps from E to F and for all continuous lattice homomorphisms $S: E \to F$ the relation $\lim_{n\to\infty} T_n g = Sg$ for all $g \in M$ implies always $\lim_{n\to\infty} T_n f = Sf$ for all $f \in E$.[1]

Remark. Usually one considers only the case F=E and S the identity on E. But in [4] Šaškin used other lattice homomorphisms implicitly, too.

Theorem. For a locally convex sequentially complete vector lattice E the following statements are equivalent:

(i) E possesses a finite Korovkin set,

(ii) E is finitely generated.

Moreover, if E is generated by k elements, E possesses a Korovkin set of at most 2k+1 elements.

Before we sketch the proof, let us give some

Remarks. (1) As the proof will show, it is enough to require E to be complete for the relative uniform convergence, e.g., we can apply the theorem to a Banach lattice E being equipped with the topology o(E,E') (i.e., the coarsest locally convex vector lattice topology compatible with the dual system (E,E')).

(2) As cited above a separable Banach lattice E with order continuous norm (e.g., a reflexive one) is generated by two elements only. It can easily be proved, that in this case E possesses a Korovkin set of three elements. Typical examples are separable $L^p$-spaces.

Sketch of the proof: (i) ⇒ (ii). Let M be a finite Korovkin set and let H be the smallest closed vector sublattice containing M. $H \neq E$ implies the existence of a continuous linear form $\mu \neq 0$ on E with $\mu(H) = 0$. The seminorm $p: f \to |\mu|(|f|)$ (for all $f \in E$) is additive on $E_+ = \{f \in E: f \geq 0\}$ and therefore the completion $E^\mu$ of $E/p^{-1}(0)$ is an AL-space. The canonical mapping S from E to $E^\mu$ maps H onto a sublattice S(H), whose closure $H_1$ is unequal to $E^\mu$.

By a well-known theorem by T. Andô [1] there exists a positive projection P from $E^\mu$ onto $H_1$. We have $T = PS \neq S$ but $T/_M = S/_M$.[2] This contradicts the assumption about M to be a Korovkin set.

(ii) ⇒ (i): (I) If A is a finite subset generating E, then $u = \Sigma_{x \in A} |x|$ is a quasiinterior point of $E_+$. Therefore $E_u = \cup_{n \in N}\{f \in E: |f| \leq nu\}$ is a dense ideal of E containing A. Since E is sequentially complete (it is enough to require E to be complete for the relative uniform convergence, in any case one has to use [5], V.6.2) $E_u$, equipped with the norm $q(f) = \inf\{\lambda > 0: |f| \leq \lambda u\}$, is an AM-space with unit u containing A. Let $H_u$ be the q-closed vector sublattice of $E_u$ generated by A. By Kakutani's representation theorem (see [5], V.8.5) $H_u$ is lattice isomorphic to $C(X)$ with a compact X. X has to be contained in $R^n$ for a suitable $n \in N$, since $C(X)$ must be finitely generated, too.

To summarize, E contains a vector sublattice G isomorphic to $C(X)(X \subset R^n)$ and dense in E (since $A \subset G$).

(II) Let $(T_n)_{n \in N}$ be an equicontinuous sequence of positive linear maps from E to F and S: E→F a continuous lattice homomorphism.

By means of a famous method in Analysis, used for example by H. P. Lotz ([3] Sec.3) and in the present situation by E. Scheffold, and which is similar to methods of Nonstandard Analysis, we can reduce the convergence problem to a "stationary" problem, i.e., we can find a positive linear map $\hat{T}$ from E to a lcvl $\hat{F}$ and a lattice homomorphism $\hat{S}$ from E to $\hat{F}$ with the property:

$$\lim_{n\to\infty} T_n f = Sf \text{ for all } f \in E, \text{ iff } \hat{T}/_G = \hat{S}/_G,$$

where G is any dense subset of E.

(III) In view of (II) we can complete our proof by the following

Lemma. Let E,F be lcvl, X compact (not necessarily contained in $\mathbb{R}^n$) and G a sublattice of E isomorphic to $C(X)$. Let M be a subset of G whose corresponding set $M_X$ in $C(X)$ contains the constant function $1_X$, so that the Choquet boundary of the linear hull of $M_X$ is equal to X. Let T be a positive linear map from E to F, and $S:E \to F$ a lattice homomorphism, and $T/_M = S/_M$. Then we have $T/_G = S/_G$.

For in our situation we can identify G with $C(X)$, and M with $M_X$. Without loss of generality F is assumed to be complete and therefore $F_{S(1_X)} = \bigcup_{n\in\mathbb{N}} \{y\in F: |y| \leq n\, S(1_X)\}$ is identifiable with a space $C(Y)$, containing $T(C(X))$ and satisfying the relation $S(1_X) = 1_Y$. Now we have reduced our problem to a classical situation and the proof can easily be finished (for the situation in mind see e.g. [7] Theorem 0.3).

Remarks. (1) The essential step (II and III together) was proved (under slight restrictions) in [8] by carrying over the proof of [2] pp. 7-8. The given proof is in part due to E. Scheffold.

(2) In the literature, several proofs of special cases of our result are not correct, because implicitly the authors use a linear lifting, which does not exist in those cases. This difficulty is avoided by the use of Kakutani's representation theorem (the representation theorem of Davies and Lotz, respectively, see [8] for a more detailed discussion).

(3) As pointed out above usually one considers only approximation to the identity operator. But for the approximation of unbounded functions, as treated in [6], one has to use other lattice homomorphisms, e.g., restriction operators.

We shall give a detailed discussion and a general approximation theorem for unbounded functions in a forthcoming paper; however, let us refer to the fact, that the theorem presented here is applicable in any case to special spaces of unbounded functions, e.g., $L^p(X,\Sigma,\mu)(1\leq p<\infty,(X,\Sigma,\mu)$ a separable totally finite measure space).

---

[1]A linear map $S:E \to F$ is called a lattice homomorphism, if it satisfies the condition $|Sx| = S|x|$ for all $x \in E$.

[2]$T/_M$: Restriction of T to M.

## References

[1] Andô, T., Banach-Verbände und positive Projektionen. Math. Z. 109 (1969), 121-130.

[2] Lorentz, G. G., Approximation of Functions. Holt, Rinehart and Winston, New York, 1966.

[3] Lotz, H. P., Über das Spektrum positiver Operatoren. Math. Z. 108 (1968), 15-32.

[4] Šaškin, J. A., Korovkin systems in spaces of continuous functions. Ann. Math. Soc. Transl. 54 (1966), 125-144.

[5] Schaefer, H. H., Topological vector spaces. 3rd ed., Springer, Berlin, 1971.

[6] Walk, H., Approximation durch Folgen linearer positiver Operatoren. Arch. Math. 20 (1969), 398-404.

[7] Wolff, M., Darstellung von Banach-Verbänden und Sätze vom Korovkin-Typ. Math. Ann. In print.

[8] ______, Über Korovkin-Sätze in lokalkonvexen Vektorverbänden. Math. Ann. In print.

Fachbereich Mathematik
Universität Dortmund
4600 Dortmund-Barop
West Germany

MONOTONE APPROXIMATION

K. L. Zeller

Let f be a (real) function of the type

(1) $f \in C[-1,1]$, $f\uparrow$

(monotone non-decreasing). We approximate f by monotone polynimials and define

(2) $E_n^*(f) = \underset{Q}{\text{Min}}\ \underset{x}{\text{Max}}\ |f(x)-Q(x)|$ $(-1\leq x\leq 1;\ \deg Q\leq n,\ Q\uparrow \text{ in } [-1,1])$.

The theory of this monotone approximation is already quite developed, see the recent report by Lorentz [1]. It is of course important to compare $E_n^*(f)$ with $E_n(f)$ (defined as in (2) but without the restriction $Q\uparrow$):

Theorem 1. (Lorentz-Zeller [2]). There exists a continuously differentiable function f of type (1) for which

(3) $E_n^*(f) \neq O(E_n(f))$

holds.

Theorem 2. (Roulier [3]). If a function f of type (1) is continuously differentiable and satisfies $f'(x) > 0$ for $-1\leq x\leq 1$, then

(4) $E_n^*(f) = O(E_n(f))$

holds.

Roulier gives a more refined result, using differences. A research team in Tübingen (R. Fahrion, F. Locher and others) investigates monotone polynomials and related problems. In the present context the following results might be interesting, since they diminish the gap between Theorem 1 and 2:

<u>Theorem 3</u>. The function f in Theorem 1 can be chosen in such a way, that it has the additional properties:

f is the restriction of an entire function.

f' has exactly one zero in [-1,1]; this zero is located at the point 0 and has order 2.

Alternatively one can place the zero at any point of [-1,1] and give it any order $\geq 2$ (resp. $\geq 1$ if at $\pm 1$). The proof uses a modification of the construction described in [2].

<u>Theorem 4</u>. Let f be a function of type (1) with $E_n(f) = O(n^{-3})$. If f' has only a finite number of zeros in [-1,1], each of them in the interior of this interval and of order 2, and if further

(5) $$E_{n-1}(f') = O(nE_n(f)),$$

then (4) holds.

The theorem can be extended to cover the endpoints $\pm 1$ and zeros of different order. For the proof we approximate (optimally) f' from below by $Q_n'$. We apply [1, Theorem 11] to $f-Q_n$ and modify $Q_n$ to become monotone: Near a zero of f' it can happen that $Q_n'$ is negative (but $\geq -2\,E_{n-1}(f')$). The conditions ensure that the length of a corresponding covering interval is $O(n^{-1})$. This makes it possible to find correcting polynomials $P_n\uparrow$ (of type Lukács-Fejér) with $\|P_n\| = O(E_n(f))$ such that $P_n'+Q_n'>0$ in the interval.

We conclude with some additional remarks. For the estimation of $E_n^*(f)$ from below Lorentz [1] mentions three methods: de la Vallée-Poussin, functionals, Bernstein. The first two are connected with the characterization of the best monotone approximation P, using points of maximal deviation of P from f and zeros of P'. A certain disadvantage lies in the fact, that the number m of points of maximal deviation with alternating signs of f(x)-P(x) is "small" if $E_n^*(f)$ is large. An elementary inequality is here

$$m \leq 2[f(b)-f(a)]E_n^*(f)^{-1} + \text{Const.}$$

The last of the methods mentioned gives estimates, which are not essentially better for monotone P than for unrestricted P (see [1]). It can, however, be complemented by other estimates and considerations (like those of Kirchberger) and seems then to be quite useful, e.g., for treating the cases where f' vanishes in an interval.

## References

[1] Lorentz, G. G., Monotone Approximation. In Inequalities, Vol. III, Oved Shisha, ed., Academic Press, New York, 1972, pp. 201-215.

[2] Lorentz, G. G. and K. L. Zeller, Degree of approximation by monotone polynomials II. J. Approximation Theory 2 (1969), 265-269.

[3] Roulier, J. A. Monotone approximation of certain classes of functions. J. Approximation Theory 1 (1968), 319-324.

Mathematisches Institut
Universität Tübingen
74 Tübingen
West Germany